Earthquakes: Prediction and Earthquake Resistant Construction

Earthquakes: Prediction and Earthquake Resistant Construction

Edited by Agnes Nolan

STATES
ACADEMIC PRESS
www.statesacademicpress.com

States Academic Press,
109 South 5th Street,
Brooklyn, NY 11249, USA

ISBN: 978-1-63989-161-0

Cataloging-in-Publication Data

Earthquakes : prediction and earthquake resistant construction / edited by Agnes Nolan.
p. cm.
Includes bibliographical references and index.
ISBN 978-1-63989-161-0
1. Earthquakes. 2. Earthquake prediction. 3. Earthquake resistant design.
4. Reinforced concrete construction. 5. Earthquake engineering.
I. Nolan, Agnes.
QE534.3 .E27 2021
551.22--dc23

For information on all States Academic Press publications visit our website at www.statesacademicpress.com

STATES
ACADEMIC PRESS

Contents

Preface

The shaking of the surface of the Earth resulting from a sudden release of energy in the Earth's lithosphere that creates seismic waves is referred to as an earthquake. They can range in intensity from weak to violent. They can destroy objects and people, and wreak destruction across the affected area. The effects of earthquakes are shaking and ground rupture, soil liquefaction, landslides, fires, tsunami, and floods. The branch of seismology which focuses on the specification of time, magnitude and location of future earthquakes within certain limits is termed as earthquake prediction. Earthquake-resistant construction is the fabrication of a building or structure that is able to withstand sudden ground shaking, and minimizing structural damage and human deaths. Ensuring proper design objectives for earthquake-resistance requires suitable construction methods. These methods can vary drastically throughout the world depending on the availability of resources, labor skills, capital, and technology in a particular area. This book is a compilation of chapters that discuss the most vital concepts and emerging trends in earthquake resistant construction. It elucidates new techniques and their applications in a multidisciplinary manner. This book aims to equip students and experts with advanced topics and upcoming concepts in this area.

The information shared in this book is based on empirical researches made by veterans in this field of study. The elaborative information provided in this book will help the readers further their scope of knowledge leading to advancements in this field.

Finally, I would like to thank my fellow researchers who gave constructive feedback and my family members who supported me at every step of my research.

Editor

1

Tunnel Vaults under Seismic Excitation

Roberto Guidotti and Alberto Castellani

Abstract

This chapter discusses factors that affect the seismic response of tunnels vaults, as the seismic ground motions, the geological condition and rock mechanics properties, and the relevance of numerical analysis, fundamental in the modeling of complex structures and processes, and in regional-scale analysis. As an example, this chapter focuses on the Laboratories of National Institute of Nuclear Physics (INFN) located in a Tunnel within the Gran Sasso mountain (Abruzzo, Italy). In addition to the L'Aquila (2009) earthquake, the chapter refers to observations reported in the literature related to the İzmit earthquake, Turkey (1999); the Chi-Chi earthquake, Taiwan (1999); and the Kobe earthquake (2004); and, as historical event, the Kern County earthquake (1952).

Keywords: tunnel vaults, seismic excitation, seismic ground motion, rock mechanics properties, numerical analysis

1. Introduction

There are two areas of concern in earthquake engineering of tunnel vaults. The first one is related to the values of the peak ground acceleration (PGA) to which the structures may survive by responding in the elastic or the plastic field. The second one is related to the fault movement, to which the structure cannot offer a valuable resistance. The chapter discusses the factors that affect the seismic response of tunnels vaults, as the seismic ground motions, the geological condition and rock mechanics properties, and the relevance of numerical analysis, fundamental in the modeling of complex structures and processes, and in regional-scale analysis. As an example, this chapter focuses on the Laboratories of the National Institute of Nuclear Physics (INFN) located in a tunnel within the Gran Sasso mountain (Abruzzo, Italy). A number of geophysics studies have been devoted to the area [1–7]. A numerical analysis on the INFN Laboratories has been presented in [8]. This chapter collects observations and references on the behavior of tunnel vaults under seismic excitation, with the aim to derive design soil motion for the INFN Laboratories, in view of future installations.

2. Factors affecting the seismic response of tunnel vaults

The response of tunnel vaults under seismic excitation depends on multiple factors. First, the earthquake motion, such as the earthquake intensity and magnitude. Second, the tunnel environment condition, as, that is, the rock condition, the overburden layer depth, the location with respect to the fault zone. Third, the structural condition of the tunnel, the presence of lining, its integrity, and the quality of the construction.

2.1. Role of ground motions

The study presented in [9] observes 71 cases of rock tunnel response to earthquake motions, with diameter varying from 3 to 6 m. Out of 72, 42 cases were damaged. The study in [9] proposes a relationship of the damage level with the earthquake magnitude, intensity and epicenter. The following observations can be drawn: (1) the tunnel has no damage when $PGA < 0.19g$ and peak ground velocity (PGV) is lower than 20 cm/s; (2) the tunnel will have minor damage when $0.19g < PGA < 0.5g$ and $20\ \text{cm/s} < PGV < 80\ \text{cm/s}$; and (3) the tunnel will be severely damaged when $PGA > 0.5g$ and $PGV > 80\ \text{cm/s}$. In these evaluations, the quantities PGA and PGV refer to "free field" conditions, that is the expected quantities at the nearby region, in plane conditions. The study in [10] provides a database containing 192 observations from 85 earthquakes worldwide. Half of the events refer to earthquakes of Richter magnitude higher than 7, and about 75% of damage is observed within 50 km from the epicenter (near-field). Among the 192 cases, no damage is observed with PGA of the horizontal components lower than 0.2 g, in agreement with [9]. In the majority of the cases, important damage has been observed for $PGA > 0.4g$.

2.2. Role of geological condition and rock mechanics properties

A study of tunnel seismic damage in Japan [11] observed that the tunnel sector with thick lining had the biggest damage percentage: 82, 38, and 16%, considering respectively thickness of 40, 30, and 20 cm. However, this observation should be revisited taking into account the nearby geological conditions. In the same study, the authors observe a damage percentage of 16% in hard rock, 40% in soft rock, 44% in joint development rock, and 61% in earth. Based on these observations, the earthquake safety of tunnel is mainly controlled by the natural condition. When the natural condition is poor, to increase lining thickness may increase earthquake forces and be detrimental to the safety of the tunnel. A more effective method could be to reinforce the surrounding rocks. The study concludes that earthquakes do not affect tunnels farther than 50 km from the epicenter.

Literature refers to peculiar applications devoted to rock mechanics. A useful database for rock mechanics properties is provided by Lama and Vutukuri [12] and has been used in [8]. In [13], a 3D dynamic analysis program for saturated porous rocks and soils is presented. The theoretical formulations incorporated in the proposed computer program are the extension of Biot's two-phase theory to nonlinear region. A numerical study shows the effects of pore water on the dynamic response of underground openings in saturated rock masses. It is shown that underground openings in saturated porous media could be significantly more vulnerable to the potential damages associated with high motions and shear failure that those in dry media. The work presented in [14] established that crack-induced stress-aligned shear wave splitting, with azimuthal anisotropy, is an inherent characteristic of most of rocks in the crust. This means that most in situ rocks are characterized by fluid-saturated micro-cracks. The evolution of such stress-aligned fluid-saturated grain boundary cracks and pore throats in response to changing conditions can be calculated, in some cases with great accuracy, using anisotropic poroelasticity (APE).

2.3. Role of fault movement

Fault movement is one of the major areas of concern in earthquake engineering of tunnel vaults. Based on the study in [15], three considerations related to the fault movement can be drawn. First, the ratio between surface displacement and sub-surface displacement has a wide range, between 0.2 and 8. The average sub-surface displacement is calculated from the seismic moment and the rupture area. The study in [15] does not mention the stiffness of the soil surface; however, the highest ratios should be assigned to the most deformable surface soils, and in particular imported backfill and reclaimed land. Second, the amplitude of displacement varies along the length of the fault, like cracks in a concrete structure or pavement. Third, the eventual movement of a fault shall be considered sub-surface movement, and not surface movement, if we consider deeply embedded tunnels. As for the amplitude of the expected displacement, Wells and Coppersmith [15] provide relations between the average displacement, in m, and the moment magnitude. Relation as those presented in [16–18] may be used to express peak ground acceleration as a function of the moment magnitude and of the distance from the epicenter. **Table 1** shows the maximum expected displacement between the fault surfaces, considering the relation in [15] between maximum expected displacement and moment magnitude and the relations in [16–18] between PGA and moment magnitude. Equation from [16–18] is provided in the Appendix A1. **Table 1** indicates that the problem regards the cases with PGA larger or equal to 0.25 g. It is at the border of the present analysis, and consequently it requires a further insight into the expectation of a fault movement.

PGA (g)	Magnitude [16, 17]	Magnitude [18]	Maximum displacement amplitude (m)
0.05	~1	~1	~0
0.15	4.6	4.6	~0
0.25	6.5	6.3	< 0.1
0.35	7.8	7.4	> 1

Table 1. Maximum expected displacement between the fault surfaces, joining the correlation of PGA and M, ([16–18], Appendix A1), and that of M and the amplitude, [15].

3. Relevance of numerical analyses in assessing the seismic response of tunnel vaults

Theoretical analyses of tunnels and lined tunnels have been proposed in the literature [e.g., 19, 20]. However, tunneling engineering is one of the areas of applied soil and rock mechanics in which the numerical methods for stress analysis are frequently adopted in practice [e.g., 21]. Their frequent use depends on several reasons related to the complex characteristics of the tunneling problems. One of the most important is the strong influence of the excavation and construction procedures, and of their technological details, on the stress/strain distribution in the rock surrounding the opening and in its supporting system. This represents a main drawback for the analytical solutions and for simplified methods of analysis, which, in most cases, cannot capture this process with a sufficient level of detail. Another important aspect of tunneling problems captured by numerical analyses is their complex geometrical nature. This includes, among other aspects, (1) the shape of the opening, (2) the presence of discontinuities in the rock mass, and (3) the presence of non-homogenous or non-isotropic layers. The extension to 3D problems is possible, provided the required amount of information and the ability to manage a more complex map of stresses.

Numerical tools are especially useful when dealing with regional-scale analysis. The study in [22] illustrates an application of the HAZUS [23] methodology to the tunnels and bridges of a highway network. The variability in the ground shaking and in the construction characteristics leads to very different probability of failure for different components (i.e., tunnels and bridges) in the network. The resulting damage levels for bridges and tunnels depend on the fragility curves used in the evaluations. They were developed for existing bridge and tunnel structural typologies in the United States. State-of-the-art fragilities with models of capacity and demand have been proposed in [24, 25].

4. Lesson learnt from direct observation of damage

4.1. The İzmit earthquake (1999)

Effects produced by faults movement are reported especially by [26, 27] for the Anatolian Motorway tunnel, Bolu tunnel, in occasion of the M_W 7.4 İzmit earthquake (1999).

4.2. The Chi-Chi earthquake in Taiwan (1999)

Effects similar to those produced by İzmit earthquake are described in [28] for the M_W 7.6 Chi-Chi earthquake of Taiwan (1999). It is reported that out of 57 galleries, 49 have suffered damage. The study is in particular devoted to the covering lining, tunnels, and design documents (see **Figure 1**). The work [29] on the Chi-Chi earthquake shows that tunnels in intensity nine areas were damaged, whereas in low intensity areas, the tunnels were undamaged. In [29], information on seven tunnels affected by the Chi-Chi earthquake are collected, and the

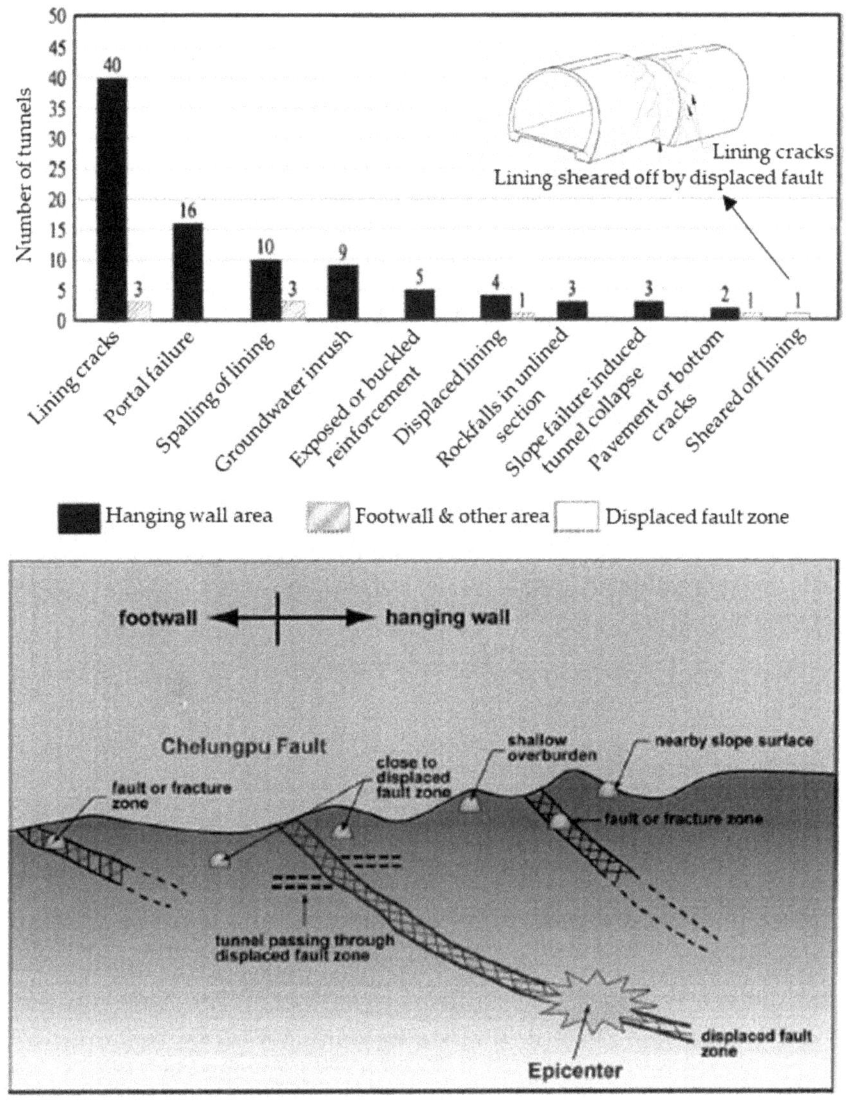

Figure 1. Number of tunnels suffering various type of damage and locations of tunnels with the respect the displaced fault zone (adapted from [28]).

intensity was adjusted according to Chinese intensity. The area intensity is a measure similar to the Mercalli scale. The results presented in [29] are in agreement with the numerical analysis reported in [8].

4.3. The Kobe earthquake (1995)

There were more than a hundred tunnels in the disaster area in the M_W 6.8 Kobe earthquake (1995), [30–34]. The damage of the tunnels has been related to the area intensity. Tunnels in the intensity 10 areas were damaged in different levels, with several tunnels major damaged for crossing fault zones. Many tunnels were damaged in the intensity nine areas, whereas only few tunnels experienced damage in intensity eight areas. No damage was reported for tunnels in intensity seven areas. The previously referred study [29] includes also information on 27 tunnels damaged during the Kobe earthquake. The study in [6] describes two cases, at depth

Figure 2. Damage at the tunnel entrance of the tunnel during the 1952 Kern country earthquake (adapted from [37]).

and shallow depth, during the 1995 Kobe earthquake: the collapse of the Daikai Underground Station in the city of Kobe [35, 36], and the damage to the Bantaki Tunnel in the mountains near Kobe. Similar damage pattern was found in the Uonuma Tunnel of the Japanese high-speed train (Shinkansen) network immediately after the 2004 Niigata-ken Chuetsu earthquake.

4.4. The Kern County earthquake (1952)

The study in [37] considers three earthquakes and, in particular, the M_W 7.3 Kern County (1952) earthquake, when a tunnel experienced damages, but just at the entrance (see **Figure 2**). The question arising therefore is whether the earthquake damage in the Bolu tunnel represents an exception, or whether the hypothesis that tunnels are affected by minor seismic risk affecting tunnels should be re-evaluated. The authors examined numerous tunnels in tectonically and seismically active areas concluding that tunnels in such areas are vulnerable not only to seismic shaking, but also to tectonic deformations. The study in [37] refers of old events where tunnels collapsed under the effects of faults. However, beside the historical relevance, these references lack quantitative data. Some useful information on faults movements causing damage are given in [38], however related to a M_W 4.0 seismic event, induced by mining in the Saar District, Germany.

5. Case study: the INFN Laboratories

The Laboratories of National Institute of Nuclear Physics (INFN) are located in a Tunnel within the Gran Sasso mountain (Abruzzo, Italy), **Figure 3**.

5.1. Ground motions during L'Aquila (2009) earthquake

During L'Aquila (2009) earthquake, acceleration records have been collected at plane conditions, at several stations on the Gran Sasso, with values of peak ground acceleration (PGA) between 0.35 and 0.5 g. A few accelerograms have been collected in the gallery, about 1400 m

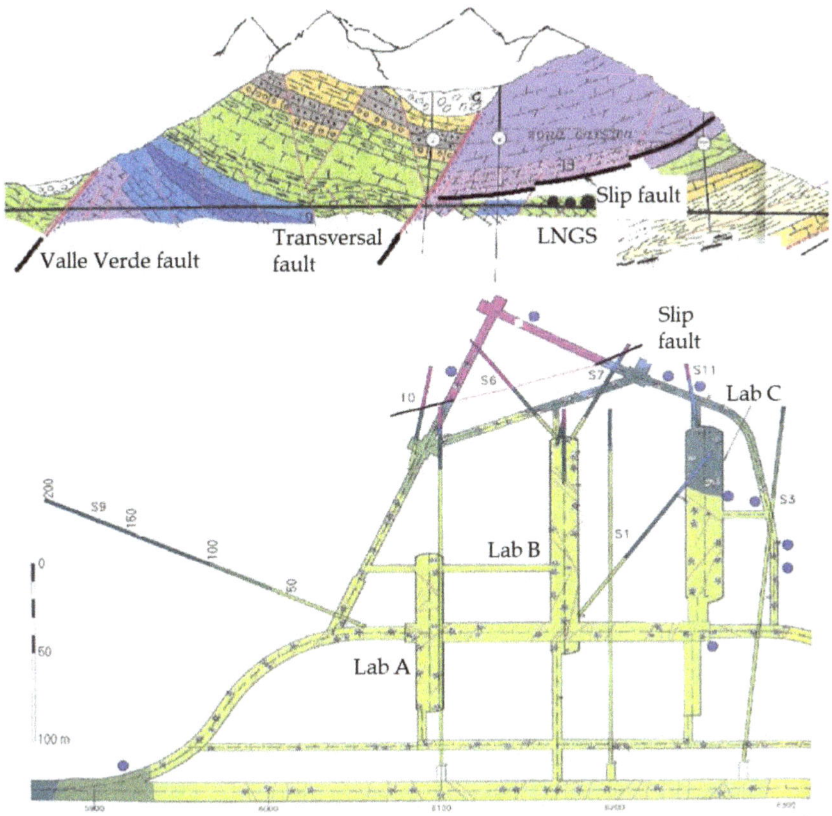

Figure 3. (top) Vertical section of the Gran Sasso mountain, and the rooms of INFN; (bottom) plant of the laboratories of the National Institute of Nuclear Physics, INFN.

below the mountain crest, with values of PGA lower than 0.1 g. This reduction confirms a qualitative result reported by Dowding and Rozen [9], considering 71 earthquakes that affected tunnels. This reduction justified also a large portion of the observations available in the literature, as discussed in this chapter. In the numerical analysis performed in [8], the accelerogram in the gallery has been reproduced, assuming as input data a representative accelerogram collected in plane conditions, with PGA equal to 0.5 g.

5.2. Geological condition and rock mechanics properties

Several reports on the geological conditions of Gran Sasso are available in the literature [1–4]. Following the L'Aquila 2009 earthquake, Amoruso et al. [4] report about significant changes in the hydrogeology of the Gran Sasso carbonate fractured aquifer. These changes are (1) the disappearance at the time of the main shock of some springs located along the surface trace of the Paganica normal fault; (2) an abrupt increase in the discharge of the Gran Sasso highway tunnel drainages and of other springs; and (3) a progressive increase of the water table elevation at the boundary of the Gran Sasso aquifer, in the months following the seismic event. The authors in [4] propose a model of the effect of the earthquake on the Gran Sasso aquifer based on historical data including seismic monitoring, spring discharge, water table elevations, turbidity and rainfall events. This model excludes the effects of seasonal recharge. The short-term hydrologic effects registered immediately after the seismic event have been ascribed to a pore pressure increase related to the aquifer deformation. Mid-term effects observed in the months following the event suggested a change in groundwater hydrodynamics. Additional groundwater flowing towards aquifer boundaries and springs in discharge areas may result from an increase in the hydraulic conductivity in the recharge area, nearby the earthquake fault zone. This increase might be attributed to fracture clearing and/or expansion. Results from numerical simulations of the pore pressure and permeability change with time are in agreement with observed field data.

5.3. Current information about faults presence and movement

Figure 4 shows current information about the presence of faults, collected during the tunnel excavation in the Gran Sasso. Discontinuities in the rock mass, or joints, are names commonly used to catalog faults during the construction phase of galleries.

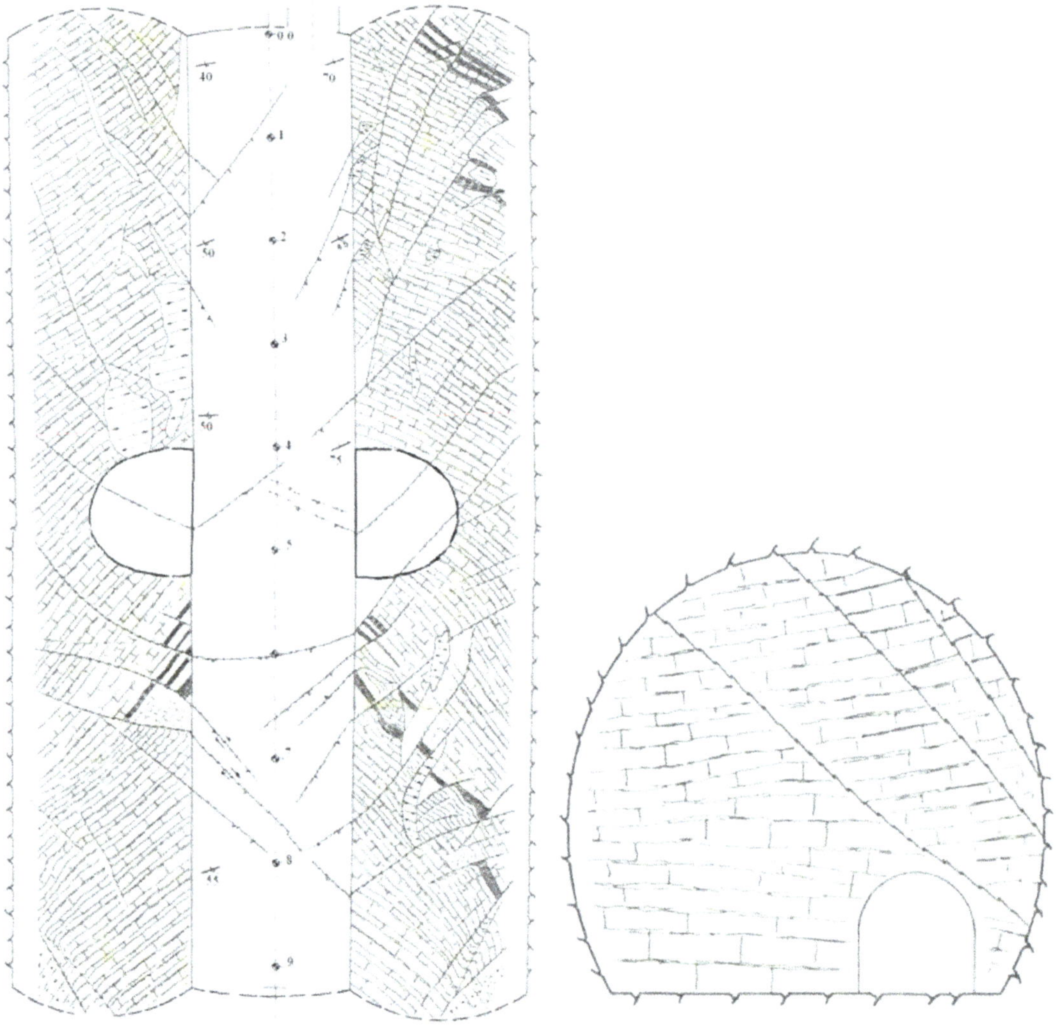

Figure 4. (left) In-plane trace of faults that pertain to the highway gallery of Gran Sasso, recorded during construction; (right) trace of faults in the vertical plane, near the tunnel entrance.

A paper focused on the Apennine geophysical condition has been prepared by Brunori et al. [5]. If a fault shows evidence of having moved at least once in the past 100,000 years, then it should be regarded as a potential source of earthquakes. Another prevailing opinion is that if the fault has moved at least once in the past 5000 years, then it should be considered a potential source of damaging earthquakes to any settlement within a radius of 50 km. Once a major fault has formed, future earthquakes are generated along the same fault line, and after years of movement, increasingly larger vertical and horizontal displacements of land may occur. With reference to the trace of faults shown in **Figure 4**, it is possible to assume two different scenarios. In the first scenario, the seismic motion is originated from a deeply embedded fault, which shows a superficial offshoot. The tunnel is concerned by the offshoot, and it is located within the epicenter area. In the second scenario, the tunnel and the fault crossing it are at some tens of kilometer from the epicenter. The seismic motion is not assigned to an energy release from the fault under object; however, the seismic motion may activate a relative motion at the sides of the fault. In the first scenario, along the fault, an energy has been cumulated, capable to activate the movement. The expected effects are more important, but the event is associated to a lower probability of occurrence, because even if the seismic event occurs, the epicenter should be exactly in correspondence to the crossing. The movement of that fault is at the origin of the earthquake. In the second scenario, along the fault the cumulated energy is not enough to activate the seismic motion, and the relative motion along the fault is an induced motion. The probability of this event is in a fair approximation linked to the probability of the earthquake motion itself, that is, 2‰ per year.

In our case, the earthquake of L'Aquila 2009 has been classified as originated at the Pettino fault, [7], (sometimes at the Paganica fault). The distance from the Gran Sasso INFN Laboratory is about 30–40 km in both cases. Therefore, it falls within the second scenario. The study in [5] focuses the analysis on the Pettino fault, a part of the Late Quaternary segmented system called the Upper Aterno fault system, which is responsible for the evolution of the L'Aquila basin, and likely, for the 1703 A.D. $M_W > 6.0$ earthquake. The Pettino fault appears, at a field survey scale, quite continuous and homogeneous along the trace. We are not aware about studies on the interactions of this fault (or Paganica fault) and the faults crossing the Gran Sasso tunnel. However, Italian seismic history reports numerous examples of cascading activation of faults nearby one to the other, following a strong earthquake. In those cases, the time delay varies from a few seconds, (?Irpinia, 1980; three shakings in 40 s), to 1 day, (Umbria-Marche, 1997), some days, (Emilia, 2012), till a week, (Calabria, 1783), or even years (Nicastro, Southern Calabria 1905, followed by Messina, 1908).

5.4. Tendons along the tunnel vaults

The lining of the tunnel and that of the Lab rooms is anchored to the rock behind by a network of tendons, **Figure 5**. According to Castellani et al. [8], the state of stress in the lining is not meaningful with respect to the existing static stress due to relaxation following the construction. However, the measure of the ovalization of the halls, expressed by a change in length of the diameters reached up to 8 cm. The risk of superficial ruptures and consequence of rocks fall

Figure 5. (top left) Tendons on one of the Laboratory rooms at Gran Sasso, recorded during construction; (top right) covering of the walls and roof. Anchor plate and tendon, one every square meter; (bottom left and right) anchor caps, showing water filtering, following the water basin movement, described by Amoruso et al. [4].

should be taken into account. For this reason, the anchors are inserted; however, according to the authors' opinion, they do not affect the overall behavior of the tunnel or the Lab rooms.

6. Conclusion

In addition to the L'Aquila (2009) earthquake, this chapter refers to observations of the response of tunnel vaults under seismic excitation reported in the literature related to (1) the İzmit earthquake, Turkey (1999); (2) the Chi-Chi earthquake, Taiwan (1999); (3) the Kobe earthquake (2004); and (4) the Kern County earthquake (1952).

Common observations are the following:

i. With respect to structures on the surface, or at shallow depth, underground deep structures subjected to dynamic waves vibrate solidly with their surroundings;

ii. Shallow galleries suffer more damage than deep galleries, with 60% of the observed damage referring to galleries of depth lower than 50 m;

iii. Galleries with rectangular cross section, excavated in open air (cut and cover structures) are more vulnerable than deep circular tunnels;

iv. Surrounding soil has the greatest importance. As discussed in [10], among 192 observations, 79% of tunnels excavated in a deformable soil have suffered damage, whereas those excavated in rock soil resulted in lower damage.

As an example, this chapter focuses on the Laboratories of National Institute of Nuclear Physics (INFN) located in a Tunnel within the Gran Sasso mountain (Abruzzo, Italy). The design acceleration at the ground floor of the Laboratory of Physics at Gran Sasso is evaluated, taking into account the ground motion attenuation in the tunnel, measured during the 2009 L'Aquila earthquake. Numerical analysis were able to reproduce such attenuation, based on local data. The survey at the Laboratory, immediately after the earthquake, confirms that the PGA at the Lab has been lower than 0.1 g, and no damage occurred, although PGA at "free field" has been around 0.5 g. This chapter confirms that similar attenuations have been pointed out in the literature of deep galleries. A few exceptions are remarked, but different conditions have been discussed between the Gran Sasso Gallery case and these exceptions. A residual risk should be investigated, connected with a possible interaction among adjacent faults (in the considered case between the Pettino and the Paganica faults).

A.1. Magnitude and peak acceleration at given distances from epicenter

Maps of seismicity are available online (e.g., earthquake.usgs.gov/earthquakes). They are expressed through the epicenters location and the measure of magnitude. In general, the most recent data available online are expressed in terms of magnitude. In order to express these data in a format comparable with for instance [9], the peak ground acceleration *PGA* needs to be related to the magnitude. Studies in [16, 17] provide one of these equations

$$\ln(PGA) = -1.101 + 0.2615M - \ln\left[\left(r^2 + 7.2^2\right)^{0.5}\right] - 0.00255\left[\left(r^2 + 7.2^2\right)^{0.5}\right], \quad (1)$$

where *PGA* is the peak ground acceleration in *g* units, *M* is the moment magnitude, *r* is the distance in *km* of the site from the epicenter. In [18], a similar equation has been proposed

$$\ln(PGA) = -1.562 + 0.306M - \ln\left[\left(r^2 + 5.8^2\right)^{0.5}\right]. \quad (2)$$

Author details

Roberto Guidotti[1] and Alberto Castellani[2*]

*Address all correspondence to: alberto.castellani@polimi.it

1 Department of Civil and Environmental Engineering, MAE Center: Creating a Multi-hazard Approach to Engineering, University of Illinois, Urbana, IL, USA

2 Structural Engineering Department, Politecnico di Milano, Milano, Italy

References

[1] Amoruso A, Crescentini L, Scarpa R. GIGS. The Interferometric Station at LNGS, LNGS Annual Reports, 1999

[2] Amoruso A, Crescentini L. Coseismic and aseismic strain offsets recorded by the Gran Sasso strainmeter. Physics and Chemistry of the Earth, Part A: Solid Earth and Geodesy. 1999;**24**(2):101-104

[3] De Luca G, Scarpa R, Abril M, Cirella A. Seismic arrays. In: Bettini A, Antolini R, editors. The Gran Sasso Laboratory 1979–1999: A vision becomes reality. Publication Assergi: INGS, 1999, 304 p. ISBN: 8886409206

[4] Amoruso A, Crescentini L, Petitta M, Rusi S, Tallini M. Impact of the 6 April 2009 L'Aquila earthquake on groundwater flow in the Gran Sasso carbonate aquifer, Central Italy. Report. Hydrological Processes. 2011;**25**:1754-1764. DOI: 10.1002/hyp.7933

[5] Brunori CA, Civico R, Cinti FR, Ventura G. Characterization of active fault scarps from LiDAR data: A case 1 study from Central Apennines (Italy). International Journal of Geographical Information Science. 2011;**0452**:R3

[6] Aiello V, Boiero D, D'Apuzzo M, Socco LV, Silvestri F. Experimental and numerical analysis of vibrations induced by underground trains in an urban environment. Turkish Journal of Earth Sciences. 2003;**12**:153-156

[7] Galli P, Giaccio B, Messina P, Peronace E, Zuppi GM. Paleoseismology of the L'Aquila faults (Central Italy, 2009, Mw 6.3 earthquake): Implications for active fault linkage. Geophysical Journal International. 2011;**8187**(3):1119-1134. DOI: 10.1111/j.1365-246X.2011.05233.x

[8] Castellani A, Canetta G, Pace S, Guidotti R. The vaults of the Gran Sasso National Laboratory Halls, during the 2009 L'Aquila earthquake. Design of an earthquake-monitoring system (in Italian). Gallerie e grandi opere sotterranee. 2012;**103**:61-75

[9] Dowding CH, Rozen A. Damage to rock tunnels from earthquake shaking. Journal of Geotechnical Engineering, Division of American Society of Civil Engineers. 1978;**104**(2): 175-191

[10] Sharma S, Judd WR. Underground opening damage from earthquakes. Engineering Geology (Amsterdam). 1991;**30**:263-276

[11] Okamoto S. Introduction to earthquake engineering. Tokyo: The University of Tokyo Press; 1973

[12] Lama RD, Vutukuri VS. Handbook on mechanical properties of rocks-testing techniques and results. Vol. II, Series on Rock and Soil Mechanics. Trans Tech Publications. 1978;**3**(1). ISBN: 0-87849-021-3

[13] Kim SH, Kim KJ. Three-dimensional dynamic response of underground openings in saturated rock masses. Earthquake Engineering and Structural Dynamics. 2001;**30**:765-782

[14] Crampin S, Chastin S. A review of shear wave splitting in the crack-critical crust. Geophysical Journal International. 2003;**155**:221-240

[15] Wells DL, Coppersmith KJ. New empirical relationships among magnitude, rupture length, rupture width, rupture area, and surface displacement. Bulletin of the Seismological Society of America. 1994;**84**(4):974-1002

[16] Ambraseys NN. Uniform magnitude re-evaluation of European earthquakes associated with strong-motion records. Earthquake Engineering and Structural Dynamics. 1990;**19**:1-20

[17] Ambraseys NN, Simpson KA, Bommer JJ. Prediction of horizontal response spectra in Europe. Earthquake Engineering and Structural Dynamics. 1996;**25**:371-400

[18] Sabetta F, Pugliese A. Estimation of response spectra and simulation of nonstationary earthquake ground motions. Bulletin of the Seismological Society of America. 1996;**86**(2): 337-352

[19] Penzien J, Wu CL. Stresses in linings of bored tunnels. Earthquake Engineering and Structural Dynamics. 1998;**27**:283-300

[20] Penzien J. Seismically induced racking of tunnel linings. Earthquake Engineering and Structural Dynamics. 2000;**29**:683-691

[21] Gioda G, Swoboda G. Developments and applications of the numerical analysis of tunnels in continuous media. International Journal for Numerical and Analytical Methods in Geomechanics. 1999;**23**:1393-1405

[22] Codermatz R, Nicolich R, Slejko D. Seismic risk assessments and GIS technology: Applications to infrastructures in the Friuli–Venezia Giulia region (NE Italy). Earthquake Engineering and Structural Dynamics. 2003;**32**:1677-1690. DOI: 10.1002/eqe.294

[23] FEMA. Multi-hazard loss estimation methodology, earthquake model, HAZUS-MH 2.1 Technical Manual. Washington, DC: Federal Emergency Management Agency; 2003. 1-699

[24] Gardoni P, Der Kiureghian A, Mosalam KM. Probabilistic capacity models and fragility estimates for reinforced concrete columns based on experimental observations. Journal of Engineering Mechanics. 2002;**128**:1024-1038

[25] Gardoni P, Mosalam KM, Der Kiureghian A. Probabilistic seismic demand models and fragility estimates for RC bridges. Journal of Earthquake Engineering. 2003;**7**:79-106

[26] Scawthorn C, Johnson GS. Preliminary report Kocaeli (Izmit) earthquake of 17 August 1999. Engineering Structures. 2000;**22**:727-745

[27] Dalgýc S. Tunneling in squeezing rock. The Bolu tunnel, Anatolian motorway, Turkey. Engineering Geology. 2002;**67**:73-96

[28] Wang WL, Wang TT, Sua JJ, Lin CH, Seng CR, Huang TH. Tunnelling in Taiwan. Assessment of damage in mountain tunnels due to the Taiwan chi-chi earthquake. Tunnelling and Underground Space Technology. 2001;**16**:133-150

[29] Xiaoqing F, Junqi L, Xiaolan Z, Runzhou L. Damage Evaluation of Tunnels in Earthquakes. In Proc. 14th World Conference on Earthquake Engineering, Beijing, China. October 12-17 2008;**8**:8-16

[30] Asakura T, Sato Y. Damage to mountain tunnels in hazard area. Soils and foundations. 1996;**36**(Special):301-310. DOI: https://doi.org/10.3208/sandf.36.Special_301

[31] Hashash YM, Hook JJ, Schmidt B, John I, Yao C. Seismic design and analysis of underground structures. Tunnelling and Underground Space Technology. 2001;**16**(4):247-293

[32] Inokuma A, Inano S. Road tunnels in Japan: Deterioration and countermeasures. Tunnelling and Underground Space Technology. 1996;**11**(3):305-309

[33] Kirzhner F, Rosenhouse G. Numerical analysis of tunnel dynamic response to earth motions. Tunnelling and Underground Space Technology. 2000;**15**(3):249-258

[34] Davis A, Lee VW, Bardet JP. Transverse response of underground cavities and pipes to incident SV waves. Earthquake Engineering & Structural Dynamics. 2001;**30**:383, 410

[35] Uenishi K, Sakurai S. Characteristic of the vertical seismic waves associated with the 1995 Hyogo-ken Nanbu (Kobe), Japan earthquake estimated from the failure of the Daikai underground station. Earthquake Engineering and Structural Dynamics. 2000;**29**(6):813-822

[36] Huo H, Bobet A, Fernández G, Ramirez J. Load transfer mechanisms between underground structure and surrounding ground: Evaluation of the failure of the Daikai station. Journal of Geotechnical and Geoenvironmental Engineering. 2005;**131**(12):1522-1533

[37] Kontogianni VA, Stiros SC. Earthquakes and seismic faulting: Effects on tunnels. Turkish Journal of Earth Sciences. 2003;**12**(1):153-156

[38] Alber M, Fritschen R. Rock mechanical analysis of a M_l = 4.0 seismic event induced by mining in the Saar District, Germany. Geophysical Journal International. 2011;**186**(1):359-372

2

Use of Polyethylene Terephthalate Fibers for Strengthening of Reinforced Concrete Frame Made of Low-Grade Aggregate

Comingstarful Marthong

Abstract

This chapter presents an experimental finding on the use of waste materials for strengthening a reinforced concrete (RC) frame prepared with low-grade materials. Compressive, tensile and flexural strength of concrete specimens made of recycled coarse aggregate (RCA) and natural coarse aggregate (NCA) were evaluated to assess RCA suitability for structural use. Results show that the strength parameter decreases in all cases. In order to extend the application of RCA concrete to structural elements, three RC frames were prepared and tested under monotonic loading. The joint regions of one of the frame were cast using Polyethylene terephthalate (PET) fiber, i.e., PET fiber-reinforced concrete (PFRC). PET fiber (aspect ratio = 25) of 0.5% by weight of concrete used in the PFRC mix was obtained by hand cutting of post-consumer PET bottles. A reference specimen was also prepared using 100% NCA and subjected to a similar loading condition. The RCA frame resulted in a brittle mode of failure and lower load carrying capacity as compared to the reference specimen. However, the presence of PFRC improved the damage tolerance and load resisting capacity and hence seismic parameters such as stiffness degradation, ductility and energy dissipation increased. This indicates that PET fiber is a viable substitute for steel fiber as a low-cost material in improving the seismic performance of frame structures.

Keywords: reinforced concrete frame, concrete mechanical strength, recycled coarse aggregate, polyethylene terephthalate fibers, monotonic loading, seismic performance

1. Introduction

Major centers of demand for aggregates are often not completely satisfied by local supplies and large quantities have to be brought in from a considerable distance. With the passage of time it is likely that both the quantities imported and the distances transported will increase with detrimental effects on cost and environmental intrusion unless alternatives are considered. Using of low grade aggregates, if technically and economically suitable, have the dual advantage of reducing the amount of quarrying required and also of reducing the amount of land required for tipping. Generally the low grade aggregates are those aggregates which lose strength generally by more than 15% upon wetting. Aggregates obtained from a demolished concrete waste are considered as a low grade aggregate. Recycling of such waste is a beneficial from the viewpoint of environmental preservation and effective utilization of resources. The processed concrete waste (recycled aggregates) either fine or coarse has been use as a replacement of the natural aggregate for a number of years. Various researchers [1–3] showed that reduction in the mechanical strength is not much prominent, when recycled coarse aggregate (RCA) replacement is up to 30%. Further, results reported [4, 5] showed that RCA concrete exhibited a similar behavior, which can be adequately used in concrete technology application. Literature revealed that most studies on utilization of RCA in concrete production have been limited to mechanical strength evaluation. For confident utilization of RCA in the construction industry, its structural behavior ought to be investigated. Some past studies concerning the behavior of beams [6], columns [7] and beam-column joints [8], RC frame structure [9] manufactured from RCA were reported. Most of their findings on their structural behavior are positive. However, due to its brittle behavior of RCA the load carrying capacity of most of the RCA specimens gets reduced. This perhaps the reasons that RCA was not prominently use in structural elements.

In all RC framed structures, the beam-column joints play an important role in the overall response of the frame structures under strong influence of seismic attack. To enhance the seismic capacity of the frame structures, steel fibers reinforced concrete (SFRC) has been incorporated in the joint region [10]. Test results revealed that SFRC enhances flexural capacity, shear strength, ductility and energy dissipation capacity. In substitution of steel fiber polymeric fibers made of nylon, aramid, polypropylene, polyethylene and polyester has been used as concrete reinforcing materials. Various experimental studies [11–13] showed PET fiber has a potential use for enhancing the mechanical properties of concrete that can replace steel fiber. The fiber content generally varies from 0.1 to 1.0%. However, a fiber content of 0.5% by weight of concrete is reported as an optimum percentage [14]. Seismic parameters such as load resisting capacity, ductility, energy dissipation capacity and stiffness of beam-column connection improve when a concrete mix incorporating PET fiber is use at the joint region [15].

Due to various negative aspect possessed by recycled aggregate for concrete production, the seismic performance of frame structure made of RCA may be lower than those of conventional concrete. Like the conventional concrete, various techniques may be employed for improving the mechanical properties of RCA concrete. One such technique on a particularly use of PET fibers for enhancing the mechanical strength and energy dissipation capacity of RCA has been reported [16]. The study reveals that a significant enhancement in tensile, flexural and energy dissipation capacity was achieved. Also, PET fiber has a potential for arresting crack formation due to fiber-bridging action inside the concrete matrix. Thus, it is expected that use of PET fiber in the joint region of an RC frame would delay the crack formation and crack growth. This would result in enhancement of displacement ductility due to delayed bond failure and hence overall improved the seismic performance of the frame structures.

Therefore, the main objectives of this study is to investigate the structural behavior of an RC frame prepared with RCA concrete and strengthened using PET fibers within the joint region and partly in both column and beams, which is the D-region as defined by ACI 318-08 [17].

2. Experimental program

2.1. Materials

Ordinary Portland Cement (OPC) of 43 grades conforming to IS: 8112 [18] was considered. The maximum size of natural coarse aggregate (NCA) was 16 mm. River sand was used as fine aggregate (FA) (0–4.75 mm size). The RCA's (5–16 mm size) are obtained from the demolished reinforced cement concrete (RCC) roof slab of 20 years old. The large pieces of the roof slab are transported to the laboratory and broken into pieces of aggregates smaller than 20 mm in size and sieved through a 16 mm sieve. The aggregates greater than 16 mm in size are further broken to a maximum size of 16 mm. All aggregates used in this study have been tested as per relevant codes [19, 20] and the physical properties are presented in **Table 1** and the particle size gradation is shown in **Figure 1**. The polymeric fiber type use for strengthening or enhancing the mechanical properties of concrete was PET fiber. The geometry of PET fiber considered was "straight slit sheet", which has the shape similar to those of steel fibers

Mix	Apparent density (kg/m³)	Bulk density (kg/m³)	Grading (mm)	Elongated particle content (%)	Water absorption (%)	Crush index (%)
NCA	2830	1560	5–16	5	0.97	5.81
RCA	1600	1130	5–16	10	5.30	12.73

Table 1. Physical property of NCA and RCA.

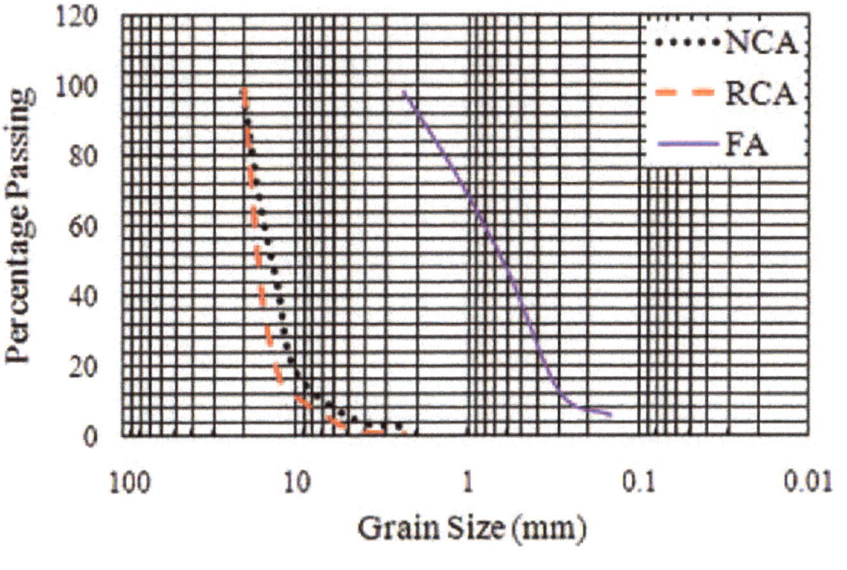

Figure 1. Particle size gradation.

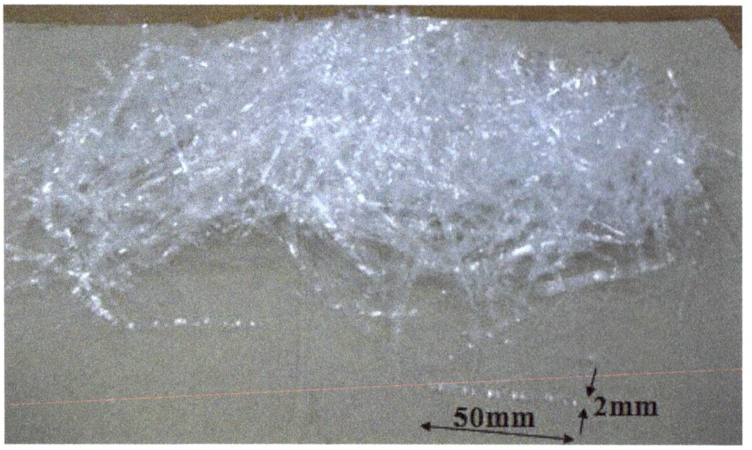

Figure 2. Straight slit sheet fiber of PET.

[21]. **Figure 2** shows the geometry of fiber use in this study, which were produced by hand cutting from post-consumer PET bottle of 1 liter capacity. The length of fiber is 50 mm, width is 2 mm and thickness is 0.5 mm with an aspect ratio (length/width) of 25. The specific gravity and tensile strength of PET fibers is found out to be 1.38 and 155 MPa, respectively.

2.2. Concrete mixture proportions

Table 2 presented the specimens and test parameters for characterizing the mechanical properties of concrete. Mix M0 is a control specimen (i.e., 100% NCA) while M100 is the mix with 100% RCA and M100-PET is the mixture of 100% RCA plus 0.5% PET fiber, i.e., PFRC All concrete mixes were prepared with the same w/c of 0.5 and the same degree of workability (slump value of 60 mm) evaluated according to IS 1199 [22]. The concrete mixes were designed for a characteristic cube compressive strength of 25 MPa which resulted in a target mean cube compressive strength of 31.6 MPa as per IS 10262 [23]. The concrete mixes were produced with 372 kg/m^3 of cement, 733 kg/m^3 of fine aggregate and 1087 kg/m^3 of coarse aggregate for a w/c of 0.5. To achieve a better workability, superplasticizer dosage of 0.5 and 0.82% by volume of water was used in the mix of M100 and M100-PET, respectively. In each sample three specimens were cast. Specimens from the

Mixes	Cube (mm)	Cylindrical (mm)	Prism (mm)	Total mixture
M0	150 × 150 × 150	100Ø × 200	500 × 125 × 125	100% NCA
M100	150 × 150 × 150	100Ø × 200	500 × 125 × 125	100% RCA
M100-PET		100Ø × 200	500 × 125 × 125	100% RCA + PET fibers
Test parameter	Compressive strength	Splitting tensile strength	Flexural strength	

Table 2. Specimen used for evaluating the mechanical properties of concrete.

mold were removed after 24 h of casting and were kept in a water tank for 28 days curing before testing.

2.3. Selection and description of RC frame

In this study, a typical full scale residential building with floor to floor height of 3.0 m and the beam with effective span of 3.6 m were considered. The full scale RC frame are then scaled down to 1/3rd size for experimental investigation. The detailing of the frame is shown in **Figure 3**. The frame has been designed following the standard code of practice [24, 25]. A cross section of 100 × 100 mm and 135 × 100 mm for column and beam elements, respectively, was considered. High yield strength deformed (HYSD) bars of 8 mm diameter (Fe 500) were used as main bars in both column and beam. Following the code provision [24] a lateral tie of 6 mm diameter mild steel bars (Fe 250) at 25 mm c/c spacing was used in the special confinement zone of the column, while the remaining part was increased to 50 mm c/c. Similarly, the shear reinforcement in beam was of 6 mm diameter bar having spacing of 25 mm c/c near the beam-column joint for a length of 225 mm and a spacing of 40 mm c/c in the remaining part. The yield stress (MPa) and ultimate stress (MPa) for HYSD bars tested as per code provisions [18, 26] were found out to be 530 and 620 MPa, while the same for Fe 250 were 285 and 450 MPa, respectively. The detailed description of the specimens is given in **Table 3**.

2.4. Casting of RC frames

Three RC frames were casted. The specimens were designated as specimen S1, S2 and S3. Specimen S1 corresponding to mix M0 was treated as the reference specimen. Keeping the

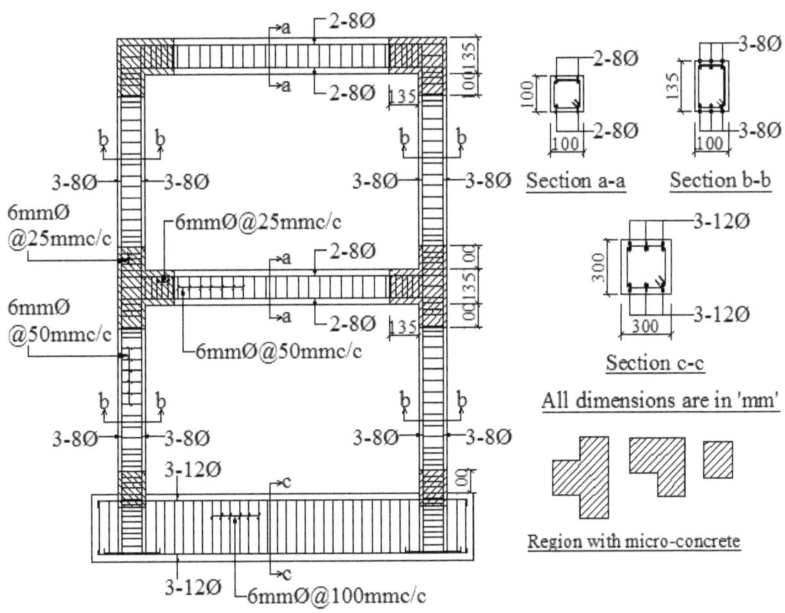

Figure 3. Reinforcement detailing of RC frame.

Beam			Column		
Span (mm)	Section (mm)	Longitudinal reinforcement	Length (mm)	Section (mm)	Longitudinal reinforcement
1200	100 × 135	2-8Ø - top2-8Ø - bottom	1000	100 × 100	3-8Ø – top bar3-8Ø – bottom bar

Table 3. Description of RC frame.

(a) (b) (c)

Figure 4. Casting of RC frame (a) M0 for S1 and M100 for S2 (b) M100 and PFRC at joint region for S3 (c) PFRC mix.

geometric dimensions, grade of concrete and steel, the amount and detailing of reinforcing bars constant as that of specimen S1 the other specimens S2 and S3 were casted using 100% of RCA concrete. However, the joint region of specimen S3 was cast with PFRC mix as per guideline [17]. **Figure 4** shows the casting of the frame.

2.5. Test set-up and loading sequence

Schematic diagram of the test set-up and the actual testing arrangement is shown in **Figure 5**. The test set-up consists of a loading frame of capacity 400 kN and a hydraulic jack of 100 kN capacity. The load was applied manually through the hydraulic jack mounted on the side of the frame. The foundation of the RC frame were held tightly in position with the help of hydraulic jacks and also with the help of mild steel plates which helped in clamping the foundation to the ground. Holes were punched into the mild steel plates which were fitted into the bolts (firmly established in the ground by concreting) with the help of nuts of appropriate size. Two dial gauges of 100 mm measuring range were used to measure the lateral displacement corresponding to the applied load. As shown in **Figure 5** the frame specimens were subjected to monotonic lateral loading at the side of the top beam. A hydraulic jack of capacity 100 kN was attached to the side of the loading frame and was used for applying the necessary lateral load to the specimens. A dial gauge was attached to the loading frame opposite to the hydraulic jack to measure the displacement undergone by the specimens

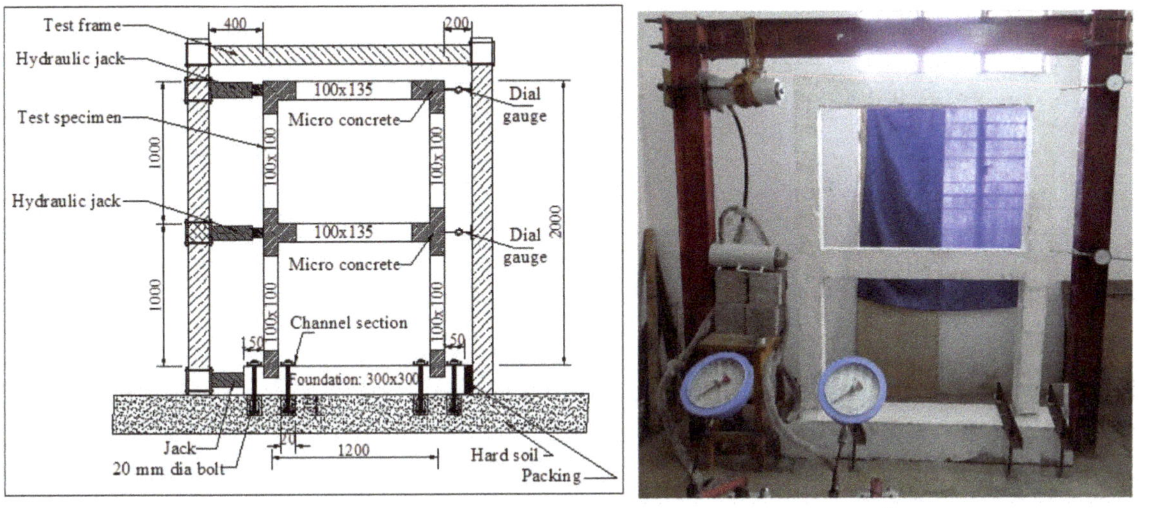

(a) (b)

Figure 5. Testing of RC frame (a) test set-up (b) actual testing arrangement.

corresponding to a particular load. A maximum displacement of 100 mm was applied in all the specimens.

3. Mechanical behavior of concrete due to addition of PET fiber

3.1. Results and discussion

At a *w/c* of 0.5, the slump measured as per IS 1199 [22] for M0 was 60 mm, while for M-100 was 40 mm, respectively. The high absorption of free water from the mixture during mixing process causes high water demand of the mix with increasing RCA contents. This shows that RCA concrete resulted in a significant effect on the workability of concrete. The addition of PET fiber in a M100 mix further reduced the workability of concrete about 58%. For a better comparison of the test results of concrete, it is essential for all concrete mixes to secure the same workability. Therefore, superplasticizer dosage of 0.5 and 0.85% by volume of water was used in M100 and M100-PET concrete, respectively.

The concrete specimen as presented in **Table 2** was used for evaluating the mechanical properties. Test parameters included are compressive strength, splitting tensile strength and flexural strength. Concrete cubes and prismatic specimens were tested for compressive strength and flexural strength [27], while a cylindrical specimen was tested for splitting tensile strength [28]. The test results presented in **Table 4** shows that all test parameters decreases in the range of 25–40% with mix of RCA concrete. However, the presence of 0.5% PET fibers in RCA concrete enhanced the compressive strength about 19%. The additions of PET fibers also enable a greater capability of resisting more tensile stress higher than the control specimens, PET fibers can restrain the crack propagation and traverse across the cracks to transfer internal force, as a result, the flexural toughness of the fibers concrete is improved which is the positive contribution of PFRC in an RC frame under seismic excitation.

Concrete Mix	Compressive strength (MPa)	Tensile strength (Mpa)	Flexural strength (MPa)	Total mixture
M0	39.25	4.55	3.58	100% NCA
M100	29.33	2.83	2.18	100%RCA
M100-PET	31.45	5.53	4.10	100% RCA + PET fibers

Table 4. Results of compressive, tensile and flexural strength of concrete specimens.

4. Behavior of RC frame due to addition of PET fiber: Results and discussion

4.1. Failure modes and load carrying capacity

Figures 6–8 presented the failure pattern of the specimens. It is observed that the initial crack formation in all the specimens mainly developed at the joint interface of beam and column. With further application of lateral load, a number of cracks were observed at the joint region and also the initial cracks widened more and more. The faster growth of cracks for specimen S2 which spreads away from the joint region reveals a brittle mode of failure as compared to specimen S1 and S3. Also as observed during experimentation specimen S2 lost its resistance after cracking whereas specimen S3 could sustain a portion of its resistance after cracking and able to resist more loads and produced a better ductility in the frame. The presence of a PFRC at the joint region for frame S3 delayed the crack formations as compared to specimen S2 and S1. Thus, **Figure 8** showed the minimum number of cracks at the joint region and wider cracks occur at the beam portion which is the desirable failure mode of frame structures.

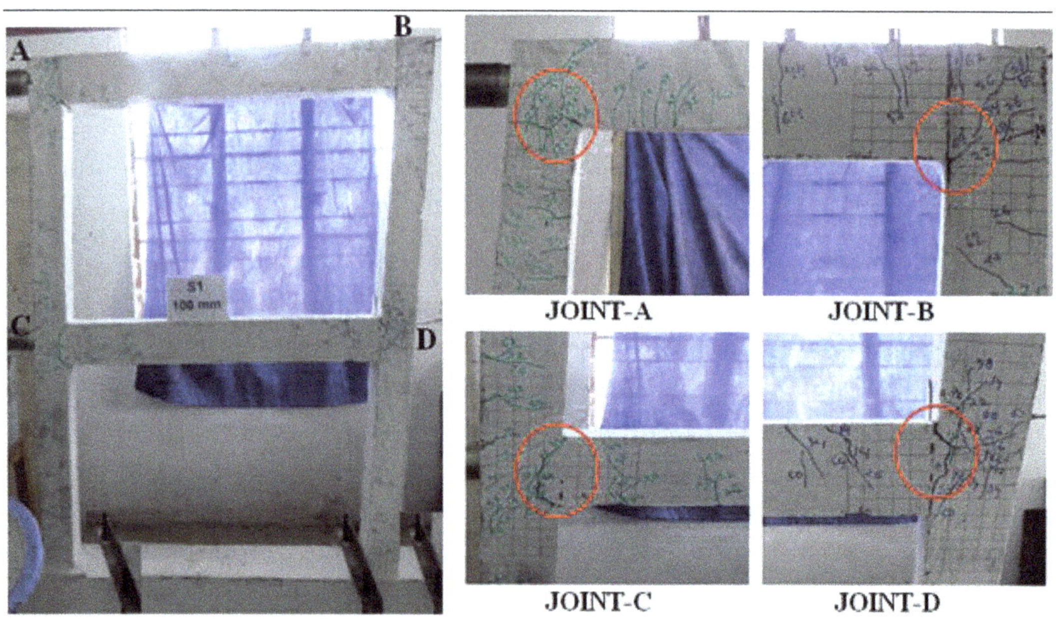

Figure 6. Failure modes of specimen S1.

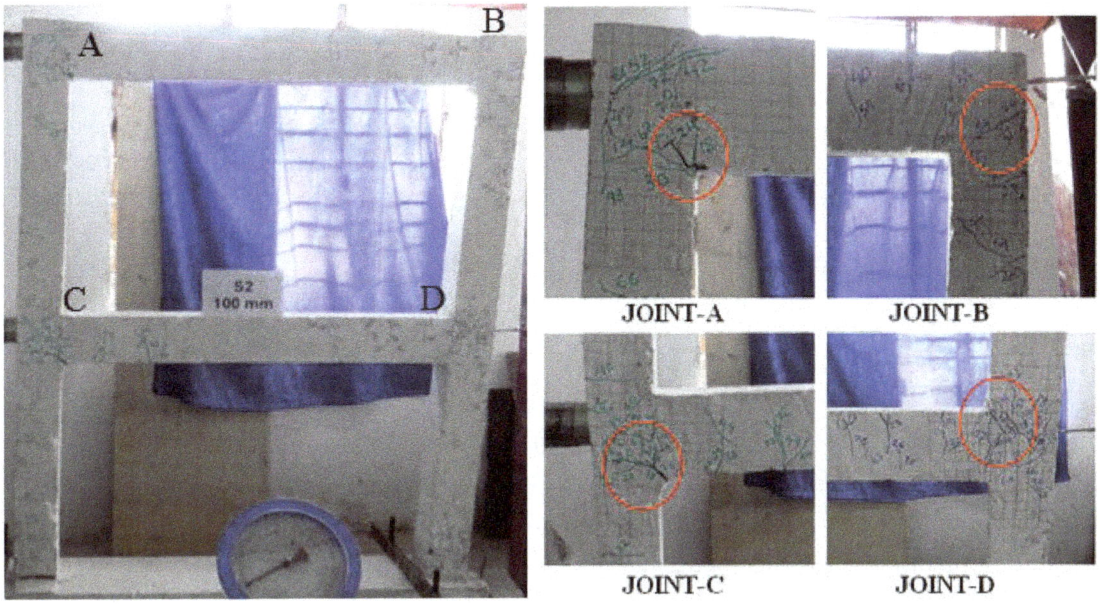

Figure 7. Failure modes of specimen S2.

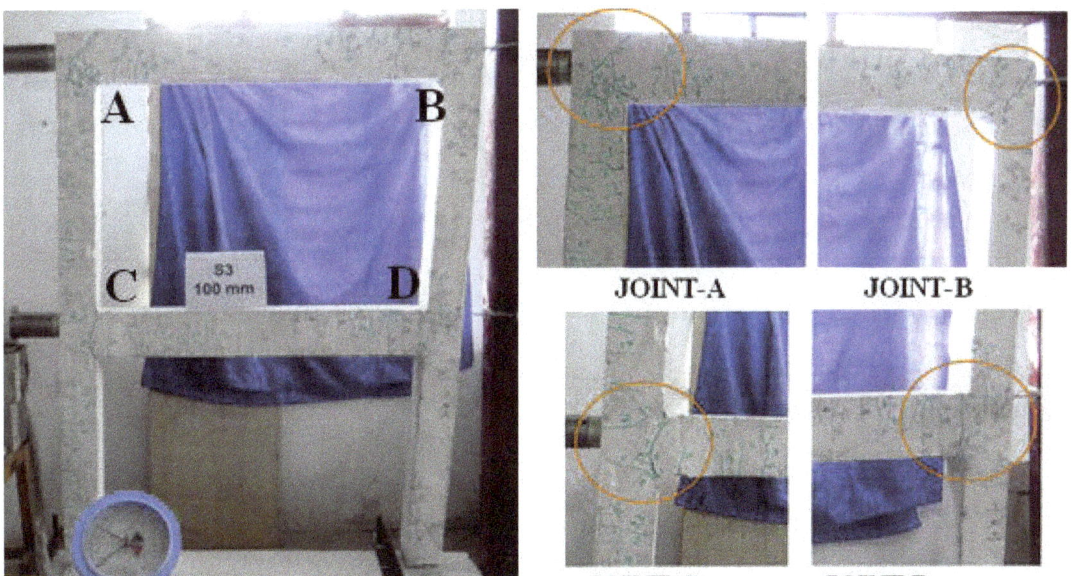

Figure 8. Failure modes of specimen S3.

The load capacity were noted for displacements starting from 1 up to 100 mm at an increment of 1 mm and the crack formations were noted for each set of displacements. **Figure 9** shows the load versus displacement curve. It can be observed that with increased in displacement, specimen S1 presented the highest load carrying capacity followed by specimen S3 and least with specimen S2. The higher load carrying capacity presented by specimen S3 at each displacement level in comparison to specimen S2 show an excellent contribution of PFRC in a RCA concrete frame which is well comparable to specimen S1. As presented in **Table 5** the addition of PFRC in an RCA concrete frame cause a strength reduction of only 6% in comparison to specimen S2 of 20%.

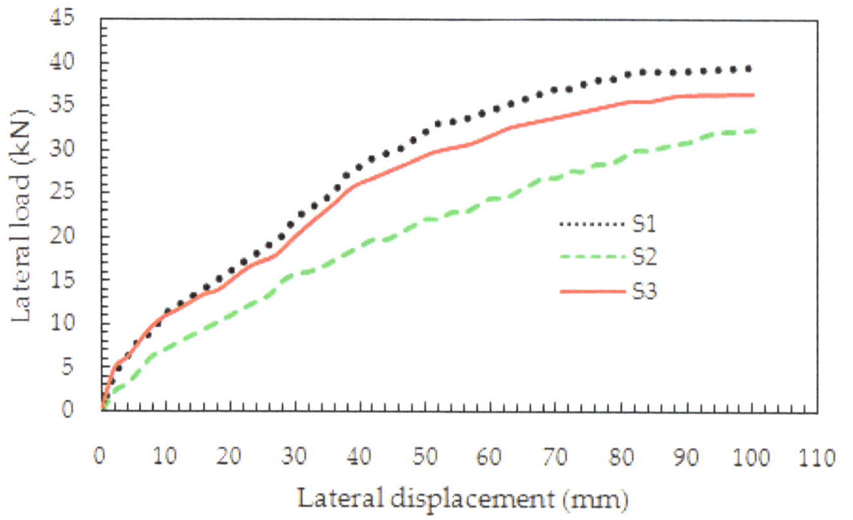

Figure 9. Load versus displacement curve of the frame.

Sl. no.	Test parameters	S1	S2	S3
1	Average load capacity, kN	39.50	31.50	36.50
	Reduction with respect to S1 (%)	–	20.25	6.00
2	Energy dissipation (kN-mm)	1304.00	825.00	1113.54
	Reduction with respect to S1 (%)	–	36.73	14.61
3	Ductility	2.00	1.26	1.65
	Reduction with respect to S1 (%)	–	37.00	17.50

Table 5. Capacity comparisons of RC frame specimens.

4.2. Evaluation of stiffness of the frames

Stiffness is an indicator of the response of a specimen and extent of strength degradation during loading. It is calculated as the slope of the line joining the peak capacity at a given displacement. The slope of this straight line is the stiffness of the specimens corresponding to that particular displacement amplitude according to Naeim and Kelly [29]. The stiffness is given in Eq. (1).

$$K = \frac{F - f}{D - d} \tag{1}$$

where, F is the maximum load of a particular specimen in the positive cycle, f is the maximum load of a particular specimen in the negative cycle, D is the displacement corresponding to the maximum load of a particular specimen in the positive cycle and d is the displacement corresponding to the maximum load of a particular specimen in the negative cycle. The present study adopted a monotonic increasing load and hence, the value of f and d are taken as zero. The drift angle is defined as the ratio of beam tip displacement to the length of the beam measured from the joint to the position of the dial gauge. The drift obtained by horizontal displacement of the beam ends are equivalent to the inter storey drift angle of a frame structure subjected to lateral loads. Drift ratio is calculated as given in Eq. (2).

$$\text{Drift ratio } (\%) = \frac{\Delta}{L} x100 \tag{2}$$

where, Δ and L are the applied displacement and the storey height of the frame measured from the top level of foundation to the top beam of the frame.

The performance of the specimen S3 due to the presence of micro-concrete at the joint region may be evaluated by comparing stiffness versus displacement with those of specimens S1 and S2. These plots are shown in **Figure 10**. Comparing these plots, similar degradations trends could be observed in all the specimens. Stiffness of the frames was gradually reduced during loading. This was occurred due to bond failure, minute cracks formed in the frame. Stiffness was getting reduced higher for specimen S2. The initial stiffness of specimen S1, S2 and S3 are 2.0, 1.1 and 1.7 kN/mm, respectively. Thus, the presence of PFRC at the joint region led enhanced the initial stiffness for frame S3 as compared to specimen S2. Further, it is also observed that the degradation in stiffness for specimen S3 is little slow with increase in lateral movement as compared to the specimen S2 which is well comparable to the specimen S1. This behavior may be attributed to the ductile properties imparted by the PET fibers at the joint region which bridge the initial crack of the joint region. The lower degradation is a desirable property in earthquake like situations. It was observed during the past earthquake that most of the RC structures failed due to sudden loss of stiffness with increasing lateral movement. Therefore, from these comparisons it can be concluded that the presence of PFRC at the joint region of the RC frame made with RCA improved the performance during seismic action.

4.3. Evaluation of energy dissipation capacity

The ability of a structural member to resist the fracture when subjected to static or to dynamic or impact loads depends to a large extent on its capacity to dissipate its energy. It was reported that the energy absorbed by a column before failure is correlated to the ductility of the column [30].

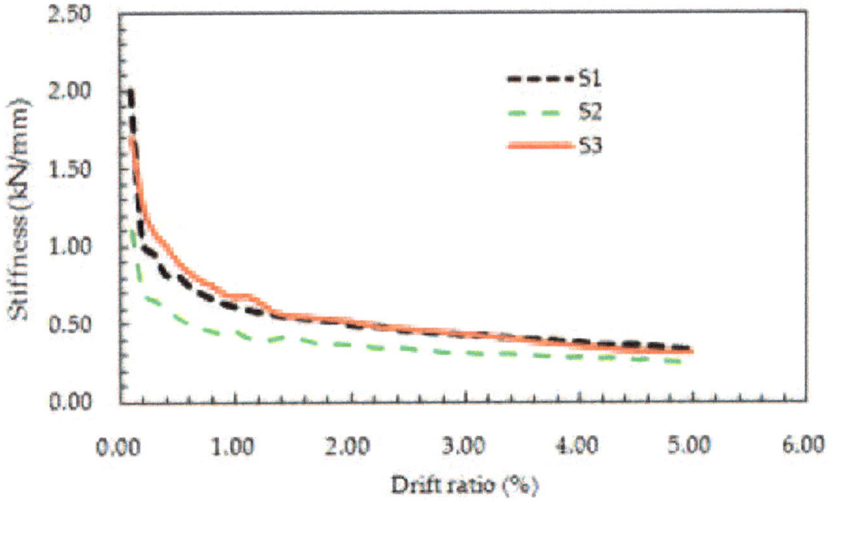

Figure 10. Stiffness degradation of the specimens.

This energy can be computed based on the area under the load versus displacement curve presented in **Figure 9** as implemented in several previous studies [31, 32]. The computed energies are tabulated in **Table 5**. RCA specimens (S2) without PFRC at the joint region presented the lowest energy dissipation capacity. The increase in energy dissipation of specimen S3 showed PFRC has a tremendous potential use in the joint region of the RC frame. The inclusion of PFRC at the joint region of an RC frame made of RCA concrete lead to an increase of energy dissipation to about 1.3 times (**Table 5**) with respect to S2. The increase in stiffness due to presence of PFRC at the joint region attracted more load corresponding to any drift angle for specimens S3, which prevent the initial crack propagations. Thus, the total area enclosed by the plot of load versus displacement was more for specimen S3 as compared to S2. This was perhaps the reason for improvement in energy dissipation. It may be noted that the energy dissipating capacity of specimen S3 as presented in **Table 5** is well comparable to specimen S1 (−15%) while for specimen S2 is 37% which demonstrated the benefit of using a recycled aggregate concrete in an RC frame strengthened with PFRC at joint region.

4.4. Evaluation of ductility of the frames

Ductility is basically the ability of a structure to accommodate deformations well beyond the elastic limit. It is the capacity to dissipate energy in hysteretic loops and to sustain large deformations. As the loads versus displacement curves for tested specimens do not have a distinct yield point, ductility capacity was determined using an idealized approximation procedure proposed by Shannag et al. [33] which has been explained in **Figure 11**. As shown in the figure, the yield displacement is calculated as the point of intersection between two straight lines drawn in the envelope curve. The first line was obtained by extending the line joining the origin and 50% of ultimate load capacity point of the curve, while the second line was obtained by drawing a horizontal line through the 80% of ultimate load capacity point. In the figure, δy represents the yield displacement. Horizontal lines drawn through the 80% of ultimate load capacity point intersect the curve at far end at points x. The abscissa of this point denoted by δu was taken as maximum displacement. The displacement ductility (μ) was calculated as the ratio of maximum displacement (δu) to the yield displacement (δy). The calculated values listed in **Table 5** clearly show a higher ductility for specimen S1. However, the presence of PFRC in

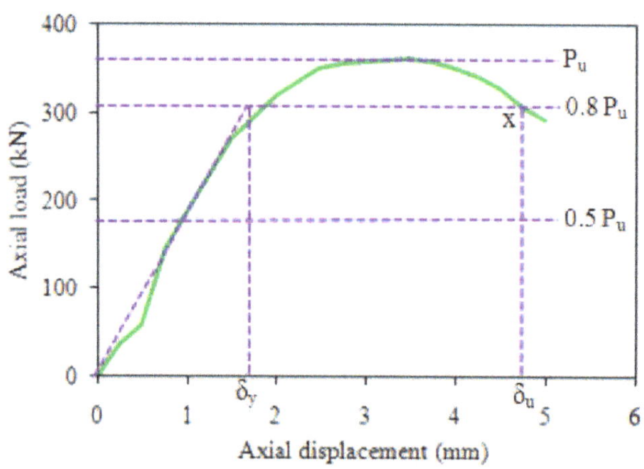

Figure 11. Procedures for ductility calculation.

specimen S3 show improved ductility value than that of specimen S2. The presence of PFRC at joint region postponed the crack development toward the beam and column region and finally it was led to higher ductility, which is well comparable to the control specimen S1 (−18%).

5. Conclusions

In the conclusion, by comparing the results obtained for the two concrete mixtures, it can be noted that, when the low grade aggregate i.e. the recycled coarse aggregate is used instead of natural coarse aggregate for concrete production, about 25–40% of mechanical strength (compressive, tensile and flexural strength) decreases. However, the presence of 0.5% PET fibers enhanced the compressive strength of RCA about 19%. The additions of PET fibers also enable RCA concrete a greater capability of resisting more tensile stress. The flexural toughness of the fibers concrete also improved which is the positive contribution of PFRC in an RC frame under seismic excitation.

In addition, on the basis of the results obtained through a monotonic loading test of an RC frame made of either RCA or NCA evaluated by means of parameter such as cracking patterns, stiffness degradation, energy dissipated and ductility. The faster growth of cracks which spreads away from the joint region reveals a brittle mode of failure for a frame made of RCA concrete. In the other hand when the joint region of an RCA frame is incorporated with PFRC the frame showed adequate structural behavior and exhibited a higher energy dissipation capacity and ductility and the lower degradation of stiffness, which is a desirable property in earthquake like situations. Therefore, it can be concluded that the, strengthening of joint region using PET fibers is found to have a significant contribution in improving the seismic performance of RCA frame structures.

Acknowledgements

Authors are thankful to the Structural Engineering Laboratory staffs of Department of Civil Engineering, NIT Meghalaya for their help in cutting the fibers and testing of specimens.

Conflict of interest

No conflict of interest to declare.

Author details

Comingstarful Marthong

Address all correspondence to: commarthong@nitm.ac.in

National Institute of Technology Meghalaya, Shillong, India

References

[1] Limbachiya MC, Koulouris A, Roberts JJ, Fried AN. Performance of recycled aggregate concrete. In: RILEM International Symposium on Environment Conscious Materials and System for Sustainable Development; 2004. pp. 127-136

[2] Rao MC, Bhattacharyya SK, Barai SV. Influence of field recycled coarse aggregate on properties of concrete. Material Structures. 2011;**44**:205-220. DOI: 10.1617/s11527-010-9620-x

[3] Xiao JZ, Li W, Fan Y, Huang X. An overview of study on recycled aggregate concrete in China (1996-2011). Construction and Building Materials. 2012;**31**:364-383. DOI: 10.1016/j.conbuildmat.2011.12.074

[4] Padmini AK, Ramamurthy K, Mathews MS. Influence of parent concrete on the properties of recycled aggregate concrete. Construction and Building Materials. 2009;**23**:829-836. DOI: 10.1016/j.conbuildmat.2008.03.006

[5] Tabsh SW, Abdelfatah AS. Influence of recycled concrete aggregates on strength properties of concrete. Construction and Building Materials. 2009;**23**:1163-1167. DOI: 10.1016/j.conbuildmat.2008.06.007

[6] Han BC, Yun HD, Chung SY. Shear capacity of reinforced concrete beams made with recycled aggregate. ACI Special Publication. 2001;**200**:503-516. DOI: 10.1002/9781118062173.ch52

[7] Chao L, Guoliang B, Letian W, Zonggang Q. Experimental study on the compression behavior of recycled concrete columns. In: Proceedings of the International RILEM Conference on the Waste Engineering and Management; 2010. pp. 614-621

[8] Corinaldesi V, Letelier V, Moriconi G. Behavior of beam-column joints made of recycled-aggregate concrete under cyclic loading. Construction and Building Materials. 2011;**2**:1877-1882. DOI: 10.1016/j.conbuildmat.2010.11.072

[9] Xiao J, Sun Y, Falkner H. Seismic performance of frame structures with recycled aggregate concrete. Engineering Structures. 2006;**28**:1-8. DOI: 10.1016/j.engstruct.2005.06.019

[10] Oinam RM, Sahoo DR, Sindhu R. Cyclic response of non-ductile RC frame with steel fibers at beam-column joints and plastic hinge regions. Journal of Earthquake Engineering. 2014;**18**(6):908-928. DOI: https://doi.org/10.1080/13632469.2014.916239

[11] Foti D. Use of recycled waste pet bottle fibers for the reinforcement of concrete. Composite Structures. 2013;**96**:396-404. DOI: 10.1016/j.compstruct.2012.09.019

[12] Marthong C, Sarma DK. Influence of PET fiber geometry on the mechanical properties of concrete: An experimental investigation. European Journal of Environmental and Civil Engineering. 2016;**20**(7):771-784. DOI: 10.1080/19648189.2015.1072112

[13] Fraternali F, Farina I, Polzone C, Pagliuca E, Feo L. On the use of R-PET strips for the reinforcement of cement mortars. Composites. Part B, Engineering. 2013;**46**:207-210. DOI: 10.1016/j.compositesb.2012.09.070

[14] Marthong C. Effects of PET fiber arrangement and dimensions on mechanical properties of concrete. The IES Journal Part A: Civil & Structural Engineering. 2015;**8**(2):111-120. DOI: 10.1080/19373260.2015.1014304

[15] Marthong C, Marthong S. An experimental study on the effect of PET fibers on the behavior of exterior RC beam-column connection subjected to reversed cyclic loading. Structure. 2016;**5**:175-185. DOI: 10.1016/j.istruc.2015.11.003

[16] Jo B, Park S, Park J. Mechanical properties of polymer concrete made with recycled PET and recycled concrete aggregates. Construction and Building Materials. 2008;**22**:2281-2291. DOI: 10.1016/j.conbuildmat.2007.10.009

[17] ACI Committee 318-08. Building Code Requirements for Structural Concrete (ACI 318-08) and Commentary (ACI 318R-08). Farmington Hills, MI: American Concrete Institute; 2008

[18] IS 432 (1). Specification for Mild Steel and Medium Tensile Steel Bars and Hard-Drawn Steel Wire for Concrete Reinforcement: Part I. Mild Steel and Medium Tensile Steel Bars. New Delhi: Bureau of Indian Standard; 1982

[19] IS 2386a. Methods of Test for Aggregates for Concrete—Part 1: Particle Size and Shape. New Delhi: Bureau of Indian Standard; 1963

[20] IS 2386b. Methods of Test for Aggregates for Concrete—Part 3: Specific Gravity, Density, Voidsr, Absorption and Bulking. New Delhi: Bureau of Indian Standard; 1963

[21] ACI Committee 544. State-Of-The-Art Report on Fiber Reinforced Concrete, ACI 544.1R-96, Re-Approved 2002. Farmington Hills, Michigan: American Concrete Institute; 1996

[22] IS 1199. Methods of Sampling and Analysis of Concrete. New Delhi: Bureau of Indian Standard; 1959

[23] IS 10262. Guidelines for Concrete Mix Design Proportioning (CED 2: Cement and Concrete). New Delhi: Bureau of Indian Standard; 2009

[24] IS 13920. Ductile Detailing of Reinforced Concrete Structures Subjected to Seismic Forces—Code of Practice. New Delhi: Bureau of Indian Standard; 1993

[25] IS 456. Plain and Reinforced Concrete—Code of Practice. New Delhi: Bureau of Indian Standard; 2000

[26] IS 1608. Mechanical Testing of Metals—Tensile Testing. New Delhi: Bureau of Indian Standard; 1995

[27] IS 519. Method of Tests for Strength of Concrete. New Delhi: Bureau of Indian Standard; 1959

[28] IS 5816. Method of Test Splitting Tensile Strength. New Delhi: Bureau of Indian Standard; 1999

[29] Naeim F, Kelly JM. Design of Seismic Isolated Structures from Theory to Practice. Canada: John Willey & Sons Inc; 1999

[30] Hadi MRNS. Behavior of FRP wrapped normal strength concrete column under eccentric loading. Composite Structures. 2006;**72**:503-511. DOI: 10.1016/j.compstruct.2005.01.018

[31] Hadi MNS. Behavior of FRP strengthened concrete columns under eccentric compression loading. Composite Structures. 2007;**77**(1):92-96. DOI: 10.1016/j.compstruct.2005.06.007

[32] Shannag MJ, Ziyyad TB. Flexural response of ferrocement with fibrous cementitious matrices. Construction and Building Materials. 2007;**21**:1198-1205. DOI: 10.1016/j.conbuildmat.2006.06.021

[33] Shannag MJ, Alhassan MA. Seismic upgrade of interior beam-column sub-assemblages with high performance fiber reinforced concrete jackets. ACI Structural Journal. 2005;**102**(1):131-138

3

Automated Model Construction for Seismic Disaster Assessment of Pipeline Network in Wide Urban Area

Hideyuki O-tani, Muneo Hori and Lalith Wijerathne

Abstract

This chapter is on rational seismic damage assessment over a wide area through the development of a module for the automated model construction of a pipeline network of lifelines using geographic information system (GIS) data. The module is assigned a functionality that can generate a simple one-dimensional line model and a two-dimensional surface model with high fidelity for the pipe shape. The source code of the module is written in object-oriented programming in order to make it easier to extend it to generate other analysis models. The module was applied to the actual GIS, and the shape of the output model was verified. Numerical analysis was performed on the output of the module, and it showed that the automatically constructed model is mechanically valid and can be used for seismic response analysis.

Keywords: automated model construction, seismic disaster assessment, pipeline network, GIS data, object-oriented programming

1. Introduction

The importance of securing the high seismic performance of the pipeline network of lifelines [1] such as water and sewerage, electricity, gas, and communication needs no emphasis. The first step toward meeting this demand is the evaluation of seismic damage. Sheer size and the complex geometry of pipeline networks make it a challenging task to evaluate seismic damage assessment in urban areas. For instance, in a populated area spanning a few tens of square kilometers, the pipeline network may be several hundred kilometers long. Seismic damage at any one point in the pipeline network could affect the services in a large supply area. Therefore, the pipeline has to be assessed at each point and carrying out this assessment manually requires an enormous amount of labor.

Several types of pipes are used in lifelines and their design life is usually 50 years or more. Therefore, the pipeline network in any place of seismic damage assessment is composed of dissimilar pipes. The structure of the ground at a location, which influences the amplification of seismic motion, is far from uniform when viewed in the scale of the length of the entire city. This nonuniformity further complicates the point-by-point seismic damage assessment of lifelines.

It, therefore, becomes necessary to assess seismic damage over a wide area by relying on a *fragility curve* [2–5] the empirical relation between the damage ratio of the pipeline network and the strong ground motion index. The curve can be improved by analyzing the effect of the localized nature of the structure and the site, by utilizing actual damage data. However, such seismic damage assessment is clearly different from the seismic-resistant design [6–8] of newly laid pipes in which nonlinear finite element analysis was used by considering the coupled effect of the soil and the structure.

The seismic damage to the pipeline network of lifelines can be assessed in a fairly rational manner and with high accuracy using the seismic-resistant design method. While the design is made only for a part of the pipeline network, assessment has to be made for the entire pipeline network of the city that is being assessed for seismic damage. It is a challenging task to construct analysis models with equivalent fidelity of those used in seismic-resistant designs, for a large pipe network.

An analysis model is essential in order to achieve a fairly rational seismic damage assessment. It does not have to be a unique or exclusive model. In the seismic damage assessment of lifelines, for instance, it would be more convenient to choose from a variety of analysis models ranging from a simple beam model to a sophisticated shell element model, which may be used according to the accuracy required for evaluation. With this objective in mind, we have aimed at developing a module that will convert the data of the pipeline network of lifelines into analysis models for the assessment of various types of seismic damage. As explained in Section 2, data on the pipeline network are available from some GIS records. The basic feature of this module is its high degree of robustness, that is, the possibility of constructing an inappropriate model is minimal.

The contents of this chapter can be summarized as follows. The module for converting data into an analysis model is described in Section 2, together with the data conversion methodology and a detailed description of the data structure of the module. Verification of the module is carried out in Section 3. In Section 4, as an illustrative example, seismic disaster assessment is conducted for a pipeline network and its connections. Section 5 gives the conclusion about the developed module.

2. Model construction module

In this section, we describe the module that converts the data recorded in the GIS into an analysis model of the pipeline network. As already mentioned, there are several options available for analysis models with varying degrees of complexity. In order to simplify the software maintenance of the module, the data recorded in the GIS are first converted to "model data." Model data refer to the essential information for the construction of the analysis model extracted from GIS, and the conversion into a specific analysis model is done by integrating the model data. The model construction module is composed of two submodules: the "GIS data conversion submodule" and the "model data conversion submodule" (**Figure 1**). Dividing the module into two submodules through the model data prevents the modifications made on one submodule from affecting the other submodule.

2.1. Source data

Six types of lifelines are considered in this study: water supply, sewage, gas, electricity, communication, and road networks. GIS created under a standard protocol is available for these six types. The GIS protocol is systematic but complex. Basically, data are recorded in a binary format in which records consisting of a fixed length header part and an undefined length content part are arranged in order; the byte order of the data constituting the header part and the content part is different for each data type. The main information of each record is compiled in **Figure 2**.

The contents of the record are divided into a shape record, in which shapes such as polylines and polygons are recorded; and an attribute record that records material properties and information such as pipeline radius, whether it refers to the inner or outer diameter, pipeline

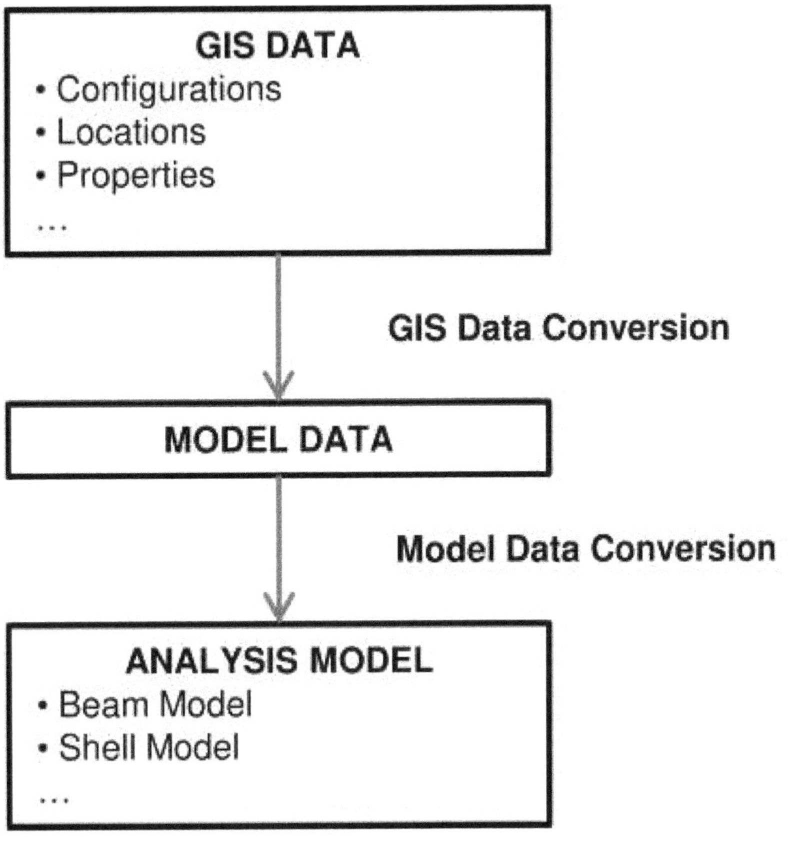

Figure 1. Schematic view of data conversion.

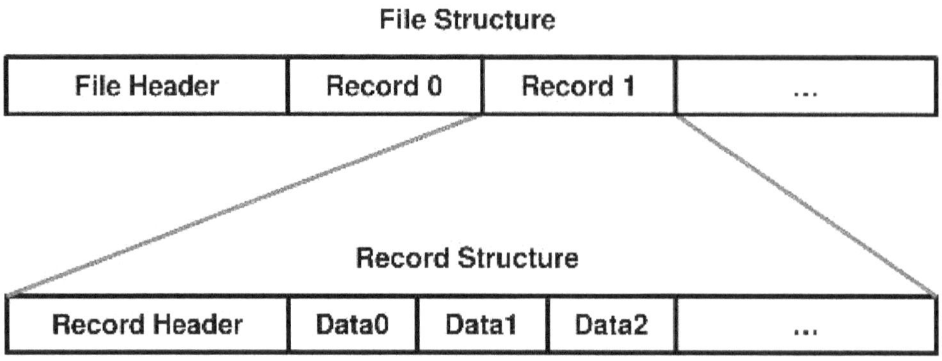

a)

Attribute	Radius
	Thickness
	Construction Date
Polyline	Coordinate set
Polygon	Coordinate set
Coordinate set	(x_0, y_0, z_0)
	(x_1, y_2, z_3)
	...

b)

Figure 2. Overview of GIS consisting binary records. (a) Typical file structure in GIS; (b) example of record contents.

thickness, etc. The record representing the shape of the pipeline network is a polyline composed of a sequence of dots. Coordinates are assigned to each point, and successive coordinates indicate the connections in the polyline. Thus, the shape of a network with branches is configured as a set of polylines. These two types of records are necessary to construct an analysis model of the pipeline network.

It must be noted here that the GIS used in this study is based on a CAD protocol and the connection of successive points in the polyline is the only information about the connectivity of the network. Even if the result of visualizing the analytical model of the pipeline appears to be satisfactory, constructing an analysis model merely by connecting these points will not be sufficient. More sophisticated data conversion ensuring that the points are mechanically connected is needed. Visualization of the connection of points is a necessary but an insufficient condition. It is believed that analysis is necessary in order to verify whether the points are connected in a manner so as to function as a pipeline network.

2.2. GIS data conversion submodule

The GIS data conversion submodule converts the GIS data into model data that can be used to construct various analysis models of the pipeline network. The model data are a set

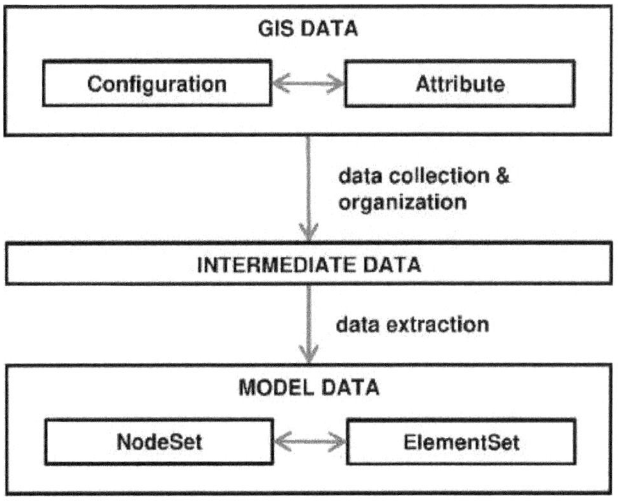

Figure 3. Main functionalities of GIS data conversion submodule.

of shape records and attribute records of the pipeline network distributed and recorded in the GIS into a single dataset. It provides a clearer description of the connectivity of pipelines than the original GIS, in order to facilitate its use as a common information source for constructing analysis models. The main functions of this submodule are to read the entries in the GIS record, extract the data necessary for the construction of the analysis model from the record, and register the extracted data as model data. These functions are summarized in **Figure 3**.

The GIS for all the six lifelines have a common protocol and share important features of the records, but the GIS of each lifeline contains its own specifications in the record details. Object-oriented programming (OOP) [1, 9] is therefore effective in creating the code of the GIS data conversion submodule. Inheritance, polymorphism, and template functions are the advantages of using OOP to handle the key functions shown in **Figure 3**. In the hierarchical structure shown in **Figure 4**, virtual classes are created as common items and concrete classes for each of the actual lifelines are created on the basis of the virtual classes. Thus, the model data conversion submodules for GIS of all the lifelines have the same appearance.

The only items common to the GIS records for all the six kinds of lifelines are the records of the polylines representing the shapes of the networks; the other items being different for each lifeline. Therefore, a common container object is created for the polyline records and separate container objects for other records. Further, an object to associate with the two kinds of objects is created in order to build an analysis model of the network using the two kinds of objects.

The following four objects are created for the container of the polyline records: NodeSet, ElementSet, Network and NetworkSet. NodeSet is a container object for the point coordinates, and the connectivity of the points is contained in ElemenSet. Network is a pair of NodeSet and ElementSet, and NetworkSet is a set of Network.

- NodeSet is a set of Node, and an entity of Node corresponds to a node in the network. Each Node entity is constructed from the polyline records, and put into a NodeSet entity without duplication. NodeSet also contains the IDs for Node.

- ElementSet is a set of Element, and an entity of Element corresponds to a record of a polyline. Each Element entity is constructed from a polyline record as a series of Node IDs, and put into an ElementSet entity without duplication. Element also contains an entity of Attribute, which is a container object associated with the property of a pipeline.
- Network is a pair of NodeSet and ElementSet, and an entity of Network is constructed from one fragmented file in GIS. Network corresponds to a part of the target pipeline network.
- NetworkSet is a set of Network, and an entity of NetworkSet corresponds to an entire pipeline network.

Figure 5 shows the construction flow of a NetworkSet entity; Segment is a straight line specified by the neighboring point coordinates in a polyline record.

2.3. Model data conversion submodule

The role of the model data conversion submodule is to construct an analysis model from model data and create input files for the program that performs the analysis using this analysis model. In this study, a model data conversion submodule that creates 1D line models and 2D surface models is presented.

2.3.1. 1D line model

1D line model is an assembly of beam elements constructed from an entity of NetworkSet, which is a model data that corresponds to an entire pipeline network. Beam elements are created by dividing an entity of Segment, which corresponds to two neighboring points in a polyline record into linear fragments of equal size. Consequently, the process of creating a 1D line model is, in contrast to the conversion process of GIS data conversion submodule shown in **Figure 5**, a process that decomposes an entity of NeworkSet into beam elements. **Figure 6** shows the decomposition process.

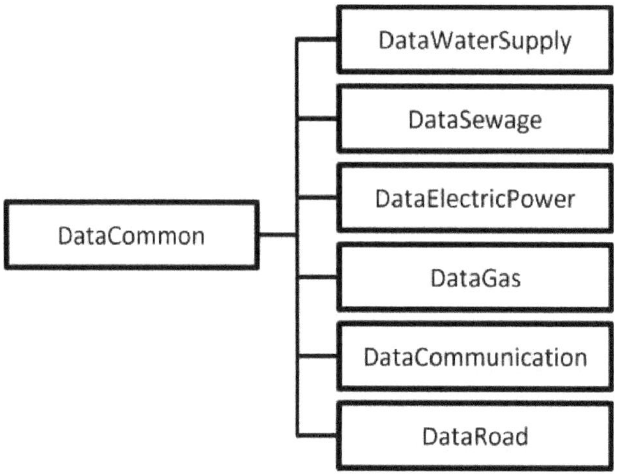

Figure 4. Data hierarchy of GIS.

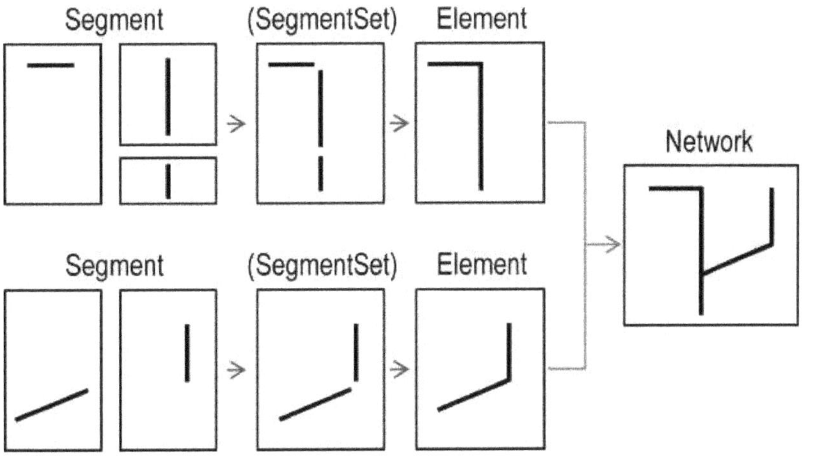

Figure 5. Flow of converting GIS to NetworkSet.

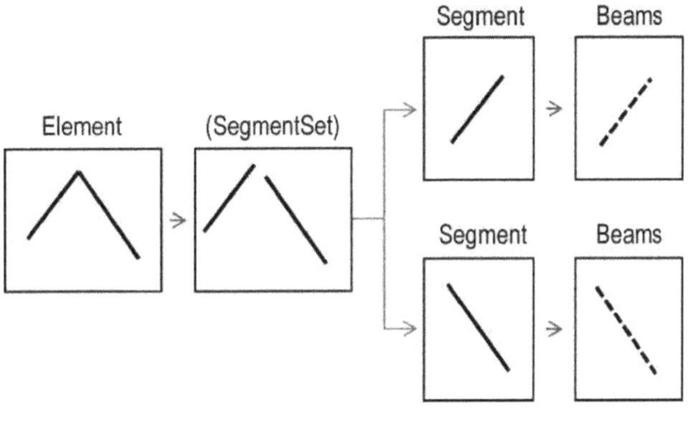

Figure 6. Decomposition flow of model data to beam elements.

It is possible to increase the transparency of the conversion process by using the same object as the model data as the container object for the 1D line model. However, to achieve high transparency, it is important that at the same time the container object has a simple structure. Therefore, the following three structures are considered as candidates for the structure of the container object of the 1D line model.

- A Beam object is created as the container object of the beam elements. Furthermore, a PipeSegment object is created as the container object to assemble the beams created from the same Segment entity. PipeSegment corresponds to Segment of the model data. In this case, the depth of the structure of the container object of the 1D line model is 5 (**Figure 7a**)).
- Without holding information about PipeSegment, as necessary, Segment is divided by creating an entity of PipeSegment. In this case, the depth of the structure of the container object of the 1D line model is 3 (**Figure 7b**)).
- The entire 1D line model is adapted to an entity of Network. In this case, the depth of the structure of the container object is 2 (**Figure 7c**)).

The first candidate features a highly transparent conversion process. Its structure is the structure of model data supplemented with the container object of the beam elements. However,

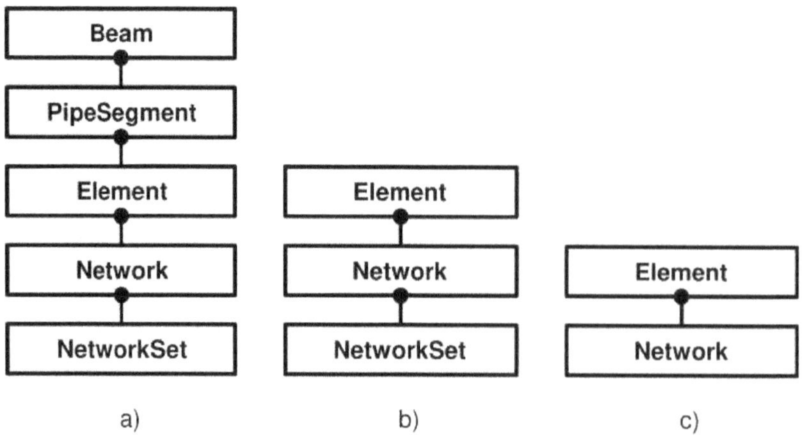

Figure 7. Candidates of layer structure for 1D line model. (a) Five layer structure; (b) three layer structure; (c) two layer structure.

the depth of its structure corresponds to 5 and it is large; therefore, its ultimate transparency is harmed. Regarding the second candidate, considering the characteristics of PipeSegment, namely that PipeSegment can be simply created by dividing Segment into the assigned number of beam elements, the number of divisions of the beam elements are allotted for each Segment. As a result, it is possible to create the necessary beam elements instead of maintaining information about the PipeSegment throughout the pipeline network. This achieves significant memory savings and simplifies the structure of the container of the 1D line model. Considering the third candidate, the structure of the container is even simpler and is premised on collectively modeling the entire pipeline network. However, it is not necessarily essential to model the entire pipeline network recorded in the GIS data. Therefore, this structure wastes memory. As stated earlier, the second candidate, or in other words, the candidate with a depth of 3 is suitable as the structure of the container of the 1D line model. Nonetheless, the names of each container are revised considering that details such as the number of divisions applied to Segment differ from the model data. **Figure 8** shows the three-layer structure of the 1D line model along with the three-level structure of the model data.

2.3.2. 2D surface model

A 2D shell model is created only for parts with conspicuous damage specified by a 1D line model. In almost all cases, locations of conspicuous damage are the junctions on pipes. The shape of the

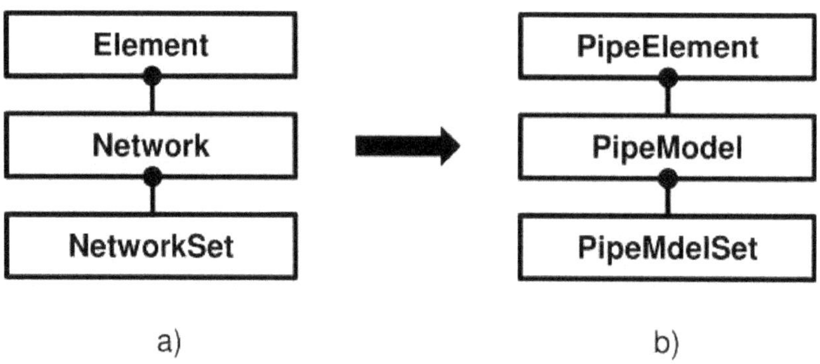

Figure 8. Three-layer structure of 1D line model together with corresponding three-layer structure of model data. (a) Model data; (b) 1D line model.

model is created based on the Non-Uniform Rational B-Spline (NURBS) surface. One merit of the NURBS surface is that it can be directly applied to isogeometric analysis that does not require mesh division. The model constructed is saved in the Initial Graphics Exchange Specification (IGES) format, which is a form that enables conversion between various CAD systems and a form that can be used by almost all finite element method (FEM) software packages. The analysis using a 2D shell model is executed based on an isogeometric analysis program or a standard FEM program.

It is straightforward to create a linear cylinder-shaped 2D surface model adapted to a beam element based on a NURBS surface. However, a junction on a pipe, the part where surfaces intersect, is not necessarily strictly represented by a NURBS curve. It may require to be approximated by a B-Spline curve.

On an actual pipeline network, there are junctions of many different kinds; however, in this study, the shape of the junction is created as the intersection of two pipes. As a result, the

a)

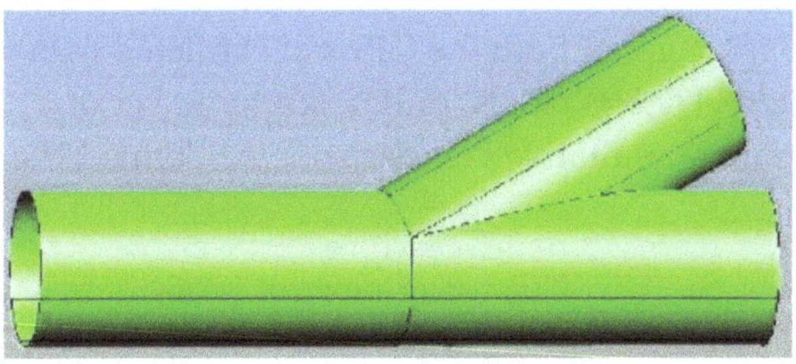

b)

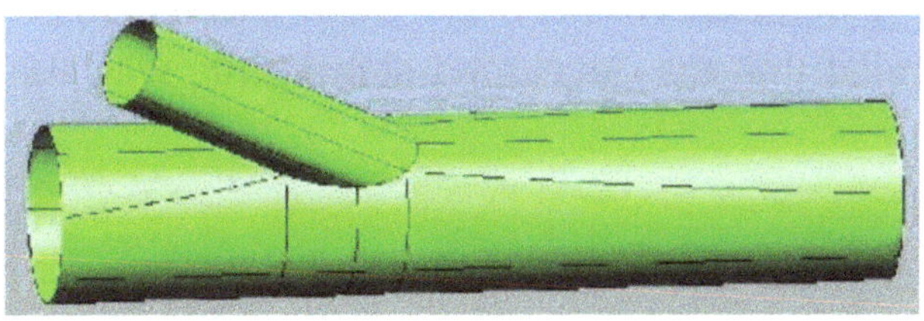

c)

Figure 9. Three types of junctions studied in this chapter: (a) elbow, (b) T-junction of equal-radius-tubes, and (c) T-junction of unequal-radius-tubes.

shapes of the junctions are the three shown in **Figure 9**, namely elbow junctions, T-junctions of pipes with equal diameters, and T-junctions of pipes with unequal diameters. Following the method of Global Curve Interpolation, the elbow junctions, and T-junctions of pipes with equal diameters can be handled as belonging to the same group.

3. Examples of constructed model

This section presents some verification tests which were conducted to ensure the correctness of the developed module for automated model construction. The target large lifelines networks in metropolis like Tokyo consist of a very large number of components and have complex configurations. The objective of these verification tests is to ensure that the developed model constructor is equipped with all the necessary logic to cope with all the complexities in lifeline networks of large metropolis.

3.1. 1D line model

The objective of the first verification test is to ensure that the developed module has necessary logics to cope with the complexities in large network. The network shown in **Figure 10a**, an actual network with spatial extents of 8.66×6.3 km^2, was used in this verification test; location and type of the network is not mentioned for security reasons.

Figure 10b shows the configuration of the data extracted by the automated model constructor. Though **Figure 10a** and **b** is visually identical, detailed comparisons are necessary to ensure the correctness of the extracted data. It is not possible to make a one-to-one comparison of the original GIS data and the extracted data; the extracted data are stored in a different format which is analysis friendly, and redundancies included in GIS data are eliminated. High resolution (8810×6376 pixels) images of both the GIS and extracted data were compared to check the accuracy. Out of the 56,172,560 pixels, only a scattered set of 2191 pixels were found to be different. To make more quantitative comparison, we defined the error index, which was estimated to be 0.0039%.

$$E = \frac{\text{Number of different pixels}}{\text{Number of pixels}}, \tag{1}$$

Further investigations indicated that these negligible differences were due to due to the antialiasing effects in image generation, confirming that all the elements were correctly extracted. This comparison confirmed that the developed submodule was able to construct a 1D line model accurately.

3.2. 2D surface model

Accuracy of the T-junctions with unequal radii pipes is demonstrated in this section as one of verification for the data conversion submodule for the 2D surface model. In the case of junctions with equal radii pipes, the intersecting curves, which are ellipses, can be exactly modeled with NURBS or accurately approximated with B-splines. On the other hand, the

a)

b)

Figure 10. Comparison of 1D line model with original GIS. (a) Original GIS; (b) constructed 1D line model.

intersecting curves of T-junction with unequal radius cannot be exactly modeled with NURBS or B-splines and produce the highest error of approximation. Hence, the error of approximating such intersecting curve is considered in this verification.

A T-junction formed by pipes of radii 2 and 1 m, intersecting at 30° angle is considered in this verification test. The surface geometry was modeled as collection of eight bivariate B-splines of third order. Varying the number of control points for approximating each of eight segments, the point wise error $\sqrt{\sum_{k=1}^{3}(x_k^{app} - x_k^{ana})^2}$, where x^{app} and x^{ana} were points on the approximated and analytical curves, was estimated. **Table 1** shows the maximum normalized errors with 6, 8, 10, and 12 control points. As shown in **Figure 11**, these errors are highly localized, and maximum error relative to radius of the smallest pipe is only 0.08%. It is thus shown that the developed submodule can generate 2D surface model with sufficient accuracy. If required, the accuracy can be further improved by increasing the order of B-splines and/or number of control points.

Number of control points per curve segments	Maximum of pointwise error (mm)
6	5.6
8	2.4
10	1.3
12	0.8

Table 1. Maximum pointwise error in B-spline approximation with different degree B-spline.

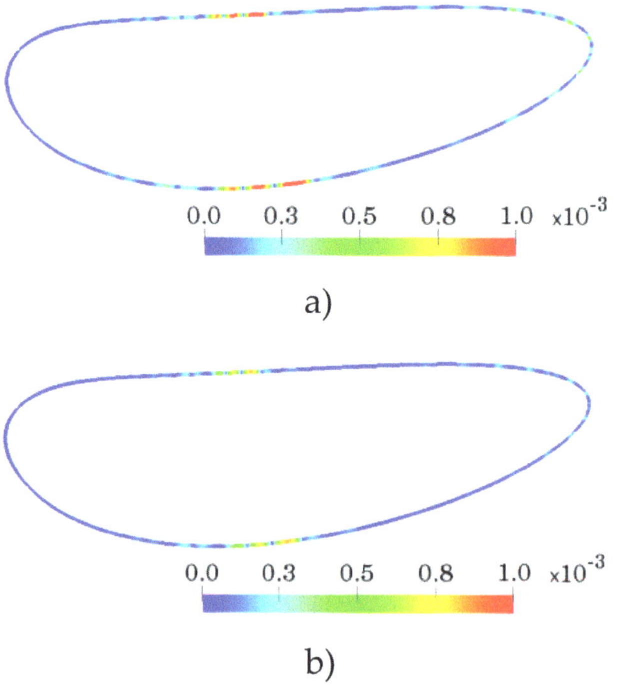

Figure 11. Distribution of error of approximation at 808 points on the intersecting curve of T-junction with 2 m and 1 m radii pipes. The intersecting curves are approximated as a collection of 8 B-spline curves, each with the given number of control points. (a) 10 control points per segment; (b) 12 control points per segment.

4. Examples of seismic response analysis

The main application of the extracted GIS data is seismic disaster assessment of buried lifeline networks. In this demonstrative analysis, small part of the network shown in **Figure 10a** subjected to a historical earthquake ground motion was analyzed to identify the damage locations. Though accuracy is high, seismic disaster assessment of an entire network with 2D analysis models is computationally expensive. Instead, first analyzing a computationally light beam element model to identify possible damage locations, and then using 2D models to analyze only the local links connected to those possible damage locations is a cost effective approach which does not sacrifice accuracy. The demonstrative examples presented in this section use this strategy to reduce computational cost.

(a) Soil property	
Shear wave velocity (m/s)	190
Density (tf/m^3)	1.9
Shear modulus (tf/m^2)	0.47
Poisson's ratio	7000
(b) Pipe property	
Outer diameter (cm)	21.63
Inner diameter (cm)	20.47
Thickness	0.58
Cross section area (cm^2)	38.4
Young's modulus (kgf/cm^2)	2.1×10^6

Table 2. Material properties used in simulation.

We made use of a general purpose finite element package, ABAQUAS, in this analysis. The input data for ABAQUS were automatically generated by the developed modules. **Table 2** shows the material properties used in the analysis; some standard values were used due to confidentiality of data. Linear analysis was used for both the 1D and 2D analysis.

4.1. 1D line model

For the 1D linear analysis with ABAQUS, two-node pipe element PIPE31 was used to model the network, and the spring element SPRING1 was used to model soil-structure interactions (i.e., the reaction force produced by surrounding soil). To represent axial, vertical, and horizontal reaction forces, three springs were included at every node of the beam element model, as shown in **Figure 12a** and **b**.

Applying the respective components of the selected input ground motion at the free ends of each spring element, quasi-static analysis was conducted for each time record of the input ground motion. Only the major portions of input ground motion were used, since quasi-static analysis was employed. Samples of the input ground motion and the induced displacement are shown in **Figure 13**. The similarities in their trends indicate that there were no fatal mistakes in automatically generating the analysis model. In this analysis, we assumed the inertial contribution from the pipes was negligible compared to confinement of surrounding soil. It is straightforward to conduct dynamic analysis if the inertial forces are significant.

While there are several ways to assess the damage, we used the maximum strain due to bending and axial deformation of pipe elements to assess the damage (i.e., maximum strain exceeds a given critical value). As shown in **Figure 14**, the distribution of maximum strain was similar to that of bending moment. This indicated that bending action of pipes was the main contributor to the induced large strain. **Figure 15** shows close-up views of the two high strain regions which are indicated in **Figure 14**. **Figure 16** shows the identified damage locations based on the maximum strain criterion.

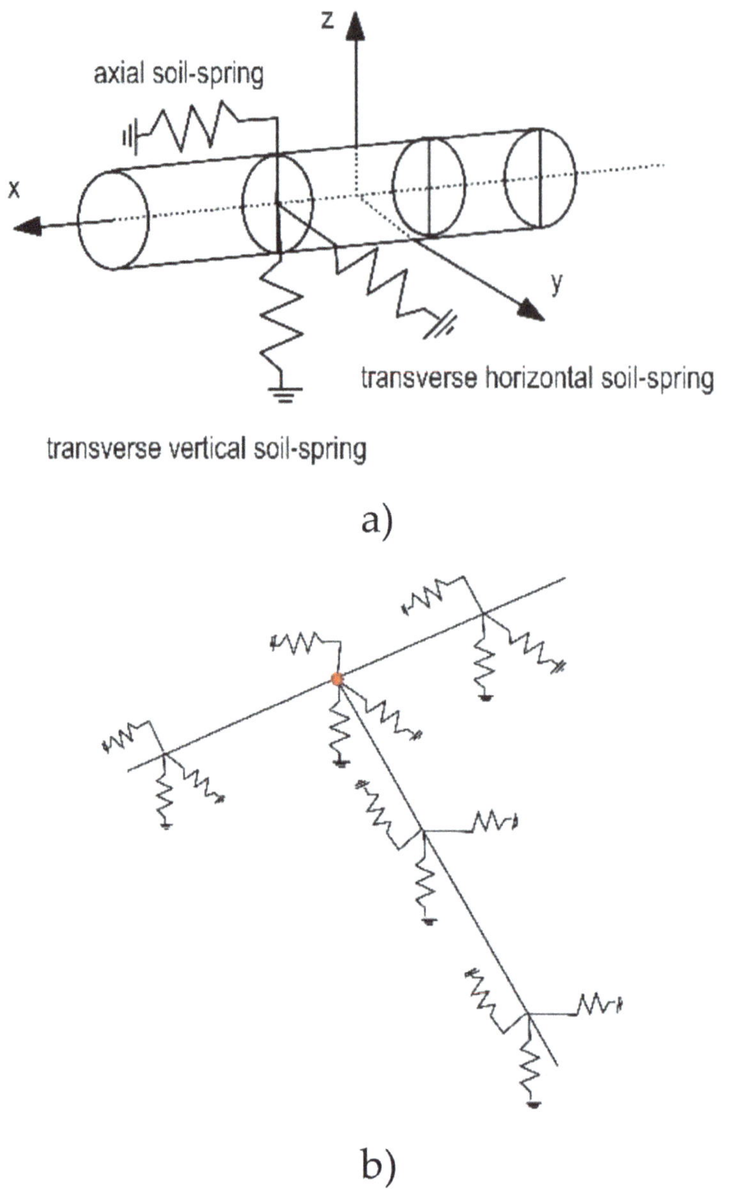

Figure 12. Soil-spring used in 1D line model: (a) shows three springs are attached to a pipe segment and (b) is a schematic view of a pipeline network attached with soil-springs.

4.2. 2D surface model for junction

Though lightweight in computation, the used beam element model mentioned earlier has a lower reliability in assessing damage states. Beam element models cannot accurately model stress concentrations at joins, hence the lower reliability. In order to increase the reliability, we model the damaged components, which are identified by abovementioned beam model, with shell elements, and the damaged status is decided based on the results of the analysis with shell elements.

We extended the data conversion submodule to automatically generate the required analysis shell element models for ABAQUS. First, the 2D B-spline surface model is generated for any pipe component identified according to the beam element analysis. Next, a mesh of sufficient spatial resolution is automatically generated based on B-spline data, and three springs are

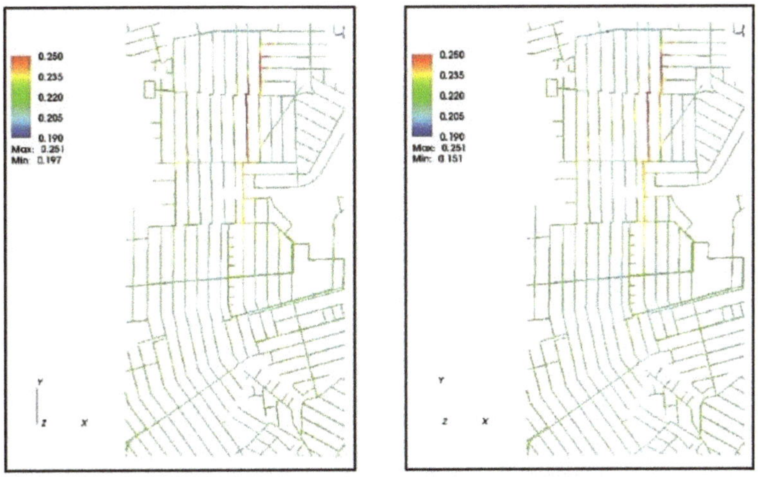

a) b)

Figure 13. Displacement distribution in part of pipeline network. (a) Input displacement; (b) induced displacement on pipes.

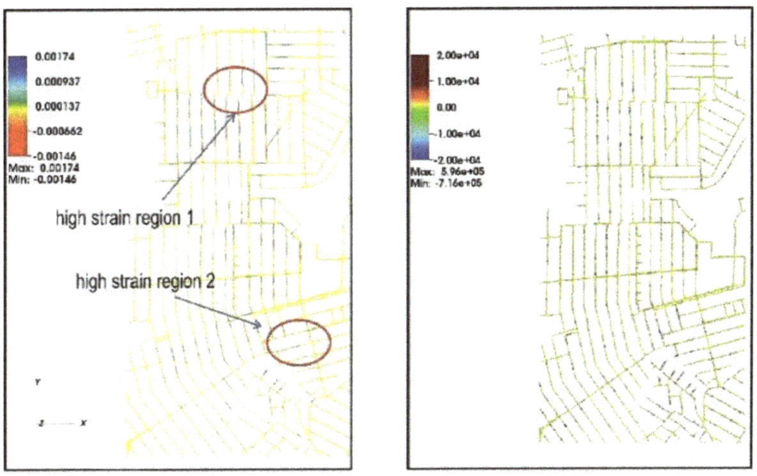

a) b)

Figure 14. Distribution of strain and bending moment in part of pipeline network. (a) Normal strain; (b) bending moment.

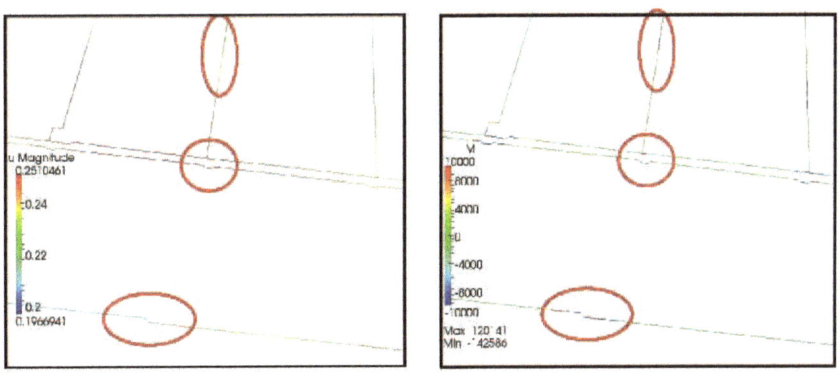

a) b)

Figure 15. Close-up view of concentration of strain and bending moment in pipeline network. (a) Input displacement (m); (b) bending moment (Nm).

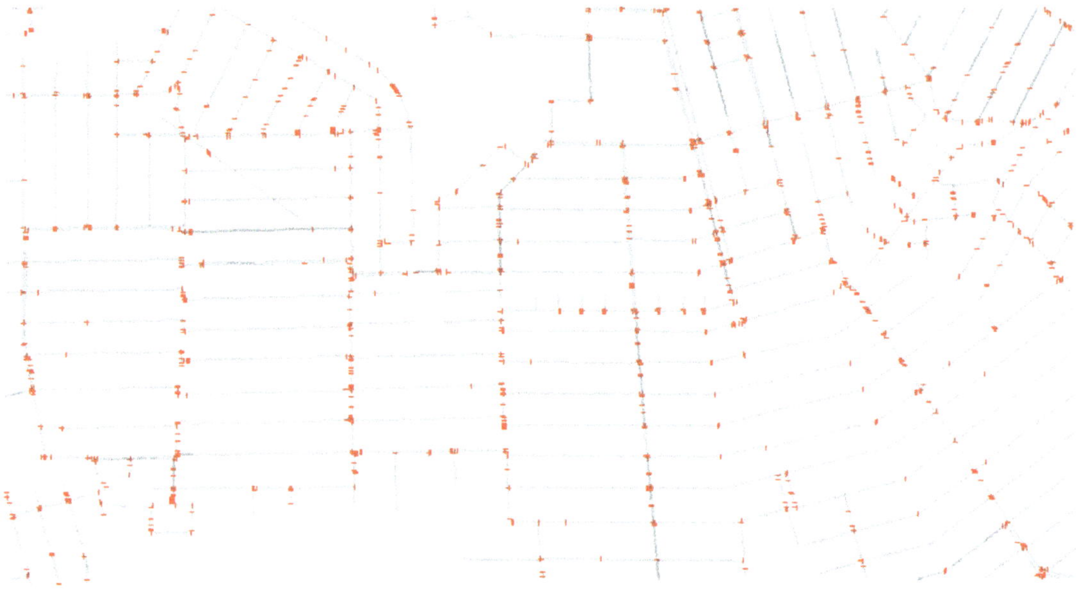

Figure 16. Example of critical points in pipeline assessed by strain.

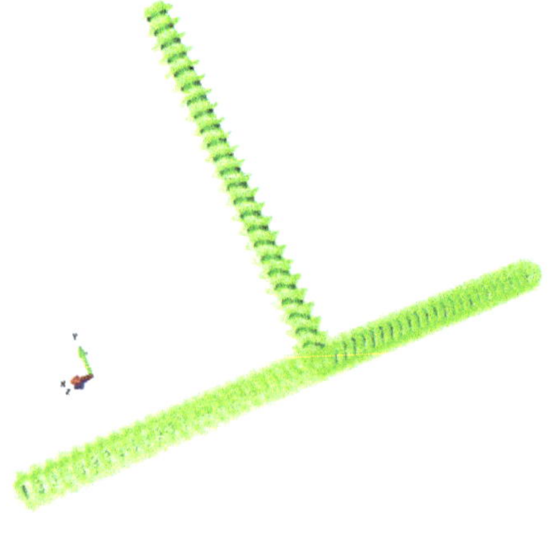

a)

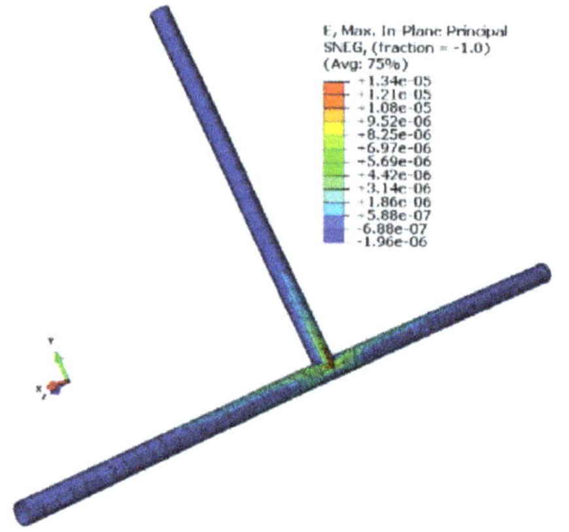

b)

Figure 17. Automatically generate shell element analysis model is shown in (a). Three springs are attached to each node to represent surrounding soil. Distribution of the maximum in-plane strain is shown in (b).

attached to each node of the shell element model to represent the reaction of the surrounding soil. The three springs are oriented in axial, radial, and tangential direction at each node (see **Figure 17a**). The spring properties are assigned according to the properties of surrounding soil. The displacements at the free ends of the springs are assigned according to the ground motion data. Including all these information, the input file for ABAQUS is automatically generated. **Figure 17b** shows the distribution of the maximum principle strain. Based on the results of this analysis, the damage state at the junction is decided.

As demonstrated in this section, the developed automated model constructor makes it straightforward to assess seismic performance analysis of a pipe network. The whole process can be automated with batch scripts. Though only a small network is used in this demonstration, analyzing a large network is straightforward.

5. Concluding remarks

A module that automatically constructs an analysis model of a pipeline network by converting the GIS data of a lifeline was successfully developed. The module functions as designed without failing to read the GIS data or output the analysis model. GIS can be used to analyze the earthquake response of a lifeline, expanding its potential power.

Acknowledgements

A large part of this chapter is based on an earlier published paper [10] "Automated Model Construction for Seismic Disaster Assessment of Pipeline Network of Lifeline."

Author details

Hideyuki O-tani[1*], Muneo Hori[2] and Lalith Wijerathne[2]

*Address all correspondence to: h.o-tani@riken.jp

1 RIKEN Advanced Institute for Computational Science, Chuo-ku, Hyogo, Japan

2 Earthquake Research Institute, University of Tokyo, Tokyo, Japan

References

[1] Hori M. Introduction to Computational Earthquake Engineering. 2nd ed. Imperial United Kingdom: College Press; 2011. 440 p. DOI: 10.1142/P644

[2] Federal Emergency Management Agency. HAZUS: Earthquake Loss Estimation Methodology. Technical Manual. National Institute of Building Sciences for FEMA; 1997

[3] Toprak S, Taskin F. Estimation of earthquake damage to buried pipelines caused by ground shaking. Natural Hazards. 2007;**40**:1-24. DOI: 10.1007/s11069-006-0002-1

[4] Alex AKK, Tang PE, Stu WPE. Lifeline earthquake engineering in a multihazard environment. In: TCLEE 2009. Reston, VA: ASCE; 2009

[5] FEMA Hazus. Available from: http://www.fema.gov/hazus [Accessed: Mar 22, 2018]

[6] Kennedy RP, Darrow AC. Seismic design of oil pipeline systems. Journal of the Technical Councils of ASCE. 1979;**105**(1):119-134

[7] Japan Gas Association. Seismic Design for Gas Pipelines. Available from: http://iisee.kenken.go.jp/worldlist/29_Japan/PART1/29_Japan_Cap5Gas_Code.pdf [Accessed: Mar 22, 2018]

[8] Constantopoulos IV, Lagasco F, Pelli F. Evaluation of dynamic site response for seismic design of pipelines. WIT Transactions on The Built Environment. 1995;15. DOI: 10.2495/SD950451

[9] Gunter CA, Mitchell JC. Theoretical Aspects of Object-Oriented Programming: Types, Semantics, and Language Design (Foundations of Computing). MA: The MIT Press; 1994. 548 p

[10] Hori M, Wijerathne L, Tanaka S, Ichimura T. Automated model construction for seismic disaster assessment of pipeline network of lifeline. Journal of Earthquake and Tsunami. 2013;**7**(3):1350039. DOI: 10.1142/S1793431113500395

4

Simulation of Broadband Strong Motion Based on the Empirical Green's Spatial Derivative Method

Michihiro Ohori

Abstract

This study sought to simulate strong broadband seismic motions beyond the corner frequencies used, for the same events, in previous studies. To correct discrepancies among the corner frequencies of events, a scaling law based on the ω^{-2} model was assumed and the spectral amplitude decay beyond the corner frequency was compensated. The observations were also corrected for location, focal mechanism, and time of occurrence. After estimating the empirical Green's tensor spatial derivative (EGTD) from 11 aftershock events, using 0.2–10 Hz band-pass-filtered waveforms, the strong motion records for the mainshock and aftershocks were simulated. In the simulation of each event, the EGTD elements were multiplied by the moment tensor elements followed by summation and then corrected in their spectral amplitude taking the corner frequency of each event into account. At the closest epicentral distance, and for most events, an acceptable agreement was found between calculated and observed waveforms. The results were also compared with the outcomes of simulations using the empirical Green's function method. As EGTD elements are determined by local underground structures, they could prove useful for future structural studies.

Keywords: empirical Green's spatial derivative method, moment tensor, ω^{-2} model, broadband strong motion, corner frequency, source time function, waveform inversion

1. Introduction

Prediction of the strong ground movements produced by large earthquakes has been demonstrated using the empirical Green's function (EGF) method, proposed by Hartzell [1] and extended by Irikura [2]. This practical technique is most effective when the focal mechanism of a small event is similar to that of a targeted event. When the focal mechanisms differ more significantly, the empirical Green's tensor spatial derivative (EGTD) method, proposed by

Plicka and Zahradnik [3], is more appropriate. This can predict the ground motion for events with diverse focal mechanisms. Using single-station inversion of waveform data of several small events whose focal mechanisms and source time functions are well determined, the EGTD elements can be estimated, with the expectation of stable and accurate ground motion prediction. However, discussion in the literature of EGTD applications has been limited [4–11].

In recent studies [9, 10], the authors sought to simulate strong broadband motions, beyond the corner frequencies, for the same events as in previous studies [6, 8]. This chapter extends these recent studies and applies the results to other stations. The scaling law based on the ω^{-2} model [12] was used to correct differences between the corner frequencies of events, assuming spectral amplitude decay at higher frequencies. The 0.2–10 Hz band-pass-filtered waveforms of 11 aftershock events were used to estimate the EGTD and then to simulate the strong motion records for main and aftershocks. The EGTD elements were multiplied by the moment tensor elements followed by summation. The spectral amplitude was adjusted by taking the corner frequency of each event into account. Agreement between the simulated waveforms, at the closest epicentral distance, and observation were acceptable.

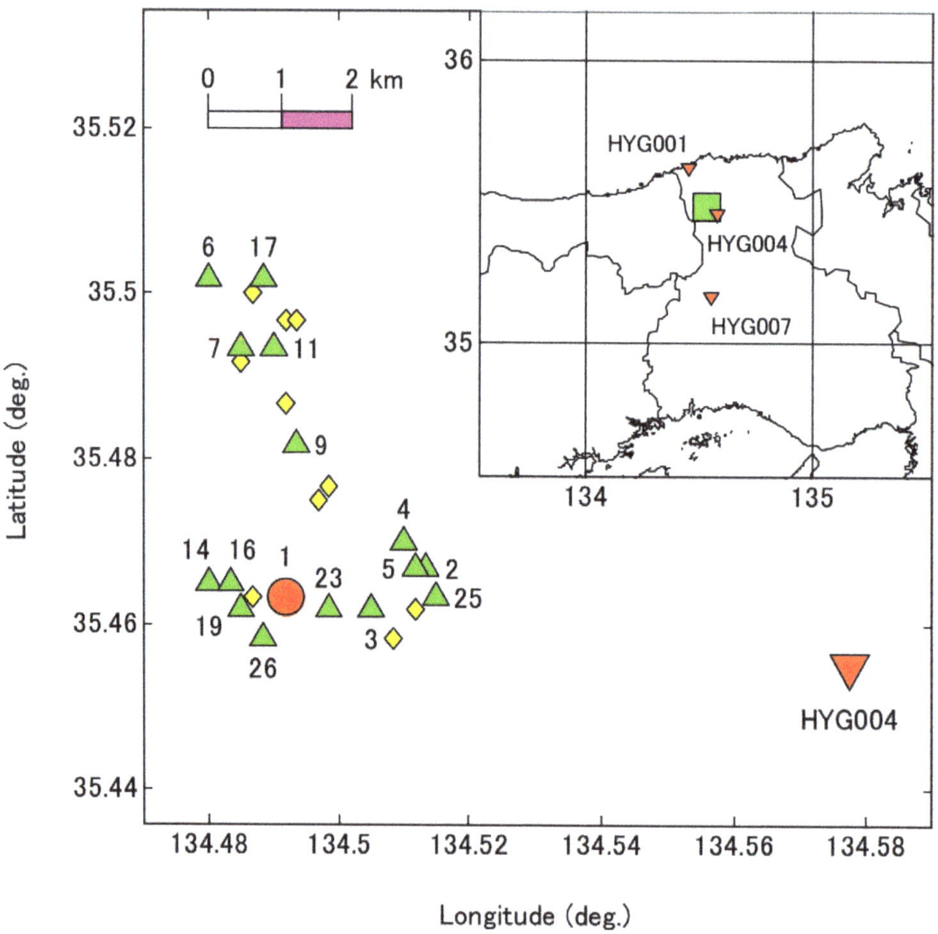

Figure 1. Map showing the epicenter locations of the mainshock and aftershocks of the 2001 Hyogo-ken Hokubu earthquake. The mainshock and 15 aftershocks analyzed are indicated by a solid circle and solid triangles, respectively. The 10 aftershocks not used in the present study are represented by open diamonds. The target station HYG004 is shown by an inverted solid triangle. The inset at the top right shows the surrounding region with the target area (solid rectangle) and the other stations, HYG001 and HYG007. (After Ohori and Hisada [8] with slight modification).

2. Locations of the epicenters and stations

Figure 1 shows epicenter locations, determined by the Japan Meteorological Agency (JMA), of the mainshock (M_J5.4) and 25 aftershocks (M_J3.1–4.7) of the 2001 Hyogo-ken Hokubu earthquake at the target station, HYG004, one of the K-NET stations operated by the National Research Institute for Earth Science and Disaster Prevention (NIED). The fault zone is roughly 4 km east-west and 6 km north-south. HYG004 is on a nearby rock site (6–10 km from the fault zone). Acceleration data for the mainshock and 15 of the aftershocks (M_J3.5–4.7) came from the K-NET. Two other stations, HYG001 and HYG007, are shown in the inset of **Figure 1**; they were also tested to verify the applicability of the EGTD method.

3. Source model

The EGTD method is most accurate if the focal mechanism and source time functions are known as accurately as possible. The source model described below, and illustrated in **Figure 2**,

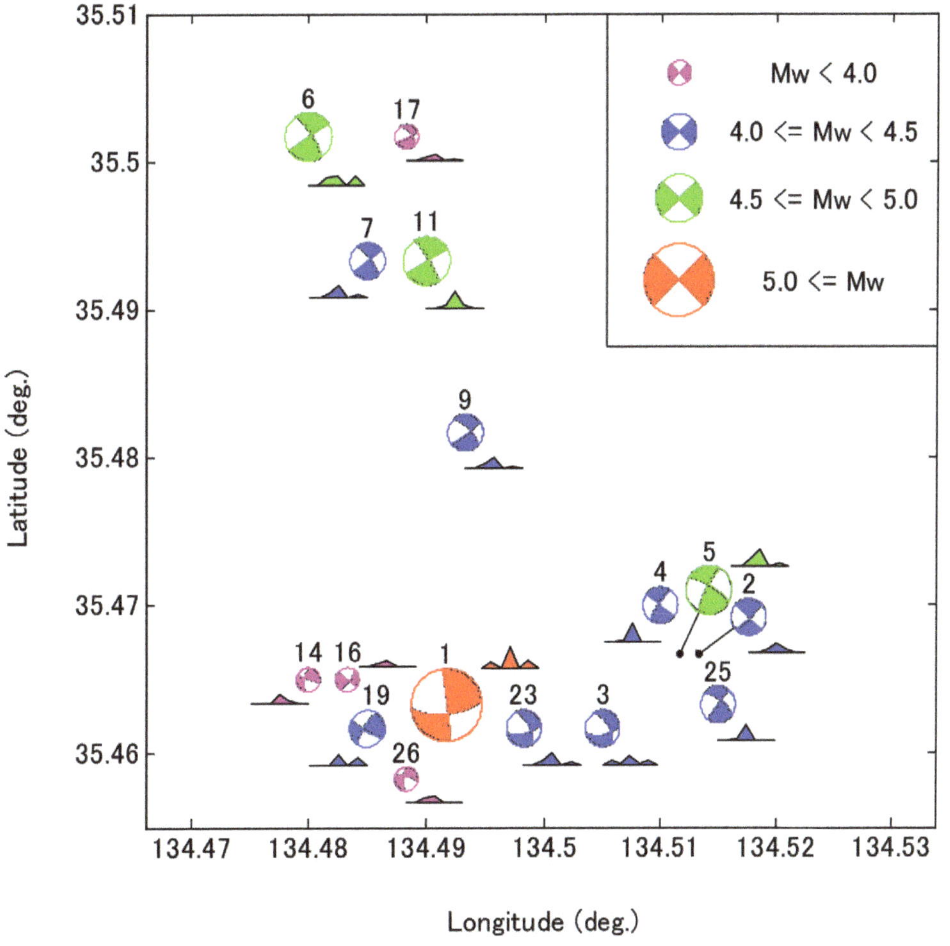

Figure 2. Map showing focal mechanisms and source time functions of 16 events (the mainshock and 15 aftershocks) from the work of Ohori and Hisada (2006). Single-station inversion for each event was performed using strong motion records recorded at the K-NET station HYG004. (After Ohori and Hisada [8]).

has been reevaluated from previous work [6]. Strike, dip, and rake of a double-couple point source, and source depth, were estimated using the grid search technique [13]. The observed acceleration records at HYG004 for the mainshock and 15 aftershocks were integrated into velocity waveform data with a band-pass filter of 0.2–1.0 Hz. A total of 5 s of data, which included the P-wave arrival and S-wave main portions, was inverted. The theoretical Green's function for the layered underground structure [14] was calculated assuming a smoothed ramp function with a rise time of 0.32 s. The searching ranges of the strike, dip, and rake angles were set within 20° of the solutions determined by the F-net of the NIED. The source depth was estimated as 8–12 km. Seismic moments, released by five sequential slips with intervals of 0.16 s, and a source time function were then determined by the least-squares method with nonnegative constraints [15]. Events 3, 17, 19, and 26 were excluded from the EGTD inversion because of a relatively large discrepancy in waveform matching between the observation data and synthesis.

4. Estimation of the EGTD

The EGTD estimation method has been fully explained by Ohori and Hisada [6, 8]. It is applicable to simulation of strong motion in a frequency range below the corner frequency. The process is summarized below. Methods for compensating the spectral amplitude decay, beyond the corner frequencies, and simulating the broadband ground motion follow.

4.1. Basic equations

Ground motion displacement $u_i(x_o, t)$ $(i = x,y,z)$ excited by a double-couple point source is theoretically expressed as the convolution of moment tensor elements $M_{pq}(x_s,\tau)$ $(p,q = x,y,z)$ and Green's tensor spatial derivative elements $G_{ip,q}\,(x_o, t \mid x_s,\tau)$:

$$u_i(x_o, t) = M_{pq}(x_s, \tau)^* \, G_{ip,q}\,(x_o, t \mid x_s, \tau) \tag{1}$$

Hereafter, $u_i(x_o, t)$, $M_{pq}(x_s,\tau)$ and $G_{ip,q}\,(x_o, t \mid x_s,\tau)$ are abbreviated as u_i, M_{pq}, and $G_{ip,q}$. Explicit expressions of M_{pq} for a double-couple point source are in the literature (e.g., see works by Aki and Richards [16]).

Considering symmetrical conditions ($M_{pq} = M_{qp}$) and no volume change [$M_{xx} = -(M_{yy} + M_{zz})$] of the moment tensor elements, Eq. (1) can be rewritten as

$$u_i = \sum_{j=1}^{5} M_j * G_{ij} \tag{2}$$

where M_j $(j = 1,2,\ldots,5)$ is defined by $M_1 = M_{xy}$, $M_2 = M_{yy}$, $M_3 = M_{yz}$, $M_4 = M_{xz}$, and $M_5 = M_{zz}$, and G_{ij} $(j = 1,2,\ldots,5)$ is defined by $G_{i1} = G_{ix,y} + G_{iy,x}$, $G_{i2} = G_{iy,y} - G_{x,x}$, $G_{i3} = G_{iy,z} + G_{iz,y}$, $G_{i4} = G_{ix,z} + G_{iz,x}$, and $G_{i5} = G_{iz,z} - G_{ix,x}$.

In the moment tensor inversion undertaken for a particular event using data of all possible components at all possible stations simultaneously, u_i and G_{ij} are given and M_j are the unknowns to be solved in a least-squares sense. Conversely, in the EGTD inversion computed for each component at each station using data from several events simultaneously, u_i and M_j are given and G_{ij} are the unknowns. While the moment tensor elements are determined from source parameters, the Green's tensor spatial derivative elements express the underground structure of the area surrounding the source and the station.

4.2. Correction of the focal mechanisms

To compensate for differences in the locations of the main and aftershocks, the focal mechanisms are horizontally and vertically rotated, as described in the literature [4, 5], such that each event can be treated as a point source at the same location. Through the horizontal rotation, shown in **Figure 3(a)**, the station azimuths of the mainshock and aftershocks are set to 90°, as

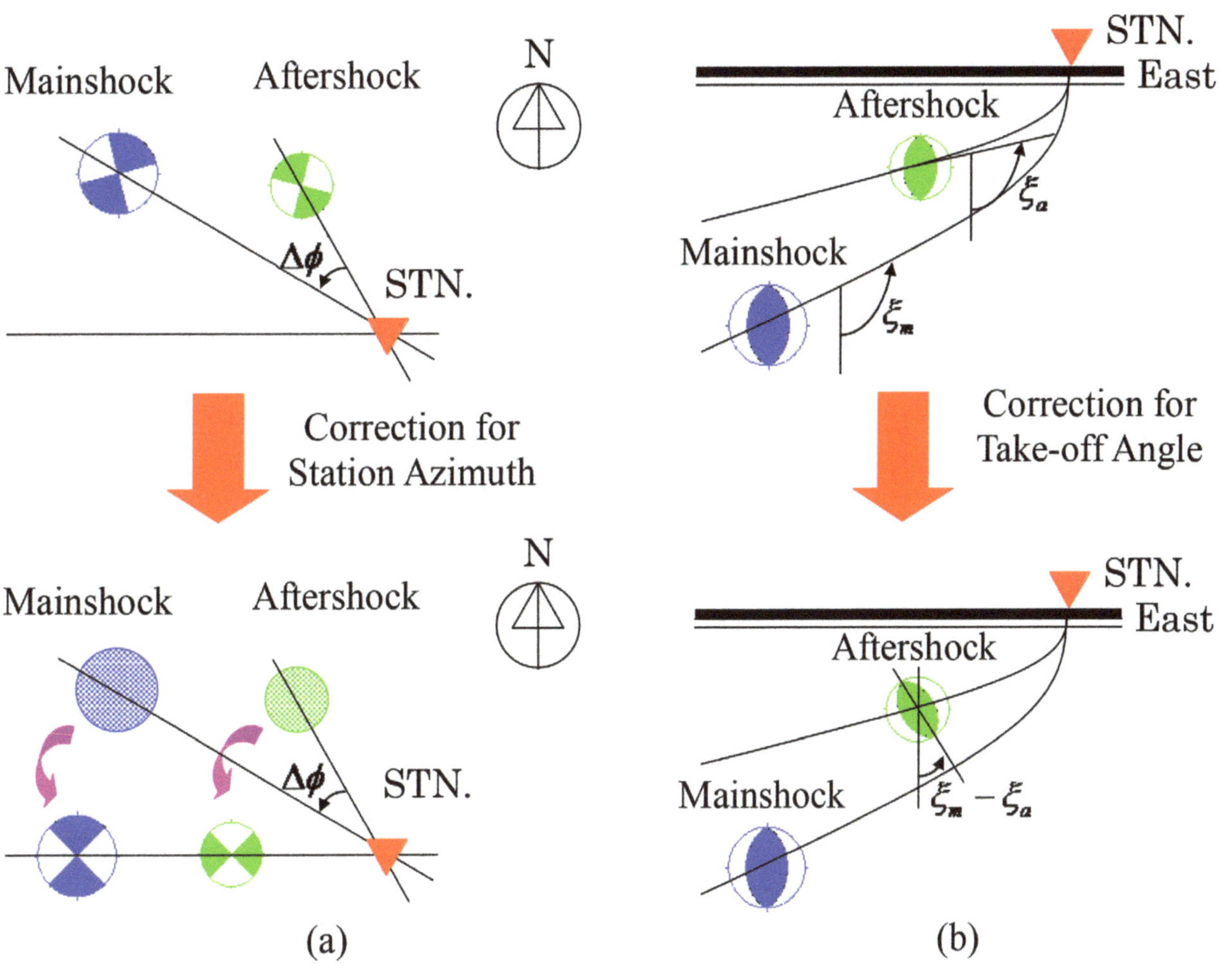

Figure 3. Schematic diagram explaining how the focal mechanisms are rotated to reduce the number of unknowns in the empirical Green's tensor spatial derivative (EGTD) and to compensate for the different locations of the mainshock and aftershocks. In the horizontal rotation (a), the station azimuths for the mainshock and aftershocks are set to 90°, measured from north, so that the number of EGTD elements is reduced to three for the radial and vertical components and two for the transverse component. In the vertical rotation (b), following horizontal rotation, the discrepancies in the take-off angles between the mainshock and aftershocks are corrected. In the top right panel, take-off angles for the mainshock and aftershocks are denoted ξ_m and ξ_a, respectively. The difference between take-off angles, ξ_m-ξ_a, is slightly exaggerated in the bottom panel for clarity. (After Ohori and Hisada [8]).

measured from north [6]. Thus, the number of Green's tensor spatial derivative elements is reduced to three ($G_{i1} = G_{i4} = 0$) for the radial component ($i = y$) and the vertical component ($i = z$) and two ($G_{i2} = G_{i3} = G_{i5} = 0$) for the transverse component ($i = x$). Through the subsequent vertical rotation, as shown in **Figure 3(b)**, the discrepancies in the take-off angles between the mainshock and aftershocks are corrected. The moment tensors are evaluated after the horizontal and vertical rotations.

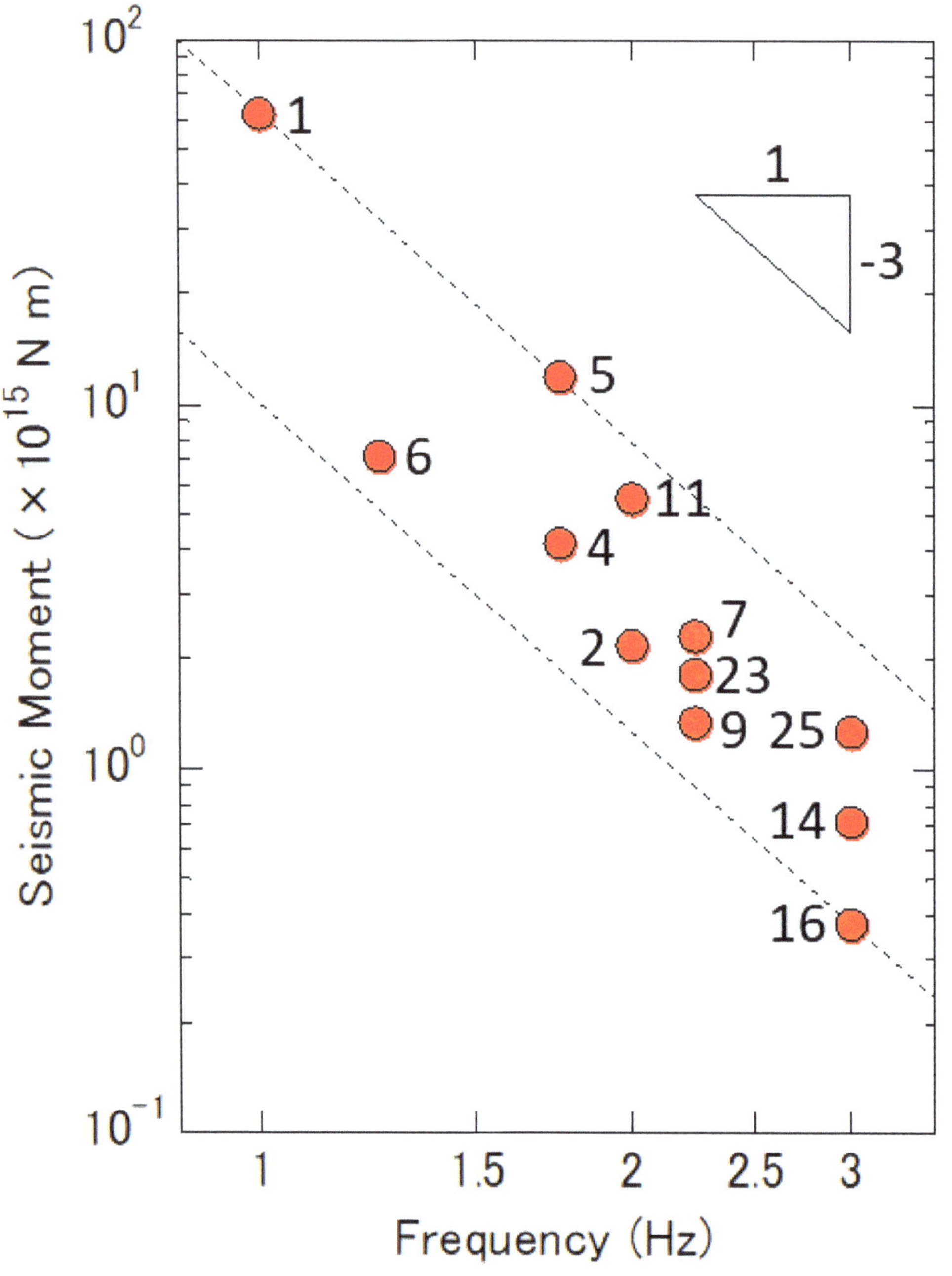

Figure 4. Relationship between the corner frequency and the seismic moment. The numeral next to solid circle represents the sequential event number in the present study. According to the scaling law, the seismic moment is inversely proportional to the cube of the corner frequency when the stress drop is constant.

4.3. Correction applied to the waveform data

Time shifts were estimated from the 0.2–1.0 Hz band-pass-filtered velocity waveforms and then applied to the observation data of aftershocks to fit their S-wave arrival time with that of the mainshock. In addition, the observation data was deconvolved to remove discrepancy in the source time function. The observed waveforms, used in the estimation of the EGTDs, were corrected such that the source time function has a constant seismic moment (1.0×10^{15} Nm, nearly equal to M_w4.0) and a single-isosceles slip velocity function with a rise time of 0.32 s.

Assuming the source spectrum obeys the ω^{-2} model [12], the corner frequency of the mainshock was about 1.0 Hz, while those of 11 aftershocks were between 1.2 and 3.0 Hz, as shown in **Figure 4**. To simulate the broadband ground motion beyond the corner frequency, the effect of the differences among corner frequencies must be removed. Again, assuming a ω^{-2} scaling law, and compensating for spectral amplitude decay beyond the corner frequency, each event could then be treated as having the same corner frequency as that of the mainshock.

4.4. EGTD estimation

Using the focal mechanisms of aftershocks, rotated and time corrected, as described above, simultaneous linear equations for each component were solved for each of the sampling data sets. No smoothing or minimization for unknown parameters was included in the EGTD estimation. **Figure 5** illustrates the transverse component elements of an EGTD for HYG004. Each element is scaled, the same as the mainshock, for an event with a seismic moment of

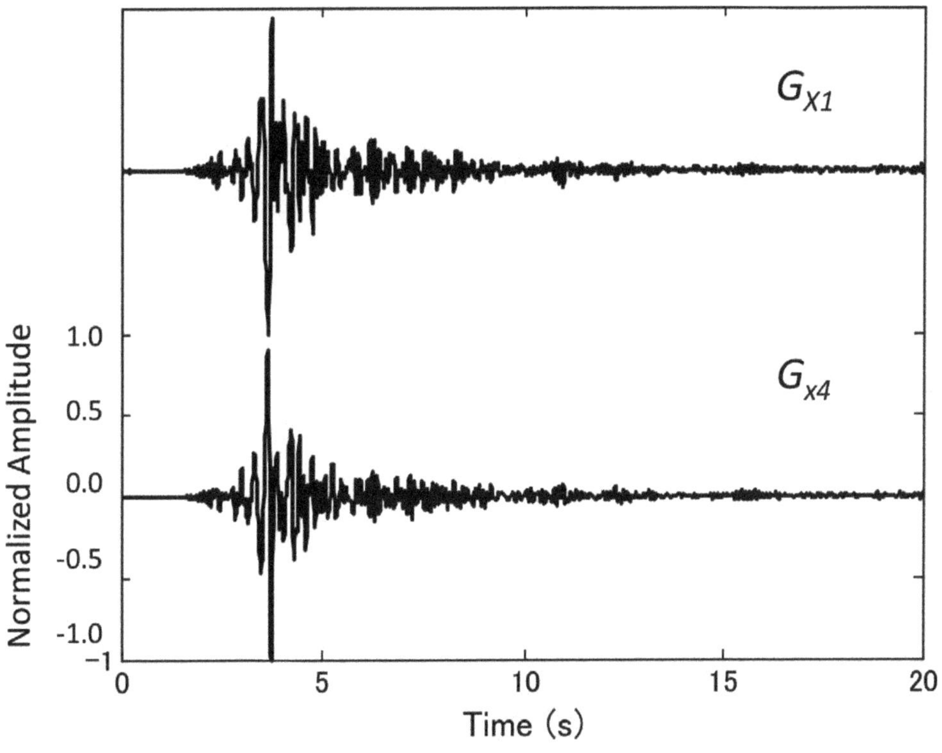

Figure 5. An example of the EGTD transverse component elements with normalized peak amplitude.

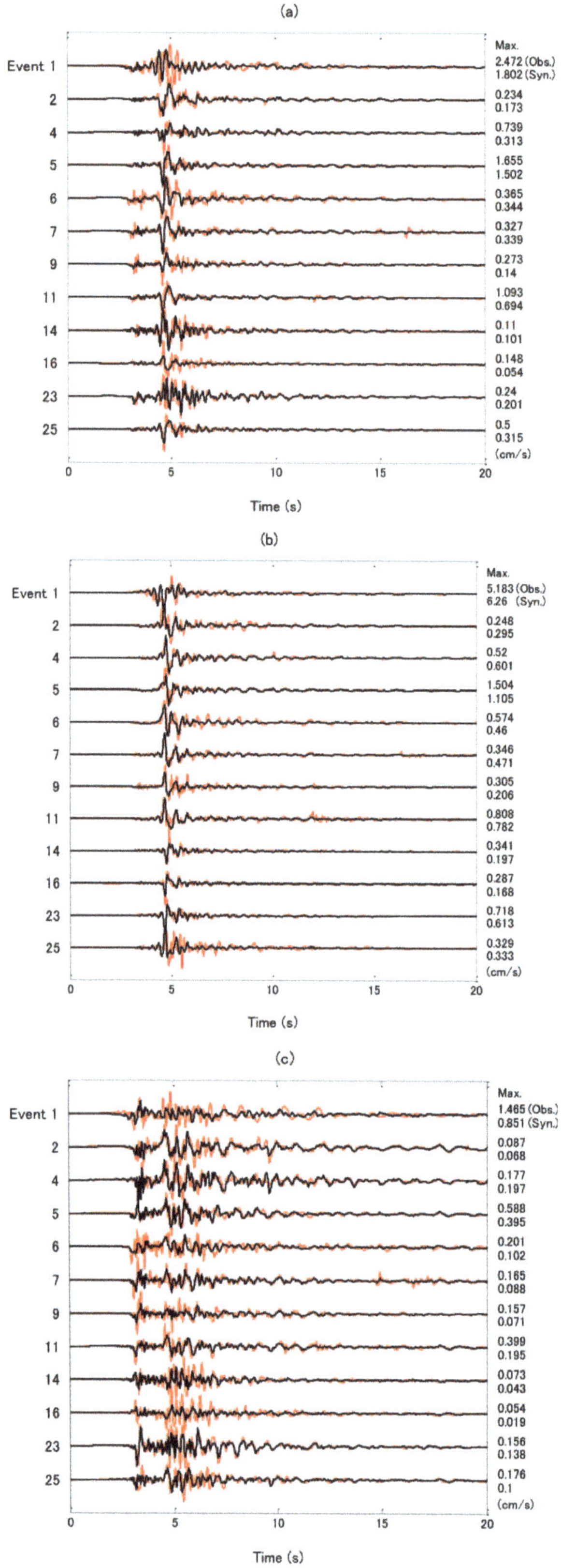

Figure 6. Comparison of 0.2–10 Hz band-pass-filtered observed velocity waveforms for HYG004 (red lines) and corresponding syntheses calculated using the EGTD (black lines). For each trace, the source time function, seismic moment, and corner frequency of each event are taken into account. Absolute peak amplitudes are given at the end of each trace: Upper numerals are observed amplitudes and lower are synthesized. Note that the top trace for the mainshock (Event 1) is not included in the EGTD inversion. (a) Radial-comp; (b) Transverse-comp; (c) Vertical-comp.

1.0 × 10^{15} Nm and a corner frequency of 1.0 Hz. As the Green's tensor spatial derivative elements are determined by the local underground structure, the EGTD elements could be useful for future structural studies.

5. Simulation of the strong ground motion using the EGTD

Figure 6 compares the observed radial, transverse, and vertical component velocity waveforms for HYG004, normalized for time, seismic moment, and corner frequency, with a 0.2–10 Hz band-pass filtering and corresponding synthesis calculated from the EGTD. The top trace for the mainshock (Event 1) is not included in the EGTD inversion. Considering the complexity of the high-frequency components, the broadband synthesis, using EGTD, acceptably reproduced the observed waveforms.

Three frequency ranges of the band-pass filter are compared in **Figure 7**: 0.2–1.0, 1.0–10, and 0.2–10 Hz. From this figure, it is evident that simulation results for HYG004 from EGTD are most accurate in a frequency range of 0.2–1.0 Hz, except that the radial components of Event 14 are somewhat overestimated. The results for a frequency range of 1.0–10 Hz are acceptable and very similar to those for a frequency range of 0.2–10 Hz. For most of the events, the maximum amplitude ratio between the synthesized velocity waveforms and observed data is between 0.5 and 1.5. On the whole, the simulation of broadband ground motion, using the EGTD method, successfully reproduced the observed waveforms.

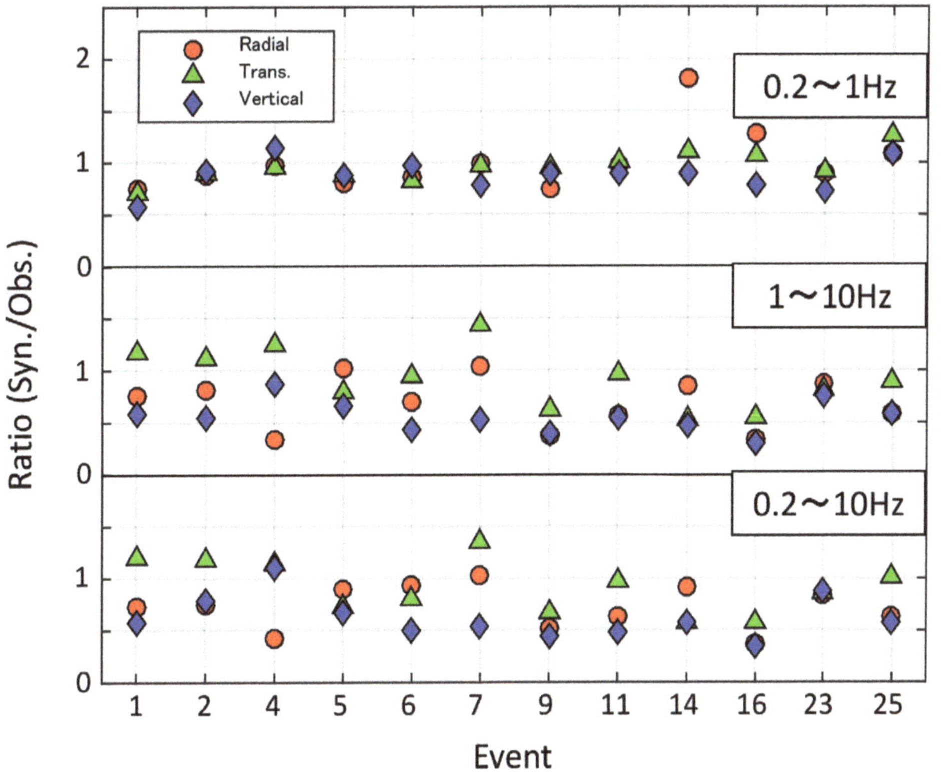

Figure 7. Comparison of the maximum amplitude ratio between the synthesis and observed data for HYG004. The frequency ranges of the band-pass filters are 0.2–1.0 Hz (upper), 1.0–10 Hz (middle), and 0.2–10 Hz (lower).

Finally, **Figure 8** shows the results for the other two stations. In this figure, because of data quality issues, only transverse components are shown. Generally, the waveform matching between synthesis and observatory data is inferior to that for HYG004 (see **Figure 6(b)**). For HYG002, the synthesis underestimated motion for Events 2, 5, 6, and

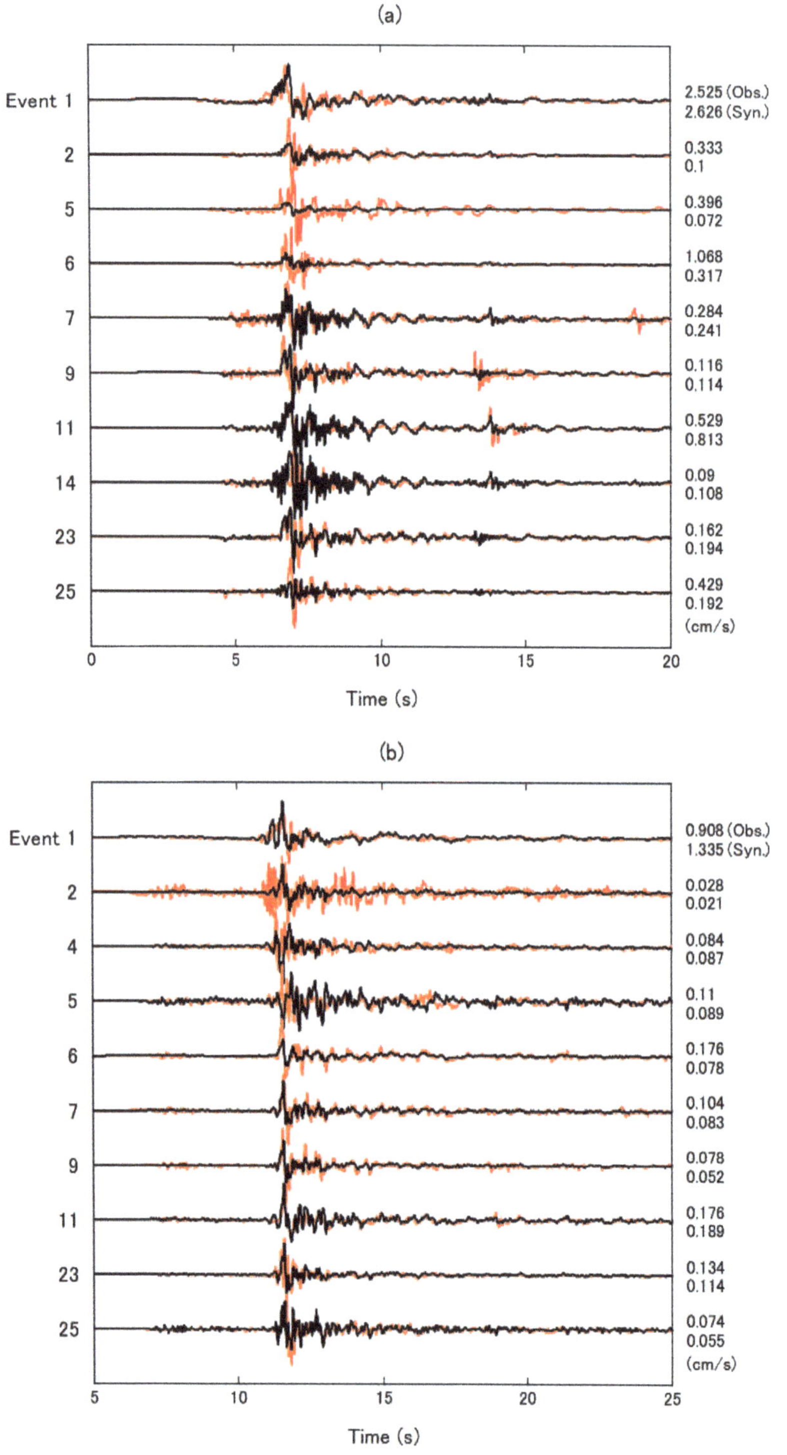

Figure 8. Comparison of 0.2–10 Hz band-pass-filtered observed velocity waveforms (red lines) and corresponding syntheses calculated from the EGTD (black lines) for transverse components of data from HYG001 (a) and HYG007 (b). Other conditions are the same as in **Figure 6**.

25. For HYG007, synthesis also underestimated motion for Event 6. On the other hand, for the mainshock (Event 1), the agreement between synthesized and observed waveforms was satisfactory.

6. Comparison of the simulation results obtained using the EGTD and EGF methods

A previous study [8] examined the accuracy of simulation results using the EGTD method by comparing them with those obtained using the alternative empirical approach—the EGF method [1, 2]. In that study, velocity waveform data with a 0.2–1.0 Hz band-pass filtering was used. This study compared the two methods over a broader frequency range. The EGF method used was almost the same as in the previous study [8], except the differences in the corner frequencies between the mainshock and each aftershock were corrected using the scaling law of the ω^{-2} model [12]. In the EGF approach, each aftershock was used in the simulation of the mainshock. To compensate for differences in radiation pattern coefficients, for each aftershock, the data were multiplied by the ratio of radiation pattern coefficients between the mainshock and the aftershock. The radiation pattern coefficients of the SV-waves were used for simulation of the radial components, while SH-waves were used for the transverse components. The source time function of the mainshock was then convolved.

Figure 9 compares the synthesized waveforms of the mainshock for HYG004 (Event 1) obtained using the EGF method, with the observation data and the synthesis obtained using the EGTD method. The EGF method is applied for each of the 11 aftershocks. The amplitude levels of the synthesized waveforms, obtained using the two methods, were mostly in agreement with the observed waveform over the full duration of the analysis. In the low-frequency range examined in the previous study [8], the EGTD method provided a better match with the data than the EGF method. In the broadband frequency range, in this present study, the EGTD method seems to give stable results but not the best match with the data. It is noteworthy that the EGF method, in the case of Events 9 and 16, acceptably reproduced the amplitude for the radial components but twice, and almost a third time, overestimated the amplitude for the transverse components. For simulation using the EGF methods in a low-frequency range, a water-level of 0.20 has best suppressed extraordinary overestimation [8]. A water-level of 0.20, as the minimum absolute value of radiation coefficient, was tested with the current data but could not suppress this overestimation. This may arise from the complexity of high-frequency source processes. The synthesized waveforms using the EGF method look different from event to event. It therefore seems that the EGTD method, by removing the dependency of the EGF method on aftershock selection, provides more stable results.

7. Conclusions

The applicability of the EGTD method to simulating near-field strong motion seismic records has been demonstrated. Previous studies [6, 8] in EGTD estimation used the low-corner

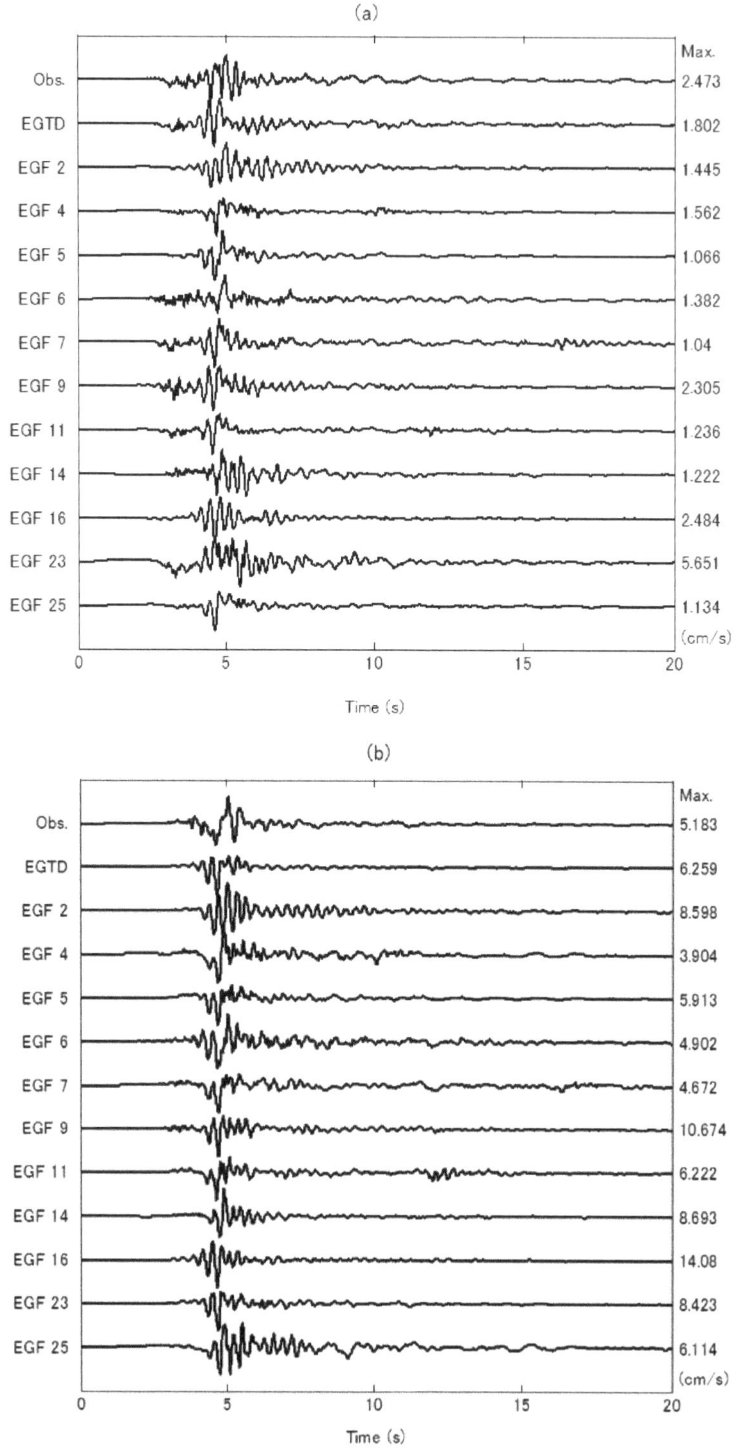

Figure 9. Comparisons of the synthesized waveforms of the mainshock for HYG004 (Event 1) obtained using the EGF method and each aftershock with the observed data and the synthesis obtained using the EGTD method. The numeral after 'EGF' represents the sequential event number of the specific aftershock used in the simulation for the mainshock. (a) Radial-comp; (b) Transverse-comp.

frequency of 1.0 Hz for the mainshock of the 2001 Hyogo-ken Hokubu earthquake (M_J 5.4) and targeted 0.2–1.0 Hz band-pass-filtered velocity waveforms. This study extended the upper limit of the target frequency range to 10 Hz, while the corner frequency of the events was in a range from 1.0 to 3.0 Hz. To correct corner frequency discrepancies, the scaling law based on the ω^{-2} model [12] was used to compensate the spectral amplitude decay beyond the corner frequency. Agreement between the observed and calculated waveforms for the mainshock was satisfactory for a long duration, and there was a good match between amplitudes. Further data accumulation and investigation enhance the applicability of the EGTD method.

Acknowledgements

The strong motion data used in this study were recorded at the K-NET stations, provided by the National Research Institute for Earth Science and Disaster Prevention (NIED) on their website. The Japan Meteorological Agency (JMA) unified hypocenter catalogue, and the F-net source parameters were also provided by the NIED on their website. This study was partially supported by Grants in Aid for Scientific Research (C) (16 K01316).

Author details

Michihiro Ohori

Address all correspondence to: ohorim@u-fukui.ac.jp

Research Institute of Nuclear Engineering, University of Fukui, Kanawa, Tsuruga, Japan

References

[1] Hartzell SH. Earthquake aftershocks as Green's functions. Geophysical Research Letters. 1978;**5**:1-5

[2] Irikura K. Semi-empirical estimation of strong ground motions during large earthquakes. Bulletin of Disaster Prevention Research Institute, Kyoto University. 1983;**33**:63-104

[3] Plicka V, Zahradnik J. Inverting seismograms of weak events for empirical Green's tensor derivatives. Geophysical Journal International. 1998;**132**:471-478

[4] Ito Y, Okada T, Matsuzawa T, Umino N, Hasegawa A. Estimation of stress tensor using aftershocks of 15 September 1998 M5.0 Sendai, NE Japan, earthquake. Bulletin of Earthquake Research Institute. 2001;**76**:51-59 (in Japanese)

[5] Ito Y. A study on focal mechanisms of aftershocks. Report of the National Research Institute for Earth Science and Disaster Prevention. 2005;**68**:27-89 (in Japanese)

[6] Ohori M, Hisada Y. Estimation of empirical Green's tensor spatial derivatives using aftershocks of the 2001 Hyogo-ken Hokubu earthquake and simulation of Mainshock (M_J5.4) strong motion. Zisin2. 2006;**59**:133-146 (in Japanese with English abstract)

[7] Pulido N, Dalguer L, Fujiwara H. Strong motion simulation on a dynamic fault rupture process and empirical Green's tensor derivatives. Programme and Abstracts of Fall Meeting of the Seismological Society of Japan. 2006;**D018**:127

[8] Ohori M, Hisada Y. Comparison of the empirical Green's spatial derivative method empirical Green's function method. Bulletin of the Seismological Society of America. 2011;**101**:2872-2886

[9] Ohori M. Simulation of broadband strong motion based on the empirical Green's spatial derivative method. In: Proceedings of Second European Conference on Earthquake Engineering and Seismology. Paper Number 1854; August 25-29, 2014; Istanbul

[10] Ohori M. Simulation of broadband strong motion based on the empirical Green's spatial derivative method. In: Earthquakes, Tsunamis and Nuclear Risks: Prediction and Assessment Beyond the Fukushima Accident. Tokyo: Springer; 2016. pp. 99-107

[11] Ohori M. Estimation of empirical green's tensor spatial derivative elements: A preliminary study using strong motion records in Southern Fukui Prefecture, Japan. In: Proceedings of 5th IASPEI/IAEE International Symposium: Effects of Surface Geology on Seismic Motion. Paper Number P109A. August 15-17, 2016; Taipei

[12] Aki K. Scaling law of seismic spectrum. Journal of Geophysical Research. 1967;**72**:1217-1231

[13] Ohori M. Source model estimation by waveform inversion using one- or two-station strong motion records, – A case for the 1990 Odawara earthquake (M_J5.1). Zisin 2. 2005; **57**:257-273 (in Japanese with English abstract)

[14] Hisada Y. An efficient method for computing green's functions for a layered half-space with sources and receivers at close depths (part 2). Bulletin of the Seismological Society of America. 1995;**85**:1080-1093

[15] Lawson CL, Hanson DJ. Solving Least Squares Problems. Englewood Cliffs: Prentice-Hall Inc.; 1974. p. 337

[16] Aki K, Richards PG. Quantitative Seismology, Theory and Methods. New York: W. H. Freeman; 1980. p. 932

5

The Earthquake Disaster Risk in Japan and Iran and the Necessity of Dynamic Learning from Large Earthquake Disasters over Time

Michaela Ibrion and Nicola Paltrinieri

Abstract

This book chapter targets how learning from large earthquakes disasters occurred and developed in Japan and Iran in the last 100 years. As research case studies, large earthquake disasters in Japan and Iran were investigated and analyzed. Normal distribution was found to be a good estimate of the magnitude distribution for earthquakes, in both the countries. In Japan, there is almost a linear correlation between magnitude of earthquakes and number of dead people. However, such correlation is not present for Iran. This lack of correlation in Iran and existence of linear correlation in Japan highlights that the magnitude of earthquakes directly affects the number of fatalities and extent of destruction in Japan, while in Iran, there is an increased complexity with regard to the factors affecting earthquake consequences. A correlation is suggested between earthquake culture and learning from large earthquake disasters in both Japan and Iran. Learning from large earthquake disasters is impacted by a multitude of factors, but the rhythm of learning in Japan is much higher if compared with Iran. For both Japan and Iran, a reactive learning approach based on past earthquake disasters needs to be constantly backed up by a proactive approach and dynamic learning.

Keywords: earthquake, earthquake disasters, Japan, Iran, learning, earthquake culture, earthquake risk, earthquake disaster risk, earthquake disaster risk management

1. Introduction

Earthquakes disasters build over time and an earthquake is a dramatic context that takes tiny units of time to bring huge human loss, injuries and massive destruction and damages [1–6].

Almost 12% of all fatal earthquakes in the world occurred along the margins of eastern side of Pacific and Japan is part of it. However, 85% of the most fatal earthquakes in the world occurred along the Alpine-Himalayan mountains belt and Iran is part of this seismic area. It was identified that both Japan and Iran belong to a particular cluster of countries that host the highest number of earthquakes and the world most fatal earthquakes [7].

The research objective of the present study targets to unveil how learning from large earthquakes disasters occurred and developed in Japan and Iran. The need for systematic learning from earthquake disasters is a challenge, which deserve particular attention.

The following section provide theoretical insights about learning from disasters. Further on, five case studies of earthquakes and large earthquake disasters occurred in Japan, were investigated and analyzed. For Iran, the case studies of 15 earthquakes and large earthquake disasters that took place over almost the same period of time, were considered. The timescale for both the countries covers more than a century-time period. As a research methodology, a historical longitudinal study with focus on earthquakes, earthquake disasters and learning from earthquake disasters was performed. A systematic comparison approach among the case studies in Japan and Iran was further implemented. Earthquake field reports, earthquake catalogs, archival documents, narratives of survivors and witnesses to earthquake disasters, cartographic materials, various academic studies and materials were consulted.

2. Learning from disasters

Disasters and crises can trigger learning and generate knowledge for communities and societies [8] and learning from disasters represents a facet of resilience, more precisely, the adaptive capacity [9, 10]. Cutter et al. [10] advised that learning from disasters occurs when disaster mitigation and preparedness are significantly improved and the resilience of communities is enhanced. Furthermore, social learning from disasters needs to be seen as result of various adaptations, actions, measures, and implementation of different policies and improvements, in order to mitigate and reduce the disaster risk. Moreover, Djalante et al. [11] emphasized that long-term planning, technical, logistical and financial preparations, implementation of legislation and policies, trainings and awareness programs, proper coordination and long-term monitoring would have their sound contributions to the learning from and adaptation to disasters. Ibrion [1] and Ibrion et al. [2, 4, 12] brought to attention that over the time the earthquake disasters in Iran have provided many lessons to be "learned" and with great potential to contribute to the socio-cultural learning and cultural memory about earthquakes and earthquake disasters. Learning from earthquake disasters is a long-term complex process, which involves various levels of participation and accountability for and requires an integrated disaster risk management and a sustainable framework such as an earthquake culture [1, 4].

2.1. Earthquake disaster risk in Japan and learning from large earthquake disasters over time

Aspects of learning from the earthquake disasters in Japan were investigated mainly through the case studies of five earthquakes and large earthquake disasters – *Great Nobi Earthquake*

1891, Great Kanto Earthquake 1923, Great Hanshin-Awaji or *Kobe Earthquake 1995, Niigata Chuetsu 2004* and *Great East Japan* or *Tohoku Earthquake 2011* – which affected various places of Japan, over a period of more than a century (**Figure 1**).

The **Great Nobi Earthquake** with Ms. (surface-wave magnitude) of 8.0 took place on 28 October *1891*, in the central part of Japan, north of Nagoya, in the Nobi Plain. This rural area and the cities of Osaka and Nagoya were terribly affected and more than 7300 people died. All the buildings that were heavily influenced by foreign architecture, especially English, collapsed. The Great Nobi Earthquake indicated the beginning of adaptation of the foreign buildings techniques, particularly European and American, to the Japanese nature and culture. This earthquake disaster was considered the "Bing Bang" for the Japanese seismic architecture and building construction in Japan. It encouraged the Japanese inventions of materials, architecture and buildings techniques designated to have seismic resistance. Almost a decade after the Great Nobi Earthquake, the seismology in Japan started to become institutionalized and attracted important funds from imperial and governmental institutions. From the beginning of the twentieth century, Japan started to be recognized worldwide not only as an earthquake country, but also as a front-runner in the production of knowledge about earthquakes [13].

On 1st September *1923*, Tokyo, Yokohama city and the surrounding towns and villages were shaken by the *Great Kanto Earthquake* with a magnitude of 7.9. The earthquake was followed shortly by a tsunami, a strong typhoon and a violent fire, which broke out all over Tokyo. The death toll and destruction, particularly from fire, were immense. It reached between 120,000–140,000

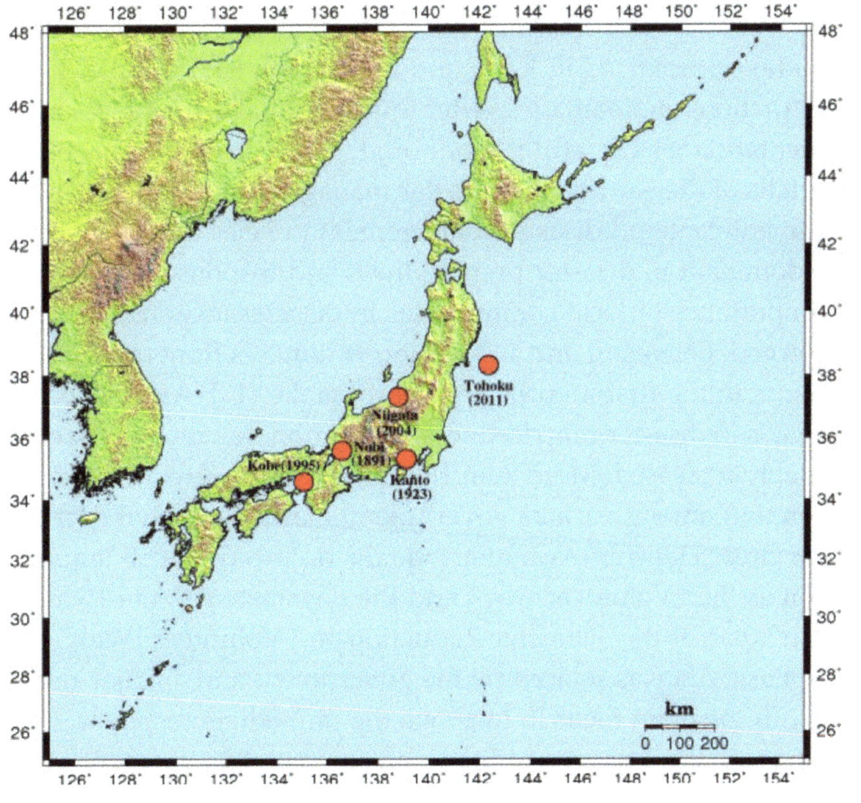

Figure 1. Japan-earthquakes and earthquakes disasters over more than a century. Source: Prof. Mohammad Mokhtari and Mr. Arash Islami, International Institute of Earthquake Engineering and Seismology (IIEES).

people and almost 2.5 million people were left homeless. The Great Kanto Earthquake contributed to the emergence of both a "culture of catastrophe" and a "culture of reconstruction", not only in Tokyo and Yokohama, but also in the whole Japan [14]. Starting from 1924 and based on the lessons from the Great Kanto Earthquake, Japan introduced and applied the first national seismic design code for the building design. Decades later, after the Fukui 1948 earthquake, in 1950, the Buildings Standard Law was adopted. Over the last half century, the building codes in Japan have been continuously revised and updated after the earthquake disasters. For instance, updates took place after Tokachi-oki in 1968, after Miyagi 1978, Kobe 1990, Niigata Chuetsu 2004. The amendment of the Building Standards Law in 1981 was particularly important because it offered the introduction of current earthquake engineering laws [15, 16]. Since 1960, the day of 1st September (the day when the Great Kanto Earthquake and a typhoon occurred) was designated as the "Disaster Prevention Day" in Japan. Every year, on this day, public awareness and educational programs are conducted all over the country, in order to commemorate and keep alive the memory about disasters in Japan [15].

An important step towards the institutionalization of disaster risk reduction and learning from disasters occurred in 1961, when the Disaster Countermeasures Basic Act was legislated. This legislative act had important consequences for Japanese disaster risk management and established the fundamental disaster prevention laws. In 1962, the national coordinating body for disaster management was established (the Central Disaster Management Council of Japan), which has role in formulating the overall policy for disaster risk management and is chaired by the Prime Minister. A special committee of the National Diet (Japanese Parliament) closely monitors disaster risk management and disaster risk reduction progress [15, 16].

On the 17 January *1995, the Great Hanshin-Awaji* known also as the *Kobe* earthquake with M_W (moment magnitude) 6.9 occurred in Kobe and its vicinity; 6437 people died and Kobe suffered massive destructions and damages. After this earthquake, many lessons were identified and their implementation was carefully monitored. The Kobe earthquake brought to attention various shortfalls of the earthquake disaster management in Japan. Until the Kobe 1995, Japan had government-centered disaster management policies and practices and the role of government was dominant in disaster preparedness and response. However, after the Kobe earthquake, the importance of local communities in disaster response and disaster management strongly emerged. Between 1 and 1, 5 million volunteers from outside Kobe contributed to help the survivors in the first months after earthquake [17]. After Kobe 1995, the role of volunteers and NGOs in disaster preparedness, mitigation, rescue, relief, reconstruction, and recovery considerably increased. More than 16 national laws were adopted in Japan in order to improve cooperation among various governmental levels and civil society organizations. The day when the Great Hanshin-Awaji Earthquake occurred, the 17 January, was declared by the government as the "Volunteer Day" and the days between the 15 and the 21 January were designated in Japan as the "Disaster Reduction and Volunteer Week" [18]. The Disaster Countermeasures Basic Act was revised by the government and a major revision was implemented in 1995, with the main focus to improve the immediate response system. In January 2001, a new position of the Minister of State for Disaster Management was created in the Cabinet Office and was fully dedicated to inter-ministerial planning and coordination. The new minister can also act on behalf of the Prime Minister for the Central Disaster Management Council [15]. Based on the lessons from Kobe earthquake disaster, the seismic engineering standards for road bridges were revised in 1996 and 2002 and their implementation closely monitored. Seismic engineering standards for railways were revised in 1998 and seismic reinforcement of elevated tracks, bridges, supporting pillars and the tunnels of Shinkansen and main artery lines were also closely monitored [16]. The Hyogo Prefectural Government with financial and technical support from the Cabinet Office organized the Disaster Reduction and Human Innovation Institution (DRI), in 2002, as a center for research and study of disaster risk management together with a museum facility in Kobe. The aims were to promote the lessons from the Kobe 1995, increase earthquake risk awareness among the younger generations and improve earthquake disaster management in Japan. The DRI museum exhibits the devastations of the Kobe earthquake disaster and reconstruction process. The Kobe's survivors were asked to convey their disaster experiences and invited to act as volunteers at the museum [15].

Just a few months before the 10th anniversary of the Kobe 1995, on 23 October *2004*, an earthquake with M_w 6.8 occurred, in *Chuetsu* region, *Niigata* Prefecture. It caused 68 fatalities and more than 4805 injuries. Moreover, another earthquake occurred in Niigata prefecture on 16 July 2007 with M_w 6.8. After the Niigata 2004 earthquake, the system of the Shinkansen trains was upgraded to avoid derails due to earthquakes. Few years after the Niigata Chuetsu 2004, the Japan Meteorological Agency (JMA) launched on 1 October 2007, the Earthquake Early Warnings (EEW) system [16].

On 11 March *2011*, an earthquake with M_w 9.0 occurred in the Tohoku region, with the epicenter near the Japan Trench off the coast of Japan. The *Tohoku* earthquake known also as the *Great East Japan Earthquake* was 45 times more powerful than the Great Kanto Earthquake in 1923, and 1450 times more powerful than the Great Hanshin-Awaji Earthquake in 1995. Less than half an hour later, the earthquake was followed by a massive tsunami, which reached to height of 40 m in some areas over the land. The nuclear reactors at the Fukushima Daiichi Nuclear Power Plant were seriously affected and uncontrolled radiation was spread in the atmosphere, water and land [17]. The number of dead people reached 20,000, but the death and destruction from the Tohoku earthquake itself was much less than previous earthquakes, because of learning from previous earthquake disasters and progress of earthquake disaster preparedness and earthquake disaster risk management [16]. Nevertheless, the magnitude of this earthquake exceeded the hazard level of any earthquake in Japan ever considered for earthquake disaster risk management and emphasized that Japan needs to be prepared for the worst scenario of hazard and disaster. "Prepare for the worst case" was recommended in Japan ([16], p. 231). The EEW system and seismic reinforcement of houses and critical infrastructures effectively reduced the number of casualties from the earthquake. After the EEW system was activated, on 11 March 2011, all Shinkansen trains were safely slowed down and stopped with no derailments or injuries to passengers on board. Moreover, because of seismic reinforcement of elevated rail tracks and bridges, none of them collapsed. All households in Japan connected to city gas service are equipped with micro-chip controlled gas meters which automatically shut down gas flow when an earthquake with a seismic intensity stronger than 5 is detected in area. Due to the learning lessons from the large-scale fire occurred in Kobe in 1995, on 11 March 2011 the gas flow to each household was stopped and the Sendai City Gas Service interrupted all the supplies from their gas factories [15]. The social learning occurred between the earthquake disasters of Kobe 1995 and Tohoku 2011 positively affected relief and reconstruction efforts after the 2011 events. For instance, gas, water, electricity were reconnected after 1 week. The

Operation Comb had as result the reopening of the railway service and roads in 4–15 days and the repair of bridges took one weak. Radios were distributed immediately and internet and phone lines were restored within 3–7 days. The construction of housing units started in 8 days and more than 100,000 people were housed in 3–4 weeks [18].

The disaster management of Japan registered many improvements, as since Kobe 1995, a government crisis center was on duty 24 h of 7 days with a stand-by emergency team. Moreover, the Japanese government took action just 4 min after earthquake and an Emergency Disaster Response Headquarters, headed by the prime minister, was organized within 30 min. Immediately after the Tohoku 2011, the National Committee for Emergency Management headed by Prime Minister was assembled. A national emergency state was declared in Japan and more than 100,000 troops of the Japan Self-Defense Forces (JSDF), both active and reserve troops, were started to be deployed and reached in affected areas within hours. JSDF reached to 107,000 people; they were fully equipped and deployed operations with 540 aircrafts and 60 vessels. They rescued 19,000 disaster victims – almost 70% of the people rescued in the cascading Tohoku disaster [16].

After the Tohoku 2011 disaster, the Prime Minister Naoto Kan acknowledged the help received from over 130 countries, 30 international organizations and 670 NGOs. This was a different attitude if compared with the situation after the Kobe earthquake in 1995, when the Japanese government displayed an extreme reluctance to receive help. Moreover, the American military forces were invited to help Japan and the operation Tomodachi (friend) took place. With the help from USA forces, on 13 April 2011, the Sendai Airport was reopened to commercial flights [19]. The Central Disaster Management Council of Japan commissioned a special Committee for Technical Investigation on Countermeasures for Earthquakes and Tsunamis in order to draw lessons from 2011 disasters. Based on the findings of this committee, the Disaster Countermeasures Basic Act was revised twice, First Amendment in 2012 and Second Amendment in 2013, in order to prepare Japan for large-scale disasters [15].

2.2. Earthquake disaster risk in Iran and learning from large earthquake disasters over time

Aspects of learning from the earthquake disasters in Iran were investigated mainly through the case studies of 15 earthquakes and large earthquake disasters – *Silakhor 1909, Salmas 1930, Torud 1953, Buyin Zahra 1962, Dasht-e Bayaz 1968, Ferdows 1968, Karzin-Qir 1972, Tabas 1978, Golbaf 1981, Sirch 1981, Rudbar 1990, Zirkuh (Qa'enat) 1997, Bam 2003, Ahar 2012* and *Shonbeh-Bushehr 2013* – which affected different places in Iran, over a period of more than a century, see **Figure 2**.

On 23 January *1909, Silakhor* earthquake with M_w 7.4 occurred in Silakhor valley. More than 8000 people died, more than 65 villages were destroyed and town of Bahrain (nowadays Dorud) and the city of Borujerd suffered damages [20]. At that time, Persia (the country name changed to Iran in 1935 under Reza Shah Pahlavi) was ruled by the Qajar dynasty and suffered the consequences of the 1907 Anglo-Russian Convention. Within the geopolitical landscape of Persia at that time, the earthquake disaster occurred in the shadow; neither relief

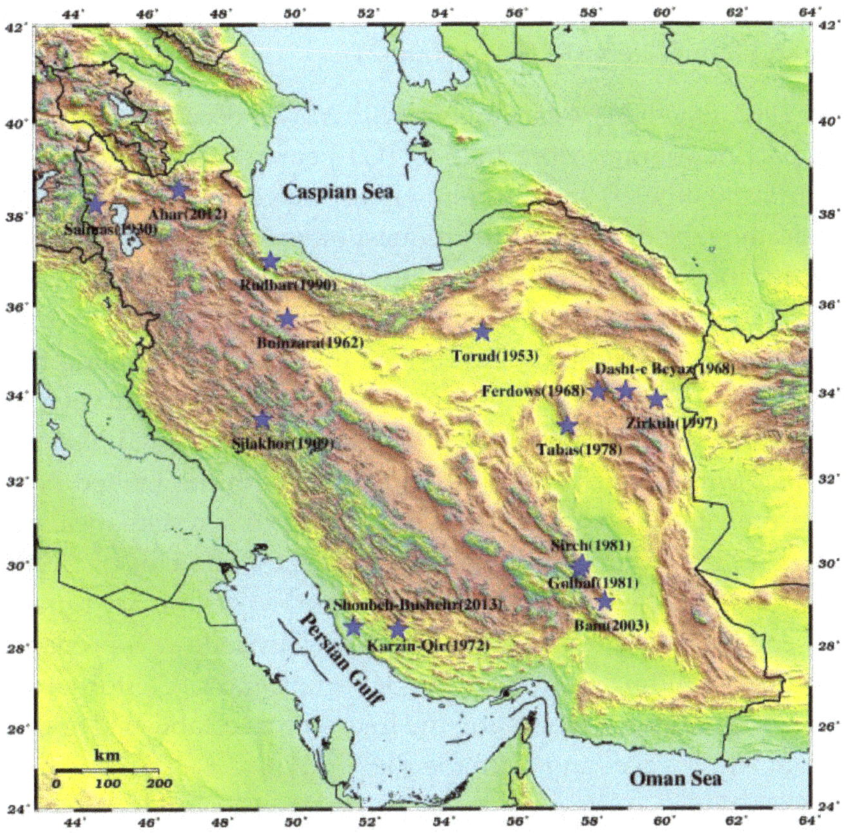

Figure 2. Iran-earthquakes and earthquakes disasters over more than a century [1].

was sent to the affected areas, nor reconstruction efforts were supported by the authorities or competing powers in Persia, Russia and England.

Salmas earthquake with M_w 7.1 occurred on 7 May *1930*, in Salmas Plain and mountains areas, northwest of Iran. The cities of Dilman and Kohneh Shar together with other 60 villages were destroyed and around 2514 people died. An alarming foreshock, which occurred in the morning before earthquake, significantly contributed to lower the number of deaths. The majority of population in the area spent the night of the earthquake outdoors and casualties occurred only among the population that did not feel the foreshock or perceived it just as a small tremor. Reaction of the local population to this warning foreshock drew attention to the existing level of earthquake risk perception in the area. The official attempts for rescue and relief in Salmas and other affected areas were impacted by a severe condition of roads after earthquake [20].

On 12 February *1953*, *Torud* earthquake with M_w 6.5 occurred in the central part of Iran on the margins of Dasht-e Kavir (literally "Great Salt Kavir") killing 930 people. Torud earthquake was investigated by the Iranian-Armenian professor Setrak Abdalian, who inaugurated the practice of earthquake geology in the field in Iran. Almost 3 years after Torud, progress towards the study of earthquake geology in Iran was registered, as in 1957, the Institute of Geophysics was established at the University of Tehran as an institution monitoring and reporting the seismic activity in Iran [20]. One of the first legal steps taken in Iran towards disaster mitigation was the establishment of the Civil Defense Organization in 1958. This organization operated under the Ministry of Interior and was supported by the Army Forces and the Red Lion and Sun Society (RLSS) of Iran [21].

On 1st September *1962, Buyin Zahra* earthquake with M_w 7.0 occurred in the southern Qazvin area, south-west part of Tehran. More than 12,000 people died, almost 3000 people were injured, Buyin Zahra and other 91 surrounding villages were destroyed. Within few days after the Buyin-Zahra earthquake disaster, at the request of governmental institutions of Iran, the United Nations Educational, Scientific and Cultural Organization (UNESCO) sent two missions to Iran in order to study the Buyin-Zahra earthquake and investigate further seismological and engineering aspects. The Buyin-Zahra 1962 marked the beginning for study of large magnitude earthquakes in Iran. Moreover, in 1962, the Geological Survey of Iran (GSI) was established in Tehran, based on a special fund project from the United Nations Development Programme (UNDP), with financial contributions from both the United Nations (UN) and the Imperial Government of Iran. GSI started to proceed to a systematical geological investigation of all parts of Iran. After the Buyin-Zahra 1962, the director of UNESCO convoked an international seminar, at the University of Tehran and a preliminary report about the first Iranian buildings code was discussed at this seminar. This code was prepared based on the United States Uniform Building Code and the San Francisco's, California Code [20]. After Buyin Zahra 1962, a Committee for Assisting the Injured was established and the board members were both from Army forces and from the RLSS [21].

Dasht-e Bayaz earthquake with M_w 7.1 occurred on 31 August *1968* in an arid area of the eastern part of Iran. More than 10,000 people died and more than 160 villages either were destroyed or suffered damages.

On the western part of the Dasht-e Bayaz fault and almost 20 h after the Dasht-e Bayaz earthquake, on 1 September *1968* and also 3 days later, two earthquakes with M_w 6.3 and 5.5 respectively, occurred in the *Ferdows* region. The town of Ferdows and many other villages were almost destroyed and more than 750 people died [20]. The first national Iranian code for the seismic resistance building design, known as the Iranian Standard ISIRI (Institute of Standards and Industrial Research of Iran), code no. 519, was approved and started to be implemented in Iran from 1969, immediately after the Dasht-e Bayaz 1968 earthquake disaster [20]. After the Dasht-e Bayaz 1968 earthquake, the Committee for Assisting the Injured in Iran established the Rescue and Relief Organization as a sub-branch of RLSS in order to conduct the emergency disaster response in Iran [21]. After the Dasht-e Bayaz 1968, a second UNESCO mission was deployed to Iran. For the first time, after an earthquake in Iran, the National Cartographic Center of Iran took aerial photos at a scale of 1:7500 for the earthquake affected areas. After Dasht-e Bayaz 1968, a systematic investigation of the large-magnitude earthquakes was deployed in Iran. A joint project dedicated to research of Iran's seismicity commenced between the Technical Bureau of the Plan and Budget Organization of the Imperial Government of Iran and the Imperial College of Science, London. Moreover, in 1971, the first research department of Tectonics and Seismotectonics was established at the Geological Survey of Iran by Dr. Manuel Berberian. In 1973, Ali Akbar Moinfar established the first strong motion network of Iran. In 1976, the satellite imagery, more precisely ERTS (Earth Resources Technology Satellite) called later Landsat, was used for the first time in Iran for active fault detection and earthquake studies [20].

Karzin-Qir earthquake, with M_w 6.9 occurred on 10 April *1972* and more than 5000 dead people and 1710 injured were registered in 50 villages [20]. After the Karzin-Qir earthquake, the role of Civil Defense Organization (CDO) of being in charge with disaster risk management in Iran increased. At that time, CDO had important roles in disaster risk management of Iran, as was responsible to assure the coordination among various governmental institutions during the disasters and crisis emergencies. Other responsibilities targeted to promote disaster preparedness and public safety in Iran. After the Islamic Revolution in 1979, in Iran, the CDO was dissolved and its functions were transferred to the Basij Mostaz'afin (Revolutionary People Union) [21].

Tabas earthquake with M_w 7.4 occurred on 16 September *1978*; the town of Tabas, was terribly affected and 85% of its inhabitants were killed as 11,000 of its 13,000 inhabitants died. Nevertheless, the total number of deaths reached to 20,000, as other 90 villages around Tabas suffered destruction or severe damage [20]. Just few months later after Tabas 1978 earthquake disaster, Reza Shah Pahlavi left Iran and the Islamic Revolution took place in 1979. Furthermore, between September 1980 and August 1988, the 8 years Iran-Iraq war occurred. Focus of Iran and the resources of country were concentrated on other dramatic events and then, far from learning from large earthquake disasters and building an earthquake culture in Iran. After the Islamic Revolution in 1979, the crisis management responsibilities were transferred to the Prime Minister's Office. However, in 1989, the position of Prime Minister was dissolved in Iran.

During Iran-Iraq war time (1980–1988), on 11 June *1981*, *Golbaf* earthquake with M_w 6.6 destroyed the town of Golbaf, situated near Kavir-e Lut, southeastern part of Iran and 1400 people were killed. On 28 July *1981*, 47 days later, the area was again severely affected by the *Sirch* earthquake with M_w 7.0. At that time, 25 villages were completely destroyed, and between 800 and 1300 people died (Berberian 2014). During the hardship of the Iran-Iraq war, various actions in Iran added to efforts towards learning from earthquakes and earthquake disasters in Iran. The Department of Tectonics and Seismotectonics, at Geological Survey of Iran, had a great contribution to the seismotectonics knowledge about Iran. Active faults for the Greater Tehran and Qazvin area were mapped and many fieldworks were conducted after each earthquake which took place in this period of time. The second revised Iranian code for seismic resistant design of the buildings, ISIRI code no. 2800 was approved in February 1988 and for the first time, a map with active faults of Iran was used for its preparation [20].

In November 1989, just few months before Rudbar 1990 earthquake disaster, based on the 24th UNESCO General Conference Resolution DR/250 and the Iranian governmental ratification, an organization named the International Institute of Earthquake Engineering and Seismology (IIEES) was established. IIEES is part of the Ministry of Science, Research and Technology and in 1999, it was approved to become a research center of Iran and has as aim to reduce earthquake risk and to improve earthquake mitigation in Iran.

Rudbar earthquake with M_w 7.3 occurred half an hour after midnight, on 21 June *1990*, in the northern part of Iran, southwest side of the Caspian Sea. The Rudbar earthquake affected both rural and urban areas. The most affected areas were the towns of Rudbar, Manjil, Lowshan and Harzehvil and many villages in Gilan and Zanjan provinces. The city of Rasht suffered also damages. More than 40,000 people were killed, more than 60,000 people were injured, and 700 villages were destroyed [20]. A major lesson brought by Rudbar earthquake disaster was the massive damage caused by earthquake-induced landslides [4]. After the Rudbar 1990, in terms of disaster risk management, the leadership was transferred to the National Disaster Task-Force (NDTF), which was part of the Ministry of Interior. The Ministry of Interior was officially assigned as the executive agency to oversee and coordinate all activities with reference to disaster prevention, mitigation, emergency relief and rescue operations, rehabilitations and reconstruction. In addition to the central Disaster Task-Force, there were also provincial, district or municipality and sub-district Disaster Task-Forces. In 1991, the National Committee for Natural Disaster Reduction (NCNDR) was established as a policy-making organization for disaster risk reduction in Iran; the NCNDR was part of the Ministry of Interior and the Minister of Interior was the head of NCNDR [21].

On 10 May *1997*, another large earthquake disaster took place after the *Zirkuh (Qa'enat)* earthquake with M_w 7.2. According to official figures, about 1568 people died and 2600 were injured. After the Zirkuh 1997, starting from 1998, the technique of InSAR (Interferometric Synthetic Aperture Radar) was used for the first time in Iran, for study of the Gowk Fault and associated earthquakes. Towards the implementation of technology for seismological studies in Iran, a Iranian-French collaboration was defined by IIEES and the first-large scale Permanent Global Positioning System (IPGN) was implemented with 25 sites in Iran, 2 in Oman and 1 in Uzbekistan [20].

Bam earthquake with M_w 6.6 occurred on 26 December *2003* and the Bam city, in Kerman province, was totally destroyed. The town of Baravat and the villages around also suffered damages. The death toll passed more than 40,000 people. The cultural landscape of Qanats, gardens of Khorma trees, Arg-e Bam and their associated cultural beliefs had a strong impact on the local earthquake risk perception before this earthquake [5, 6].

After Bam 2003, the authoritative command of disaster risk management in Iran was given to the President of Iran. General Policies on Natural Hazards Risk Mitigation and Disaster Management were prepared by the Expediency Council of Iran in 2005. However, many of these policies are yet pending to be implemented in Iran. Moreover, in 2008, the National Disaster Management Organization (NDMO) with national, city and provincial structures was established in Iran for disaster risk management and risk mitigation. Many difficulties were registered for coordination of various organizations, lack of expertise for implementations of different plans and policies, lack of regular updates and supervision and especially, the allocated budget was not sufficient. The Integrated Disaster Risk Management (DRM) in Iran suffers of many shortcomings as many institutions and organizations are working individually with no effective collaboration among them. Moreover, DRM laws and regulations, policies and plans did not take in account local conditions and community based preparedness and participation to disaster reduction and disaster risk management. Effectiveness of many earthquake disasters plans and policies has not been satisfactory in Iran, as many of them were not applied in practice, properly monitored and without adequate legal enforcements for their application [22].

On 11th August *2012*, two earthquakes-*Ahar* earthquakes- with a magnitude of M_w 6.4 and 6.3 respectively, occurred at an interval of 11 min, in northwest part of Iran, northeast side of Tabriz. Ahar and Varzeqan towns suffered damages, 300 people died and more than 3000 suffered injuries. Rescue and relief reached the area very late [20].

On 9th April *2013*, at 16.22 local time, *Shonbeh-Bushehr* earthquake with M_w 6.4 occurred in Dashti district, south-west part of Bushehr province and 37 people died and 850 were injured. The Building and Housing Research Centre (BHRC) as part of the Ministry of Housing and Urban Development is conducting periodical revisions of the Iranian code of practice for the seismic resistance design of buildings in Iran. However, the implementation of the seismic building codes is still on long way to become in practice in Iran.

3. Discussions

All the above earthquakes and earthquake disasters in Japan and Iran, over more than a century-time period, shall be considered as important wake-up calls towards the learning from large earthquake disasters and building an earthquake culture. Their death toll (**Figure 3**) have required long-term sustainable strategies for earthquake disaster risk reduction in Japan and Iran. The disaster of Shonbeh-Bushehr 2013, Iran, with 37 people dead, was not presented in **Figure 3**.

A probability density function (PDF) of the number of the dead people in Japan and Iran was plotted and a wide scatter is seen for Iran case, while a narrower scatter with lower mean value is observed for Japan (**Figure 3**, right). It should be noted that for the Tohoku 2011, only 6% of 20,000 death people were taken in account for plotting of the PDF, as according to estimations, the majority of death was caused by the tsunami drownings and only 6% of causes of death were caused by collapse of buildings and fires after earthquake [18]. Moreover, with reference to Great Kanto 1923, the death toll was not consider for the PDF, as there is a significant uncertainty about the number of dead people associated with earthquake. In fact, the majority of death was caused by secondary hazards such as fires, tsunami, landslides, drowning in the water [14].

In Iran, there is no linear correlation between magnitude and number of dead people, as even a medium magnitude earthquake caused one of the highest number of death, such as the

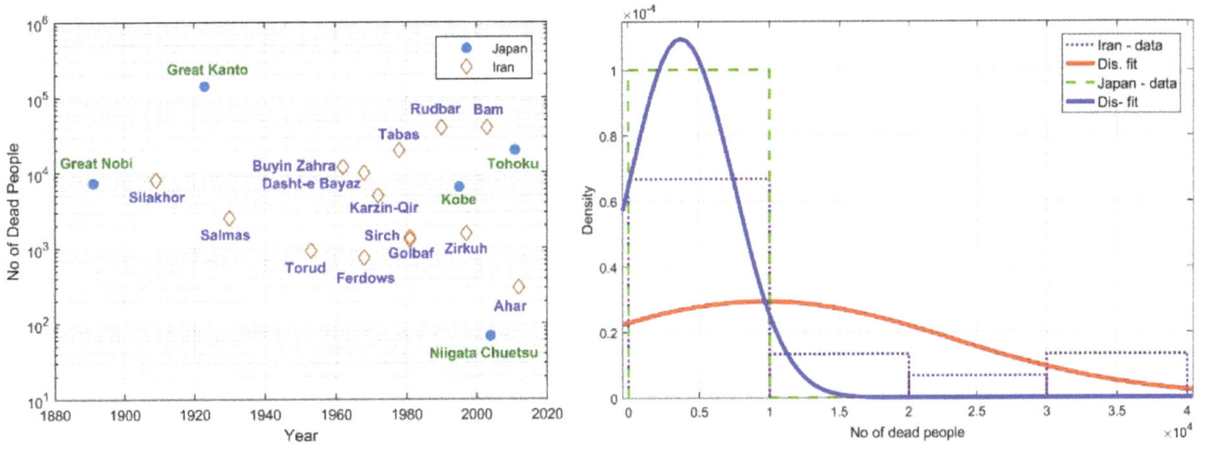

Figure 3. Earthquakes and earthquake disasters in Japan and Iran over more than a century: The number of dead people versus year (left) and the probability density function (PDF) of the number of the dead people (right).

earthquake in Bam in 2003. However, in Japan, there is almost a linear correlation between the magnitude of earthquakes and the number of dead people (**Figure 4**). This lack of correlation in Iran and existence of linear correlation in Japan highlights that the magnitude of earthquakes directly affects the number of fatalities and extent of destruction and damage in Japan, while in Iran, there is an increased complexity with regard to the factors affecting earthquake consequences.

In Iran, it is seen that all the 15 earthquakes are less than 7.5 magnitude, while in Japan, the magnitude is much higher and can reach 9 (**Figure 4**).

The probability density function (PDF) of the earthquakes' magnitude in Iran and Japan, over more than a century- and the probability paper plot of the Normal distribution are shown in the **Figure 5**.

A normal distribution was derived from the magnitude distribution of earthquakes in both Iran and Japan, as illustrated in the probability plot in **Figure 5**. The mean values of the normal distributions are 6.88 for Iran and 7.72 for Japan. The variance for Iran is 0.146 and for Japan is 0.817. This indicates a wider scatter of magnitude for Japan. It is also interesting to observe that while an earthquake with magnitude of 8 or higher is towards the tail of the distribution with low probability in Iran, such high magnitude earthquake has a high probability to occur in Japan (**Figure 5**). Despite the lower mean value for the earthquakes in Iran, this did not lead to a lower number of fatalities. This further confirms the lack of linear correlation between magnitude and number of dead people in Iran and the complexity of

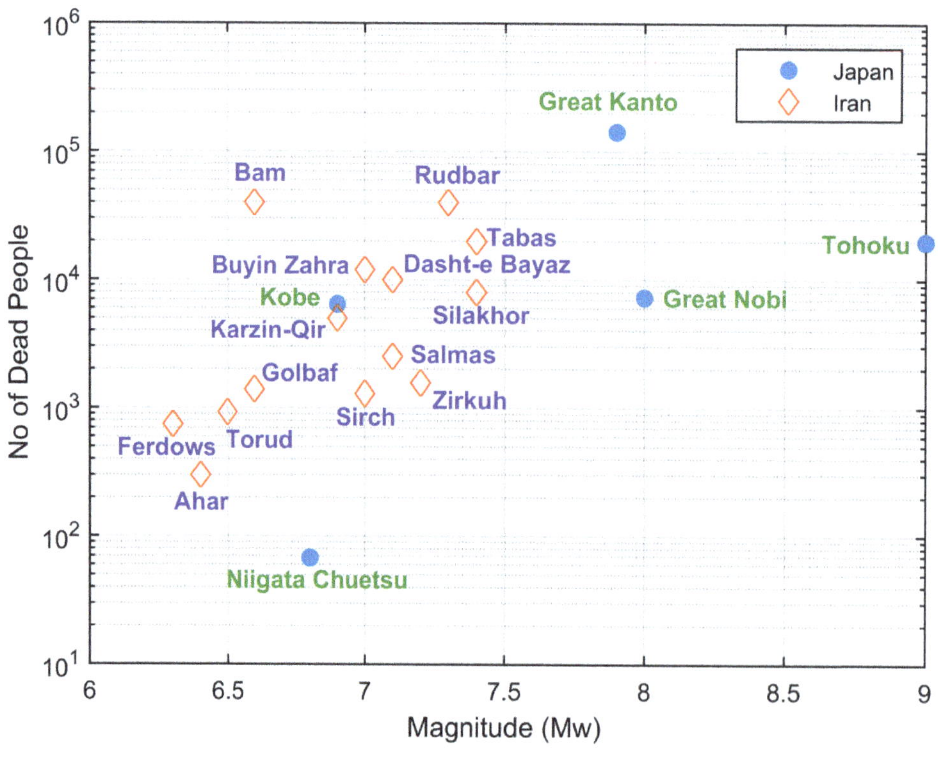

Figure 4. Earthquakes and earthquake disasters in Japan and Iran over more than a century and the number of dead people versus magnitude.

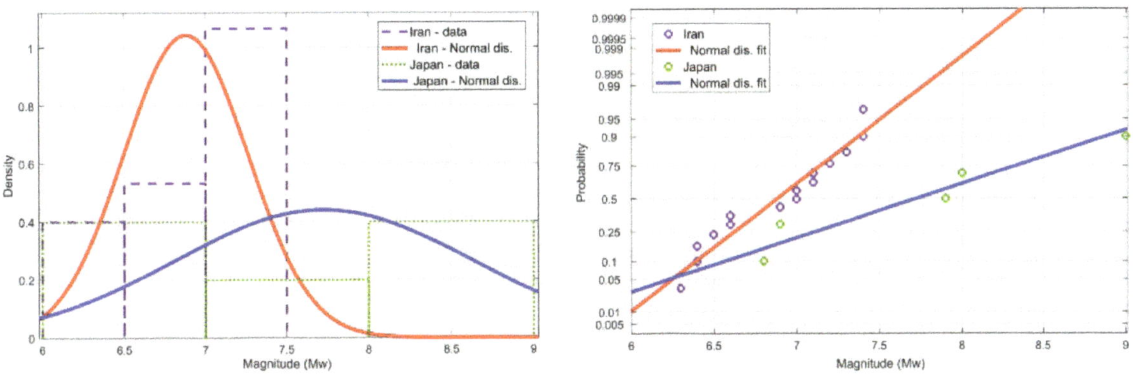

Figure 5. Probability density function (PDF) and the probability paper plot of the normal distribution of the earthquakes' magnitude in Iran and Japan.

factors concurring in an Iranian earthquake disaster. From the Japanese perspective, Japan has learned the lessons from earthquake disasters over time, but the magnitude and seismic intensities of earthquakes has high impact on the number of fatalities.

Japan long-term experience of earthquake disasters has encouraged a culture of prevention [15], a seismic culture [23] and a culture of Lessons-Learned [16]. Over the time, Japan learned how to prepare for and to respond to earthquake disasters and how to foster an earthquake culture, and to learn from disasters. Over the last century and in the present century, learning from earthquake disasters, building of the earthquake culture and earthquake disaster risk reduction were not among the national priorities of Iran. The earthquake culture in Iran is still far from becoming a reality of the present time [1].

Learning from large earthquake disasters in Japan and Iran have been impacted by various factors, **Figure 6**. As a remark, the landscape of factors that have influenced and impacted learning from large earthquake disasters in Japan and Iran, as presented by **Figure 6** is non-exhaustive and other factors can be further identified. Some of the factors are non-existent for Iran, like Earthquake Early Warning (EEW), museums and monuments about preservation of memory from earthquake disasters, accountability of governance and decision makers for earthquake disaster risk reduction (EDRR).

The impact of these factors on the learning in Japan and Iran was not constant over time and further introspection is required about their dynamism over the time.

In **Figure 6**, it can be observed for the case of Japan that science and technology has had both a negative and positive impact on the learning from large earthquake disasters. After March 2011 cascading disasters, in Japan, the trust of society in government and especially, the confidence in science and technology and overreliance on scientific information were dramatically shaken. Many voices from the Japanese society expressed that the Tohoku 2011 disasters was seen as divine punishments for strong beliefs of the Japanese people about their supremacy and control over nature through science and technology. The Tohoku 2011 disasters were believed to be tough lessons for the Japanese nation, because its people had supposedly become arrogant and disrespectful towards the power of nature. A politician declared that "this disaster is divine punishment for our selfishness". At a conference of the Seismological

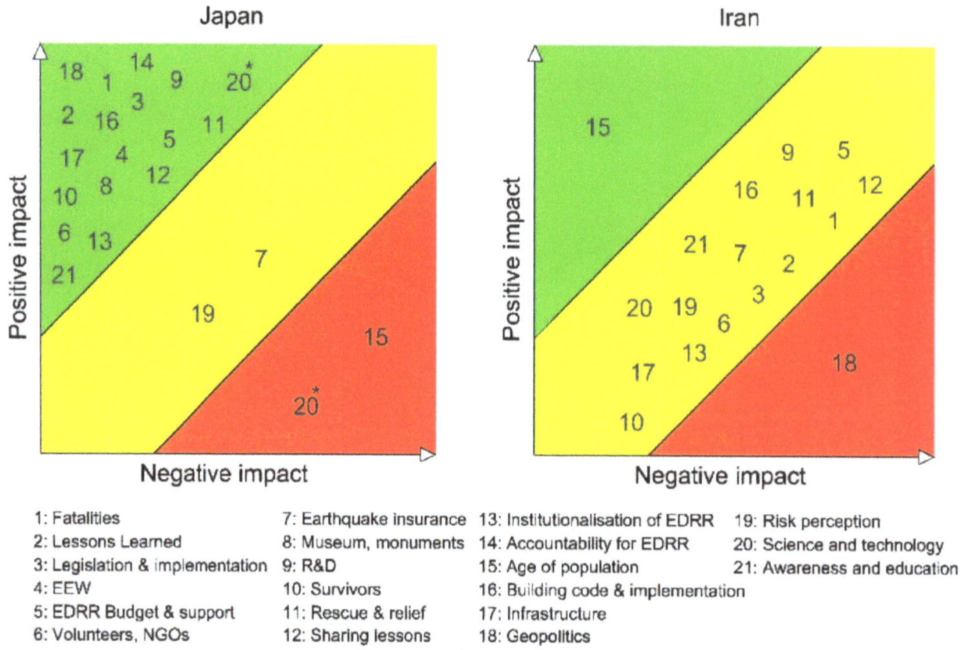

Figure 6. Overview of a non-exhaustive landscape of factors that have affected learning from large earthquake disasters in Japan and in Iran.

Society of Japan, there were many Japanese earthquake researchers who categorized the Tohoku 2011 disaster as a "failure of seismology". The reasons why the Tokohu 2011 events were categorized as "unforeseen" (souteigai, in Japanese) were seen as unacceptable ([17], p. 73). Moreover, the myth of nuclear safety in Japan was seriously shaken among society. After the Tohoku 2011, in addition to scientific knowledge, the importance of local knowledge about disasters and disaster preparedness started to strongly revive in Japan [17].

The geopolitics in Iran and in the Middle East area, the complex and dynamic power structures in Iran had a strong negative impact on the learning from earthquake disasters and building an earthquake culture. Over more than a century, the earthquake disasters were frequently overshadowed by other disasters in Iran, such as: the 8 years Iran-Iraq war, intervention of foreign powers in Iran (Russia, UK, USA), political turmoil (Mosadeq 1953, Iranian Revolution 1979). Despite very rich oil and gas resources of Iran, the country focus and support have been shifted further from mitigation and preparedness for earthquake disasters and development of an earthquake culture [1]. One of the critical issues faced nowadays by the Japanese society is the aging of population, which can have serious negative impact on disaster preparedness. Japanese population is aging with the fastest rate in the world. According to extrapolations from existing trends, one in three people will be over 65 years old by 2050 [24, 25]. Moreover, Maki [24] highlighted that together with aging, massive migration of young population after the Tohoku 2011 disasters, and depopulation in many areas of Japan will jeopardize disasters preparedness, response and resilience in Japan. On the positive side, Iran has a higher percentage of young population compared to Japan. According to demographics, the median age for Iran is 29.4 and for Japan is 46.9 [26].

In Japan, a few years later after the Niigata 1964 earthquake, more precisely from 1966, the Earthquake Insurance Law was ratified and started to be implemented, but the earthquake insurance is still not well developed in Japan. In Iran, despite many earthquake disasters, the insurance penetration ratio is very low. Laws, regulations and policies in Iran, in the last decades, did not encourage development of a catastrophe insurance pool [27]. Earthquake disasters in Iran brought many lessons to be "learned" and with great potential to contribute to the social-cultural learning and social-cultural memory [1, 4, 12]. However, from the investigated case studies of large earthquake disasters, only few lessons were consistently oriented towards the earthquake disaster risk reduction in Iran [4]. In Iran, an institutionalization of earthquake disaster risk occurred in the last half century, but legislation, policies and practices need to be supported by adequate planning and budget, implemented, and continuously monitored [21, 22]. Moreover, the JICA report [21] drew attention to important shortcomings of the earthquake disaster risk reduction in Iran. The critical areas included were not limited to the implementation of legislation and policies in Iran, the lack of national, regional, urban and rural disaster prevention and disaster risk management plans, the funds and budgets not properly allocated for earthquake disaster mitigation, preparedness and emergency stages, lack of implementation of seismic codes and seismic retrofitting. In addition, measures and tasks related to disaster preparedness and disaster risk management were not clearly identified, defined and explained, at both national and regional levels.

Tohoku 2011 cascading disasters highlighted for Japan that "Preventive investments pay, but be prepared for the unexpected", as it is important to understand that "the risks from natural hazards can never be completely eliminated" [16]. Furthermore, the Tohoku 2011 cascading disasters (earthquake, tsunami and nuclear catastrophe) highlighted for Japan the absolute requirements for a dynamic learning from earthquake disasters, a sustainable earthquake culture and the necessity of continuous improvement towards an integrated earthquake disaster risk management.

Over time, a profound institutionalization of earthquake/disaster risk reduction efforts occurred in Japan. In addition, the required legislation, policies, practices and required budget exist and are being implemented and carefully monitored by the government and Parliament. Earthquake disaster risk reduction, which is mainstreamed in the national policies and Japan disaster risk management, is always part of the national political agenda. Moreover, security of financial resources and the required budget for earthquake disaster risk reduction is continuously assured [15]. Worldwide, an effective and sustainable way of learning from earthquake disasters and mega-disasters was shown by Japan and its status of earthquake nation [13, 16]. After a necessary adaptation to the Iranian local context, the Japanese culture of continuous learning from past earthquake disasters and continuous improvements of earthquake disaster preparedness can serve as a model for Iran [1, 4, 12].

4. Conclusions

An investigation of earthquakes and large earthquake disasters in Japan and in Iran over a period of more than a century brought to attention that a normal distribution was found to be a good estimate of the magnitude distribution for earthquakes in both Japan and Iran. In Japan, there is almost a linear correlation between magnitude of earthquakes and number of dead people. However, in Iran, there is no linear correlation between magnitude and number of dead people, as even a medium magnitude earthquake caused one of the highest number of death. This can be an indicator that the magnitude and seismic intensity of earthquakes directly affects fatalities and damages in Japan, while in Iran there is an increased complexity in the factors affecting earthquakes consequences.

There is a strong correlation between the earthquake culture and learning from large earthquake disasters. The learning from large earthquake disasters in both Japan and Iran requires existence of a dynamic system such as the earthquake culture. Nevertheless, the existence and development of the earthquake culture is conditioned by the dynamic learning from earthquake disasters over time.

Learning from large earthquake disasters in both Japan and Iran has either encountered medium or less impact from different factors or has been influenced in a negative and positive way. However, the rhythm of learning from earthquake disasters in Japan is much higher comparative with Iran. Over more than a century, Japan learned how to mitigate, prepare for and respond to earthquakes and earthquake disasters and foster an earthquake culture. Improvements may be still needed, but a steady progress towards the integration of the earthquake disaster risk management has occurred in Japan. On the other hand, in Iran, integration of earthquake disaster risk management is still lacking a structured approach and is requiring urgent and critical measures and actions.

Dynamic learning from earthquake disasters, a sustainable earthquake culture and an integrated earthquake disaster risk management can positively contribute not only to the earthquake disaster risk reduction, but also towards tsunami risk disaster reduction and nuclear safety in Japan and in other countries at risk around the world.

In Iran, an institutionalization of earthquake disaster risk reduction took place in the second half of the last century and there has been a favorable foundation and factors that can contribute to the building and development of an earthquake culture. However, after more than a century-time period of dramatic experiences of many large earthquake disasters, the learning from large earthquake disasters is slow and an earthquake culture is still pending to become a reality in Iran. Earthquakes and earthquake disasters contributed to status of Japan as an earthquake nation with a mature earthquake culture, but for Iran, this did not happen until present time.

Learning from the Japanese culture of dynamic learning from large earthquake disasters is recommended for Iran and other countries at seismic risk. Moreover, the Japanese earthquake culture can inspire adaptations and adoptions to the Iranian context and specificity of place.

A proactive approach to earthquake disaster risk reduction and improvement of the earthquake disaster risk management is required for both Japan and Iran. A reactive approach to learning based on the past earthquake disasters needs to be constantly supported by a proactive approach and dynamic learning. Future and even present earthquake disaster risk in Japan and especially, in Iran might well exceed the proportions of the seismic risk from the last century.

Acknowledgements

Prof. Amir R. Nejad, NTNU, is greatly acknowledged and thanked. The financial support from NTNU Publishing Fund, Trondheim, Norway is acknowledged and thanked.

Conflict of interest

Authors declare there is no conflict of interest.

Author details

Michaela Ibrion* and Nicola Paltrinieri

*Address all correspondence to: mibrion5@gmail.com

Faculty of Engineering, Department of Mechanical and Industrial Engineering, Norwegian University of Science and Technology (NTNU), Trondheim, Norway

References

[1] Ibrion M. Earthquake culture: a significant element in earthquake disaster risk assessment and earthquake disaster risk management. In: Svalova V, editor. Risk Assessement. Rijeka, Croatia: InTech; 2018. ISBN:978-953-51-5702-1

[2] Ibrion M, Mokhtari M, Parsizadeh F, Nadim F. Timescape of the earthquake disasters in Iran: The intricacies of earthquake time and earthquake disaster risk reduction. Geografiska Annaler: Series A, Physical Geography. 2015;**97**(1):197-216. DOI: 10.1111/geoa.12093

[3] Ibrion M, Parsizadeh F, Pakdaman Naeini M, Mokhtari M, Nadim F. Handling of dead people after two large earthquake disasters in Iran: Tabas 1978 and Bam 2003 – survivors perspectives, beliefs, funerary rituals, resilience and risk. International Journal of Disaster Risk Reduction. 2015;**11**:60-77. DOI: 10.1016/j.ijdrr.2014.11.006

[4] Ibrion M, Mokhtari M, Nadim F. Earthquake disaster risk reduction in Iran: Lessons and "lessons learned" from three large earthquake disasters; Tabas 1978, Rudbar 1990, and Bam 2003. International Journal of Disaster Risk Science. 2015;**6**(4):415-427. DOI: 10.1007/s13753-015-0074-1

[5] Parsizadeh F, Ibrion M, Mokhtari M, Lein H, Nadim F. Bam 2003 earthquake disaster: On the earthquake risk perception, resilience and earthquake culture – Cultural beliefs and cultural landscape of Qanats, gardens of Khorma trees and Arg-e bam. International Journal of Disaster Risk Reduction. 2015;**14**(4):457-469. DOI: 10.1016/j.ijdrr.2015.09.011

[6] Ibrion M, Lein H, Mokhtari M, Nadim F. At the crossroad of nature and culture in Iran: The landscapes of risk and resilience of seismic space. International Proceedings of Economics Development and Research. 2014;**71**(8):38-44. DOI: 10.7763/IPEDR. 2014. V71.8

[7] Bilham R. The seismic future of cities. Bulletin of Earthquake Engineering. 2009;**7**:839-887. DOI: 10.1007/s10518-009-9147-0

[8] Folke C, Hahn T, Olsson P, Norberg J. Adaptive governance of social-ecological systems. Annual Review of Environment and Resources. 2005;**30**:441-473. DOI: 10.1146/annurev.energy.30.050504.144511

[9] Carpenter S, Walker B, Anderies JM, Abel N. From metaphor to measurement: Resilience of what to what? Ecosystems. 2001;**4**:765-781. DOI: 10.1007/s10021-001-0045-9

[10] Cutter SL, Barnes L, Berry M, Burton C, Evans E, Tate E, Webb J. A place-based model for understanding community resilience to natural disasters. Global Environmental Change. 2008;**18**:598-606. DOI: 10.1016/j.gloenvcha.2008.07.013

[11] Djalante R, Holley C, Thomalla F. Adaptive governance and managing resilience to natural hazards. International Journal of Disaster Risk Science. 2011;**2**(4):1-14. DOI: 10.1007/s13753-011-0015-6

[12] Ibrion M, Mokhtari M, Parsizadeh, F, Lein H, and Nadim F. Towards a culture of resilience and earthquake disaster risk reduction in Iran – "Lessons-Learned" from earthquake disasters. In: Proceedings of the 7th International Conference on Seismology and Earthquake Engineering. Tehran: IIEES; May 18-21, 2015. paper 00159

[13] Clancey G. Earthquake Nation, the Cultural Politics of Japanese Seismicity, 1868-1930. Berkeley: University of California Press; 2006. p. 331. ISBN: 0-520-24607-1

[14] Schencking JC. The Great Kanto Earthquake and the Chimera of National Reconstruction in Japan. New York: Columbia University Press; 2013. p. 400. ISBN: 978-0-231-16218-0

[15] Nishikawa S. Incorporating science and technology for disaster risk reduction – The Japanese experience. Planet@Risk. 2015;**3**(1):95-106

[16] Ranghieri F, Ishiwatari M. Learning from Megadisasters, Lessons from the Great East Japan Earthquake. Washington: The World Bank; 2014. p. 391. DOI: 10.1596/978-1-4648-0153-2

[17] Yamori K. A historical overview of social representation of earthquake risk in Japan: Fatalism, social reform, scientific control and collaborative risk management. In: Joffe H, Rossetto T Adams J, editors. Cities at Risk: Living with Perils in the 21st Century, Advances in Natural and Technological Hazards Research. 33. Dordrecht: Springer Science + Business Media; 2013. pp. 73-91. DOI: 10.1007/978-94-007-6184-1

[18] Leng R. Japan's civil society from Kobe to Tohoku, impact of policy changes on government – NGO relationship and effectiveness of post-disaster relief. Electronic Journal of Contemporary Japanese Studies. 2015;**15**(1):1-23

[19] Carafano JJ. The Great Eastern Japan earthquake, assessing the disaster response and lessons for the United States Report, Heritage Foundation, Douglas and Sarah Allison Center for Foreign Policy studies; 2011. Available from: http://www.heritage.org/research/reports/2011/05/the-great-eastern-japan-earthquake-assessing-disaster-response-and-lessons-for-the-us [Accessed: December 5, 2016]

[20] Berberian M. Earthquakes and Coseismic Surface Faulting on the Iranian Plateau, A Historical, Social and Physical Approach. Amsterdam: Elsevier; 2014. p. 776. ISBN: 978-0-444-63292-0

[21] Japan International Cooperation Agency (JICA) and Centre for Earthquake and Environmental Studies of Tehran (CEST), Tehran Municipality. The Study on Seismic Microzoning of the Greater Tehran Area in the Islamic Republic of Iran. Final Main Report, SSF-JR-00-186, November 2000. Japan: Pacific Consultants International and Oyo International Corporation; 2000

[22] JICA and Tehran Disaster Mitigation and Management Center (TDMMC). The Comprehensive Master Plan Study on Urban Seismic Disaster Prevention and Management for the Greater Tehran Area in the Islamic Republic of Iran. Final Main Report, GE-JR-04-039, December 2004. Tehran: Pacific Consultants International and Oyo International Corporation; 2004

[23] Pierotti P. History and praxis. In: Ferrigni F, Helly B, Mauro A, Mendes Victor, L. Pierotti P, Rideaud A, Teves Costa P, editor. Ancient Buildings and Earthquakes, the Local Seismic Culture Approach: Principles, Methods, Potentialities. Bari: Edipuglia; 2005, p. 29-51

[24] Maki N. How resilient is Japan? Response and recovery lessons from the 1995 Kobe and the 2011 Tohoku disasters, A STS Forum on the East Japan Disaster; 2013. Available from: https://fukushimaforum.wordpress.com/workshops/sts-forum-on-the-2011-fukushima-east-japan-disaster/manuscripts/session-1/how-resilient-is-japan-response-and-recovery-lessons-from-the-1995-kobe-and-the-2011-tohoku-disasters/ [Accessed: January 10, 2017]

[25] The Social Impact Investment Landscape in Japan (SIILJ). Report published by the Japan National Advisory Board and submitted to the G8 Impact investment Taskforce July 2014. 2014. Available from: http://www.socialimpactinvestment.org/reports/Japan%20NAB%20FINAL.pdf [Accessed: August 5, 2016]

[26] CIA (Central Intelligence Agency). The World Facts Book, Middle East, Iran; December 19, 2016. Available from: https://www.cia.gov/library/publications/the-world-factbook/geos/ir.html [Accessed: January 12, 2017]

[27] Spence R. Saving lives in earthquakes: Successes and failures in seismic protection since 1960. Bulletin of Earthquake Engineering. 2007;**5**:139-251. DOI: 10.1007/s10518-006-9028-8

6

Finite Element Models of Elastic Earthquake Deformation

Sui Tung, Timothy Masterlark and
Daniel Sai Huen Lo

Abstract

The Earth's surface deforms in response to earthquake fault dislocations at depth. Deformation models are constructed to interpret the corresponding ground movements recorded by geodetic data such GPS and InSAR, and ultimately characterize the seismic ruptures. Conventional analytical and latest numerical solutions serve similar purpose but with different technical constraints. The former cannot simulate the heterogeneous rock properties and structural complexity, while the latter directly tackles these challenges but requires more computational resources. As demonstrated in the 2015 M7.8 Gorkha, Nepal earthquake and the 2016 M6.2 Amatrice, Italy earthquake, we develop state-of-art finite element models (FEMs) to efficiently accommodate both the material and tectonic complexity of a seismic deformational system in a seamless model environment. The FEM predictions are significantly more accurate than the analytical models embedded in a homogeneous half-space at the 95% confidence level. The primary goal of this chapter is describe a systematic approach to design, construct, execute and calibrate FEMs of elastic earthquake deformation. As constrained by coseismic displacements, FEM-based inverse analyses are employed to resolve linear and nonlinear fault-slip parameters. With such numerical techniques and modeling framework, researchers can explicitly investigate the spatial distribution of seismic fault slip and probe other in-depth rheological processes.

Keywords: FEM, earthquake, deformation, inverse model

1. Introduction

With the wealth of geological and geodetic information accumulated around seismogenic zones over the past decades, we are posed to ask: in what way we can unify and take advantage of these data to study the earthquake hazard of those areas? The rate of interseismic creeping/slow slip [1, 2], coseismic slip [3], and afterslip [4] are usually estimated with fault dislocation models that predict surface deformation from in-depth fault slip motions. Customary analytical (Okada) solutions analyze rectangular slip in an isotropic half-space [5] and serve as a good initial approximation for inferring fault behaviors which are critical for assessing regional strain accumulation related to seismic hazard [6, 7]. However, the more we study, the more we find that the shallow part of the crust (especially the upper crust) is not as simple as, or even far beyond, a uniform half-space (**Figure 1**) [8]. The major shortcomings of an Okada solution rest on its assumptions of homogeneous crust (HOM) and a rectangular fault dislocation [5] which are inadequate according to *in situ* geological observations [9]. Failure of simulating the realistic crustal domain could induce fundamental uncertainties in predicting faulting-induced displacements, which could propagate into the interpretations of related earthquake studies [10]. For instance, we found that ignoring the heterogeneous crust (HET) in deformation models could yield to considerable prediction errors when simulating seismic deformation of the 2015 Gorkha, Nepal earthquake (**Figure 1**) [3]. This can be explained by the lateral and vertical material variations across the epicentral area, which poses a technical challenge for conventional analytical solutions (**Figure 1**). The importance of HET has also been suggested by many other colleagues. Hearn and Bürgmann [11] develop a finite element model for the 1999 Izmit Turkey earthquake and show that the GPS-recovered seismic moment is up to 40% greater for models incorporating depth-dependent shear modulus than it is for uniform elastic half-space models. The corresponding Coulomb stress change in the lower crust is ~300% larger than a model domain using a homogenous shear modulus. They conclude that models of co-seismic ruptures and postseismic viscoelastic relaxation associated with large strike-slip earthquakes should

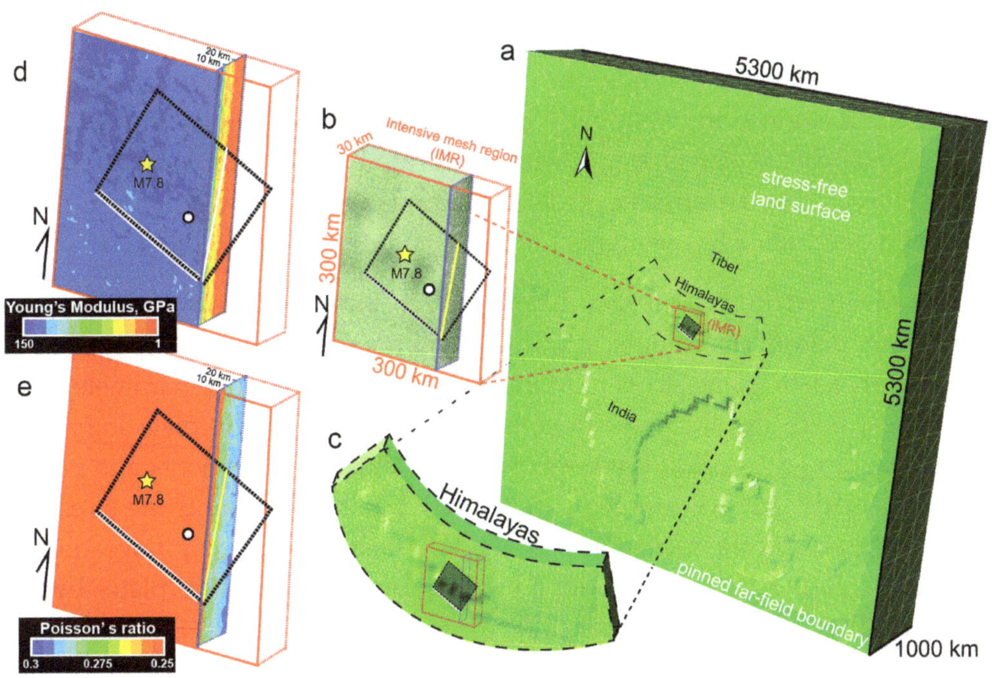

Figure 1. (a) Topography-shaped FEM domain of the 2015 M7.8 Gorkha, Nepal earthquake. (b and c) Nearfield meshes are refined within the central Himalayas [7]. Domain spatial distribution of (d) Young's Modulus and (e) Poisson's ratio are derived from CRUST2.0 [34].

incorporate depth-dependent elasticity, particularly for the triggering studies beneath the lower crust. Williams and Wallace [12] build finite element models (FEMs) in conjunction with a New Zealand-wide seismic tomography model to assign elastic properties. They find that these heterogeneous models typically require ~20% less slip than homogeneous models where the slip is deep or there is reasonable geodetic coverage above the slipping region. In cases where the slip is shallow (and mostly offshore) and there is little geodetic coverage directly above the slipping region, the heterogeneous models can predict significantly larger amounts of slip (~42%). These changes in the predicted magnitude of slip have important implications for quantifying slip budgets accommodated by slow slip at subduction zones worldwide [12]. The sensitivity of fault-slip solutions to HET is also demonstrated by Trasatti et al. [13] for the 2009 L'Aquila earthquake, showing up to 20% of discrepancy between Okada and FEM-based heterogeneous solutions which reveal new fault-slip features near the epicenter. Tung et al. [3, 10] also find the co-seismic GPS displacements are significantly better recovered by a HET model than a HOM model at the 95% confidence level. This model uncertainty is generally larger than those inherited in the geodetic measurements. The advantages of using FEMs over analytical solutions for simulating fault deformation are also exemplified among other earthquake studies [3, 10, 13–25].

With the advancement of computation power, FEM and large data acquisition techniques such as space geodesy, remote sensing, and imaging, we are now able to study the seismic activities on large-scale tectonic plates across continents with unprecedented detail and precision. For finite elastic deformation, elastoplastic analysis over a large domain, based on the Hellinger-Reissner and the Hu-Washizu functionals, 3D solid enhanced assumed strain formulations are among the most efficient and stable finite elements [26–35]. For high accuracy FE solutions over complicated domains of curved boundary, however, we could also use quadratic solid elements such as 10-node and 20-node tetrahedral elements, 20-node and 27-node hexahedral elements, etc. [36, 37]. Such methods are demonstrably useful for simulating a variety of complex science and engineering systems. Nonlinear contact problem of a hip joint is analyzed using T4, T10, H8 and H20 elements [38]. Fluid-saturated, inelastic, pressure-sensitive porous solid medium subjected to dynamic large deformation is analyzed by the mixed theory formulation using solid quadratic H27 elements [39].

Following the advanced numerical simulations, much work has been done recently on the deformation, stress distribution, faults, ruptures, dynamics, and wave propagation of tectonic plates by FEMs. An elastic plane stress FEM incorporating realistic rock parameters was used to calculate the stress field, displacement field, and deformation of the plate interactions in the eastern Mediterranean [40]. A 3D FE model of ~3000 hexahedral elements and nodes is set up by Lu et al. [41] for the surface topology, major active fault zones and the stress field of the Chinese continent to study the mechanism of the long-distance jumping migration over active seismogenic areas. Shear zones are identified over regional-scale tectonic plates by 2D FEMs of faults and boundaries of tectonic plates [42]. By means of cascaded FE simulations, glacial isostatic adjustment is extended to investigate the relationship between glacial loading/unloading and fault movement due to the spatial–temporal evolution of stresses [43]. Lithospheric pressure and density fields are determined by novel FEM-based gravity inversion which is implemented within the open-source *escript* modeling environment [44]. Sophisticated 3D

FEMs highlight that surface processes acting on normal-fault-bounded mountain ranges may sustain fault slip for millions of years even after regional extension has stopped [45]. By combining several datasets, a FE module was developed to estimate the gravitational potential energy of the lithosphere and calculate stresses acting on the real (non-planar) geometry of African plate [46]. A 3D FEM is also employed to simulate the original co-seismic Coulomb stress patterns through space and time as modified by post-seismic viscoelastic flow [47]. FEM-derived solutions are also integrated into the regularized linear inversion of InSAR data over the volcano surface to image the 3D deformation field and pressure distribution [48, 49].

Finite element generation is an important step for advancing 3D large-scale numerical modeling, as almost three quarters of the overall analysis time is devoted to mesh generation and the related geometrical analysis. A comprehensive account of various mesh generation techniques is described and discussed in the textbook "Finite Element Mesh Generation" by Lo [50]; and in general, unstructured meshes are generated by the Delaunay triangulation, the advancing-front approach and the quadtree/Octree techniques etc., whereas structured meshes of hexahedral elements can be synthesized by some mapping and sweeping processes. Transition quadrilateral and hexahedral elements [51] and universal connection hexahedral elements [52] have also been developed for adaptive refinement analysis. However, in conjunction with the popular mesh generation methods mentioned here, other techniques could also be employed for specific applications to broad-scale earthquake problems. A full waveform inversion method that incorporates seismic data on a wide range of space-temporal scales on both crustal and upper-mantle structure is developed with the multi-grid FE scheme [53]. Furthermore, a non-conforming octree-based scheme on a fictitious domain for the numerical modeling of earthquake induced ground motion of realistic surface topology of the Earth's crust was presented by Restrepo and Bielak [54]. Other interdisciplinary examples are the adaptive multi-material grids generated from image data for biomedical fluid–structure simulations [55], and the conformal finite element/volume meshes derived from 3D measurements of the propagation of small fatigue cracks [56].

2. Data

2.1. Seismic tomography

The propagation of earthquake waves is a function of rock material properties within the crustal layer that hosts the waves [57]. These material properties alter the traveling velocities of the P and S wave subjected to the local elastic rock properties. A tomography model refers to a velocity model that describes a 3D distribution of P-wave velocity *Vp*, S-wave velocity *Vs*, which is interchangeable with the spatial distribution of elastic moduli, namely, Young's Modulus *E* and Poisson's ratio *v*, as formulated by [3, 10]:

$$E = \frac{\rho V_s^2\left(3V_p^2 - 4V_s^2\right)}{V_p^2 - V_s^2} \;\&\; v = \frac{V_p^2 - 2V_s^2}{2\left(V_p^2 - V_s^2\right)} \tag{1}$$

In general, E increases as a function of depth, while v decreases in the deeper crust and converges to 0.25 for mantle rocks [58] (**Figure 1**). It is noteworthy that the homogeneous elastic half-space assumptions, which are commonly used in the models of earthquake deformation [5], assume that these moduli are uniform in space and generally conflict with the tomographic observation. Tung et al. [3, 10] quantifies the implications of ignoring seismic tomography in elastic deformation models for the 2015 M7.8 Gorkha, Nepal earthquake (**Figure 1**) and the 2016 M6.2 Amatrice, Italy earthquake. In these examples, over-simplifying the relatively weak materials near the surface translates to substantial prediction errors of InSAR and GPS signals. This underpins the necessity of modeling seismic deformation within the HET domain of FEMs which are one of the few existing methods capable of simulating 3D crustal rock heterogeneity.

2.2. Geodetic data

2.2.1. GPS data

The time series of Earth positioning are collected by thousands of GPS receiver stations using radio-wave signals from the constellations of Global Positioning System (GPS) satellites (**Figure 2**). Generally, these data provide a 3D displacement field of a station location with uncertainties close to 1 mm depending the atmospheric noise and other data processing errors [59]. Some GPS stations sample the ground positions continuously, while others are re-visited periodically through multiple surveying campaigns [11, 60]. Furthermore, some of the former become able to provide real-time or near-real-time data feed with automatic data processing procedures and web-based data sharing platforms, such as EarthScope-PBO-UNAVCO, USGS-NEIC and NSF-Cascadia Initiative [19]. Continuous GPS sites record systematic positioning data and generally require more considerations such as sustainable power supply, data logging protocols and

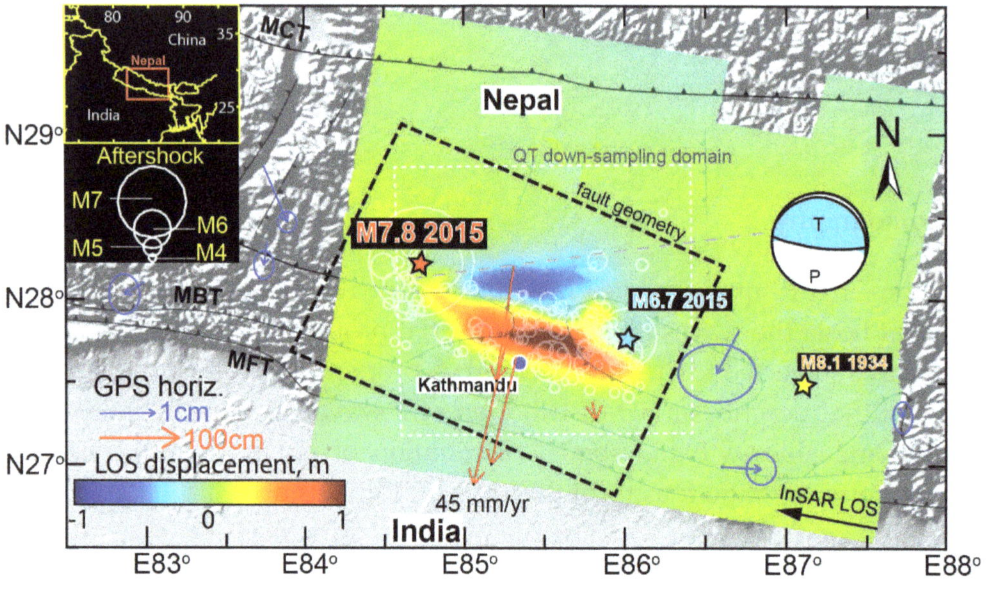

Figure 2. Coseismic deformation of the 2015 M7.8 Gorkha, Nepal earthquake mapped by InSAR and GPS data [7].

secure station design, while campaign-style measurements rely on more labor-intensive surveying strategies and are of lower temporal resolution over a longer time period. The technical details of GPS survey implementation are far beyond the scope of this work. For our purposes, we mainly focus on GPS data to constrain the three-component displacement field before and after an earthquake at given GPS stations (**Figure 2**).

2.2.2. InSAR data

In 1993, Massonnet et al. [61] presented the first Interferometric Synthetic Aperture Radar (InSAR) image to map the displacement field of the 1992 Landers earthquake. This image is derived from the changes in the phase distribution of radar scenes acquired respectively from two separate satellite passes over the epicentral area. An InSAR image unfolds the LOS displacement field that represents the difference between the positions of surface location at the time of the second and first satellite passes. By comparing the differential radar phase arrivals recorded before and after the earthquake, the spatial distribution of phase interference estimates the displacements parallel to the look direction of the satellite in unwrapped InSAR images (**Figure 2**). No information is available about the displacement between the two image acquisition times. The technical details of InSAR processing are far beyond the scope of this chapter, but aspects relevant to signal modeling are described here. The process of unwrapping InSAR data is to integrate the spatial phase data to map the line-of-sight (LOS) displacements. For our purposes, the obtained displacement field refers to that induced by earthquake dislocations [61]. FEMs are designed to predict these unwrapped phase data and hence characterize seismic sources [3, 10]. Moreover, InSAR observations are susceptible to artifacts caused by atmospheric noises and mismodeled orbital effects [62]. The former can be avoided via reducing the temporal baseline separation between two satellite passes, while the latter can be accounted for with linear inverse methods, as discussed in Section 4. Due to these artifacts, each pixel of an InSAR image is not completely independent so that a data covariance matrix is involved to empirically weight each pixel [63]. Alternatively, geospatial reduction techniques such as quadtree decomposition may be applied to filter unwanted signals, account for covariance and improve computational efficiency of matrix inversion.

2.3. Topography and bathymetry

Unlike the conventional HOM assumption [5], our FEMs and corresponding meshing regimes are capable of calculating the fault deformation over surface topography [3]. This surface is configured as a stress-free surface because we assume that there are only minimal normal stress variations and shear resistance. It is well known that the shape of such free surface affects deformation predictions, especially for tsunami modeling studies [64]. We can visualize this aspect by considering how the calculated deformation field would be affected by the limiting case of a vertical cliff near the rim of continental shelf. In this case, the ground surface is orthogonal to the assumed flat surface of an HOM domain. Matsuyama et al. [65] underlines the importance of including non-uniform topography and bathometry in fault deformation model to assess the tsunami hazard and coastal impact upon tsunamigenic events. Subjected to the ongoing tectonic movements and irregular structural settings, seismogenic/tsunamigenic zones usually attain a variable topography or bathymetry, which can be well accommodated by our FEMs for better accuracy of source characterization and tsunami wave predictions (**Figure 1**) [3, 66].

3. Model configurations

3.1. Domain and mesh configurations

Since FEMs are designed to simulate the crustal body of the seismogenic zone in a scale of few tens to thousands of kilometers, one of the initial decisions is to define a particular model coordinate system and units. A FEM is an assembly of numerous finite-volume elements stitched together to form a broader modeling domain (**Figure 1**). Those elements may attain different degrees of freedom (DOF) and geometry. For instance, a linear (p = 1) 4-node tetrahedral T4 element having 4 vertices attains a DOF of 4, while a linear (p = 1) 8-node hexahedral H8 element comprises 8 vertices and inherits 8 DOF. The latter could be further improved by the enhanced assume strain to be very competitive in regular simple geometry and structural shell problems. Furthermore, DOF applied to the solution variables may, for example, have 3 displacement components (DOF = 3) plus an additional pore pressure DOF. The meshing schemes of these elements are generally divided into two main categories, namely, structured meshing and free meshing. The former requires the meshes to be created according to a certain degree of uniformity. The element orientation, volume and nomenclature are defined in a structured manner, which is favorable for low-level modeling and solving procedures. On the contrary, the latter loosens all these criteria to let mesh "fill up" the model domain with the least number of elements. The choice of element type and meshing scheme heavily depends on the nature of the problems researchers are going to resolve. When earthquake slip is along a complex fault curvature, tetrahedral elements are preferred with regards to their smaller interior angles and thus ability to effectively tessellate a sharply-turning geometry such as listric faults and abruptly-changing topography (**Figure 1**). The prediction differences between the tetrahedral and rectangular elements become negligible when the fault is planar and the surface is flat. Given the same element-edge length and constant model domain, the tetrahedral mesh aggregation usually contains more elements than the rectangular aggregation as the volume of an individual tetrahedral element is smaller than that of a rectangular element. Hence, the computational time is longer for the former. Similarly, the free meshing algorithm allows more efficient and flexible tessellation of complex geometry than the structured approach, requiring more computational power. The modeling accuracy could be further boosted by incorporating quadratic elements (p = 2) instead of linear elements (e.g., T4 and H8). Quadratic Tetrahedral T10 element, which is one of the most versatile elements for both flexibility and accuracy, can fill up most complicated domains using an automatic mesh generation scheme, while the corresponding hexahedral H20 element provides another accurate formulation for simple geometry. As expect, using quadratic elements not only substantially improves simulation accuracy but also increases the number of domain nodes and hence computing time.

This leads the researcher to cautiously consider a fundamental trade-off problem between the FEM approximations and the limitations of the available computing resources. There might be cases in which differences between 1D model and a 2D model are negligible for a smaller study area. However, the computational time of model configuration and execution of a 3D domain is at least several tens to thousands times longer than that of a 2D domain, depending on the adopted meshing scheme and element/seed size. A 3D domain subjected to tetrahedral random meshing with the largest number of elements is compensated by a maximum flexibility of simulating the tectonic and lithospheric environment. For a given size of domain space, a large number of smaller elements translates to larger solution matrix of algebraic operations that may become numerical unfeasible when the computing time is too long or the calculation process is non-accomplishable. Alternatively, a small number of larger elements satisfies a smaller matrix problem that only requires nominal computing facilities, but at a cost of losing precision to resolve the equations of elasticity. Thus, apart from a general adaptive refinement analysis [3, 10], a common approach is to tessellate the near field region with a relatively small element size which gradually increases near the far-field boundaries [3, 10, 62, 67]. This radially-decaying meshing strategy satisfies the need for a refined resolution of nearfield areas expected with a relatively higher strain gradient (**Figure 1**), while the far-field boundary conditions are still connected numerically through large elements between the deformation source (i.e., the earthquake fault(s)) and the outer lateral surfaces exhibiting relatively low strain gradients. When installing the heterogeneous distribution of rock material into the FEM domain, elements of similar elastic properties (similar values of E and v ideally with respect to their integration points) are grouped into discrete element sets such that the entire FEM is a representation of multiple element sets (**Figure 1**) [3, 10]. As such, the resolution of rock heterogeneity is controlled by the element size as well as the discretization of elastic parametric values. From our modeling experiences, we default the number of element set to be about 100 for describing both regional and local crustal material variations (**Figure 1**).

3.2. Governing equations of elasticity

The governing equations regulate the physical behavior of a system. The governing equations for the elastic materials in a heterogeneous domain are [5, 58]:

$$\frac{\partial}{\partial x_j}\left[G(x)\left(\frac{\partial u_i}{\partial x_j}+\frac{\partial u_j}{\partial x_i}\right)\right]+\frac{\partial}{\partial x_i}\left[\lambda(x)\left(\frac{\partial u_k}{\partial x_k}\right)\right]\delta_{ij}=0\ \ \&\ \ \delta_{ij}=\begin{cases}0, i\neq j\\ 1, i=j\end{cases} \tag{2}$$

where x is a spatial component of coordinate axes $\boldsymbol{x}$; u refers to the corresponding displacement; G and λ are respectively the shear modulus and Lame's parameter; δ is the Kronecker delta; component indices i and j span over orthogonal axes 1, 2, and 3 for a 3D domain such that x_1, x_2, and x_3 are equivalent to Cartesian coordinates x, y, and z. The subscript k represents summation over all these three components. These equations describe elastic behavior in a domain comprising a spatial distribution of isotropic elastic material properties G and λ which can be derived from E and v [68]. Noting that when the elastic properties are taken outside of the spatial derivatives, along with appropriate initial and boundary conditions, Eq. (2) is reduced to the Navier formulation [69] and becomes a description of a HOM space that is commonly assumed in deformation models [5]. However, some seismogenic zones such as subduction margins require localized complexity, or a distribution of material properties, $[G(\mathbf{x}), \lambda(\mathbf{x})]$ or $[E(\mathbf{x}), v(\mathbf{x})]$. As we incorporate seismic topographical data (e.g., CRUST2.0 [8]) into the calculation of rock material (**Figure 1**), Eq. (2) describes elastic behavior of a 3D elastic domain inheriting a spatial distribution of isotropic elastic properties. If researchers want to include elastic anisotropy, they can replace scalar elastic moduli by tensors. FEMs are, so far, the best mathematical tool that satisfy these elastic equations over arbitrary crustal domains.

3.3. Loading conditions and kinematic constraints

The loading conditions can be viewed as the impulse that triggers the fault model to deform. For our purposes of simulating fault-slip deformation and consistency with analytical solutions, the loading conditions are assigned with a set of kinematic constraints developed by Masterlark et al. [70]. The fault discontinuity in FEMs is meshed with multiple node pairs which consist of two overlapping nodes sharing the same initial geographic location. A quasi-static fault slip is applied to these node pairs by locally offsetting these two node members, node n1 and n2 of each pair along the rake, θ_{rake}. The loading condition specifies the subfault dislocation, Δm, of each node pair through three equations of motion along the orthogonal axes:

$$m^{n1}_{strike} - m^{n2}_{strike} = \Delta m_{strike} = \Delta m \times \cos(\theta_{rake}) \tag{3}$$

$$m^{n1}_{dip} - m^{n2}_{dip} = \Delta m_{dip} = \Delta m \times \sin(\theta_{rake}) \tag{4}$$

$$m^{n1}_{normal} - m^{n2}_{normal} = 0 \tag{5}$$

where m^{n1}_{strike} and m^{n2}_{strike} refer the along-strike motion of node n1 and n2 respectively; m^{n1}_{dip} and m^{n2}_{dip} are the along-dip motion of node n1 and n2 respectively; m^{n1}_{normal} and m^{n2}_{normal} denote the fault-normal motion of node n1 and n2 respectively. The null fault-normal displacements of Eq. (5) ensure the footblock and hanging block are welded together along the fault-normal axis throughout the entire calculation. Non-slipping node pairs are constrained with $\Delta m = 0$ equivalent to a welded condition. For the FEM of the 2016 M6.2 Amatrice, Italy earthquake [10], 665 node pairs are assigned to assemble the fault so that the loading conditions and kinematic constraints contain ~2000 equations of motion. The predicted earthquake deformation with these equations are proved consistent with half-space analytical solutions.

4. Model calibration

The primary purpose of seismic source characterization is to resolve the spatial and temporal distribution of fault dislocations during earthquakes. Fault deformation models reveal fundamental elastic behavior of fault slip to interpret the observed quasi-static earthquake displacements. Geodetic data that map the surface deformation of an earthquake, are used to quantify the slip directionality, θ_{rake} and magnitude, Δm of each subfault node (**Figure 3**) [63, 70]. These

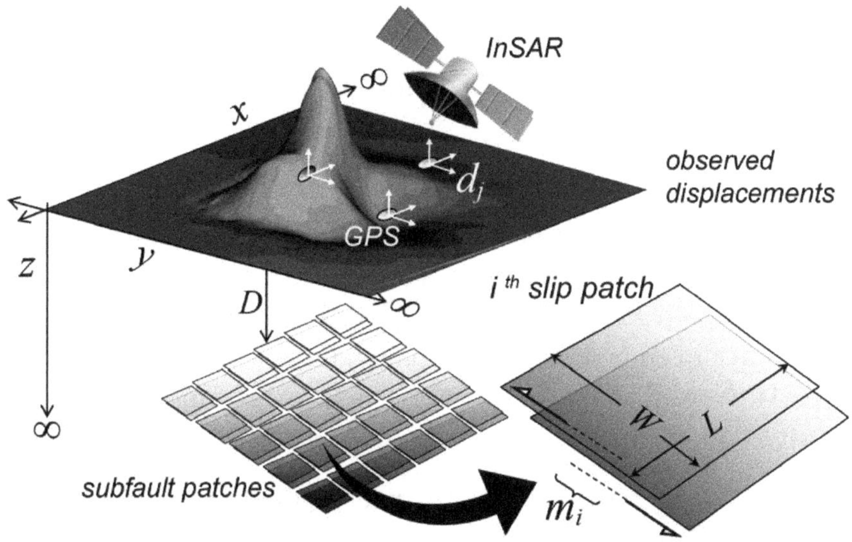

Figure 3. Linear inversion of coseismic displacements observed by GPS and InSAR data for a subfault-slip distribution.

data are regarded as the calibration targets, while the slip parameters are deemed as calibration parameters. To implement, the calibration procedure involves both forward modeling and inverse modeling [71]. The former yields predictions with respect to a particular input slip pattern, where the latter calibrates the pattern retroactively by fitting the geodetic observations [71]. The forward model is an essential component for inverse modeling as it signifies the deformation field induced by each individual subfault of an earthquake rupture (**Figure 3**) [3, 10]. Characterizing an earthquake source from the coseismic deformation field first involves finding the fault location, orientation and eventually the detailed slip distributions (**Figure 3**) [10]. The choice of calibration parameters strongly depends on the availability of data constraints and background information. Well-studied earthquakes having more data warrant more calibration parameters than those having relatively sparse information. There are some calibration parameters (e.g., Δm) which change linearly with the calibration data (e.g., $\boldsymbol{u}$), while others (e.g., fault strike, dip and depth) vary nonlinearly with the geodetic data. For instance, doubling the slip magnitude doubles the ground movement observed in the data, whereas doubling the depth or dip of an earthquake fault of a constant slip magnitude does not double the resulting surface deformation [5]. The former could be easily calibrated via matrix inversion methods [71], while the latter necessitate exclusive sampling of a multidimensional nonlinear parameter space [10]. In particular, the latter require exponentially more samples when the loading entity has to be adequately characterized by a large number or DOF of nonlinear parameters. The number of random samples required to explore a suite of nonlinear parameters, thus, depends on the degree of nonlinearity and the algorithmic efficiency of stochastic optimization. A rule of thumb could start from bisecting the parametric space. For instance, a model of 7° of freedom may requires $2^7 = 128$ samples for each iteration of stochastic sampling [10]. While experience plays an important role in selecting a specific suite of nonlinear parameters, the calibration process and its solution convergence quantitatively control the precision of these calibration parameters [10]. The details are described in the following paragraphs.

4.1. Forward model

The predicted three-component displacement, $\boldsymbol{d}_i$, for location i due to the slip, $\boldsymbol{m}_j$ of subfault, j, is:

$$\boldsymbol{d}_i = \boldsymbol{G}_{ij}\, \boldsymbol{m}_j \tag{6}$$

where $\boldsymbol{d}_i$ is a three-component displacement vector $[d_{x,i}, d_{y,i}, d_{z,i}]^T$; $\boldsymbol{m}_j$ is a two-component slip vector $[m_{dip\_slip,j}, m_{strike\_slip,j}]^T$; the superscript T denotes the matrix transpose operator; $\boldsymbol{G}_{ij}$ is the unit-slip-displacement Green's function between location i and subfault j based on the FEM predictions of unit fault slip embedded in a domain having a specified distribution of material properties (**Figure 1**). Note that the deformation is a linear function of slip and nonlinear function of the fault location, dip and strike as well as the distribution of domain materials.

The LOS displacement, $\boldsymbol{d}_{LOS,i}$ for the ith InSAR pixel is a linear combination of contributions from the fault slip and a plane shift:

$$\boldsymbol{d}_{LOS,\,i} = \boldsymbol{d}_i\, \boldsymbol{V}_i^{\mathrm{T}} + p_1 x_i + p_2 y_i + p_3 \tag{7}$$

where V_i is the line-of-sight (LOS) unit vector of the ith InSAR pixel; p_i are coefficients of a plane to account for the plane-shift displacements attributed to errors in modeling orbital effects [62]. The generalized matrix formulations for Eqs. (6) and (7) are respectively:

$$\boldsymbol{d} = \boldsymbol{G}\, \boldsymbol{m} \& \boldsymbol{d}_{LOS} = \left[\boldsymbol{G}\, \boldsymbol{V}^T, \boldsymbol{x}, \boldsymbol{y}, \boldsymbol{1}\right] \left[\boldsymbol{m}, p_1, p_2, p_3\right]^T \tag{8}$$

where $\boldsymbol{d}$ is the column vector of displacement data; **G** is the integrated Green's function matrix; $\boldsymbol{m}$ is the slip vector; $\boldsymbol{d}_L$ is the InSAR data column vector; $\boldsymbol{V}$ is the LOS unit column vector; $\boldsymbol{x}$ and $\boldsymbol{y}$ are the pixel location column vectors; ***1*** is a unity vector; The complete data vector that includes both GPS and InSAR data could be constructed by appending the matrices given in Eq. (8).

4.2. Inverse model

The common goal of inverse model is to estimate the calibration fault parameters based on the observed seismic data. While recognizing that a forward model is the linkage between the calibration data and the calibration parameters, inverse models step forward to optimize the calibration parameters and minimize the prediction errors against the calibration data. As mentioned above, those linear and non-linear calibration parameters are analyzed differently based their relations with the earthquake deformation. Our FEMs primarily contribute to the calculation of the Green's function matrix, $\boldsymbol{G}$ (Eq. (8)), depending on the characteristic model configurations that may include nonlinear calibration parameters. Variations of the nonlinear calibration parameters typically exert asystematic and non-systematic influence on $\boldsymbol{G}$, whereas those of linear calibration parameters multiply matrix entries with a constant factor.

4.2.1. Linear inverse analysis

With the consideration of both strike-slip (*ss*) and down-dip (*dd*) component, a complete Green's function matrix becomes $\boldsymbol{G} = [\boldsymbol{G}_{dd}, \boldsymbol{G}_{ss}]^T$ and has dimensions of 2 M x N, given there are M dislocating nodes-pairs (within the fault-slip region) and N displacement data points. Similarly, the dislocation vector has dimensions of 2 M so that $\boldsymbol{m} = [\boldsymbol{m}_{dd}, \boldsymbol{m}_{ss}]^T$. Given $\boldsymbol{m}$ is a vector representing M dislocating nodes-pairs and $\boldsymbol{d}$ is a vector of N displacements, the under-determined problem of linear inversion (when 2 $M > N$) always poses non-unique solutions of fault-slip models (**Figure 3**). Elastic dislocation problems for multiple slip patches generally have both over- and under-determined aspects - a given patch influences all data, and each datum constrains all patches. A physical solution can be hence resolved by simultaneously [71]: 1) estimating the slip distribution that minimizes misfit to geodetic data, 2) damping spurious solution oscillations, and 3) accounting for the uncertainties of geodetic data. To do so, first, we pre-multiply Eq. (6) by a weight matrix, $\boldsymbol{W}$ to account for the geodetic data uncertainties [71]:

$$\boldsymbol{WGm} = \boldsymbol{Wd} = \boldsymbol{G}_w\boldsymbol{m} = \boldsymbol{d}_w \quad \& \quad \mathbf{W}_{\mathrm{ij}} = \begin{cases} 1/\sigma_{\mathrm{j}} & \text{if i} = \text{j} \\ 0 & \text{if i} \neq \text{j} \end{cases} \tag{9}$$

where $\boldsymbol{W}$ is an N x N weight matrix formulated from the reported 1-sigma uncertainties of geodetic measurements, σ. σ_j is the uncertainty of the jth element of $\boldsymbol{d}$. Alternatively, $\boldsymbol{W}$ can be derived from data covariance matrix, $\boldsymbol{C_d} = (\boldsymbol{W}^T\boldsymbol{W})^{-1}$ through Cholesky decomposition. Neglecting the data uncertainties implies that $\boldsymbol{C_d}$ and $\boldsymbol{W}$ are an identity matrix, for which the uncertainties and weights are unity for all data. Second, we can reconfigure Eq. (9) using second-order Tikhonov Regularization to damp the null space of the data kernel (smooth the fault-slip distribution, $\boldsymbol{m}$) by:

$$\left(\boldsymbol{G}_w{}^T\boldsymbol{G}_w + \beta^2\boldsymbol{L}^T\boldsymbol{L}\right)\boldsymbol{m} = \boldsymbol{G}_w{}^T\boldsymbol{d}_w \& \boldsymbol{L} = \begin{bmatrix} \boldsymbol{L}_{dd} & 0 \\ 0 & \boldsymbol{L}_{ss} \end{bmatrix} \tag{10}$$

where $\boldsymbol{L}$ is a 2 M x 2 M matrix of damping. For curved fault configurations, $\boldsymbol{L}$ is literally substituted by the global conductance matrix, $\boldsymbol{L}_{GCM}$ referring to finite element approximation of Laplacian operator, $\nabla^2\mathbf{m} = 0$ [10, 17]. This global conductance matrix, $\boldsymbol{L}_{GCM}$ is the only mathematical tool to impose Laplacian regularization over slip locations of irregular fault geometries, for example, associated with the 2015 M6.2 Amatrice, Italy earthquake [10]. Using such Laplacian operator for smoothing allows us to conveniently impose and test Dirichlet (null; x = 0) and/or Neumann ($\partial\mathbf{m}/\partial$x = 0) specifications along the boundaries of the rupture surface [10]. With desired geodetic points, $\boldsymbol{d}$, the Green's function matrix, $\boldsymbol{G}$, and the global conductance matrix, $\boldsymbol{L}_{GCM}$ and weight matrix, $\boldsymbol{W}$ all essential components of the slip inversion are ready in Eq. (10) to solve the least-squares solution of $\boldsymbol{m}$ (**Figure 3**):

$$\boldsymbol{m} = \left(\boldsymbol{G}_w{}^T\boldsymbol{G}_w + \beta^2\boldsymbol{L}_{GCM}{}^T\boldsymbol{L}_{GCM}\right)^{-1}\boldsymbol{G}_w{}^T\boldsymbol{d}_w \tag{11}$$

where β is the regularization parameter controlling the tradeoff between minimizing misfit, $\boldsymbol{e}_w{}^T\boldsymbol{e}_w$ (given $\boldsymbol{e}_w = \boldsymbol{d}_w\text{-}\boldsymbol{G}_w\boldsymbol{m}$) and satisfying $\nabla^2\boldsymbol{m} = 0$ [71]. Some other workers recast the linear inversion through a Bayesian method to arrive at a physical solution, $\boldsymbol{m}$ [2, 72]. A Markov-chain Monte Carlo (MCMC) method is used to sample numerous combinations of subfault slips, $\boldsymbol{m}_i$, and smoothing coefficient, β_i (and possibly relative data weights, α_i) to create posterior probability density functions which in terms provide estimates of the above parameters

[72]. The uncertainty of the calibration parameters is characterized by the parameter covariance matrix, C_m [71]:

$$C_m = \left(G_w{}^T G_w\right)^{-1} G_w{}^T d_w G_w \left(G_w{}^T G_w\right)^{-1} \tag{12}$$

Eqs. (11) and (12) provide a mechanism for providing estimates of central tendency and uncertainties for linear calibration parameters, in a way that accounts for the data uncertainties. From the 2015 M7.8 Gorkha, Nepal earthquake, the HOM domain without considering heterogeneous rock properties in calculating **G** significantly degrades the fidelity of predicted GPS displacements beyond the data uncertainties (**Figure 3**), suggesting that a HET FEM domain is necessary for improving model predictions [3].

4.2.2. Nonlinear inverse analysis

This procedure is specially designed for nonlinear deformational parameters such as fault location, width, length, dip and strike to quantify the geometry and location of earthquake rupturing faults. As such, nonlinear inverse analyses are always conducted before the inverting for linear fault-slip parameters [10]. The solutions of those nonlinear parameters then later influence the accuracy of the linear slip solutions. For instance, uncertainties in fault dip propagate into the magnitude of subfault slip components such that a larger dip mistakenly resolved by the nonlinear analysis gives rise to larger slip magnitude predicted by the linear solutions. The nonlinear inverse method constitutes perturbing a nonlinear parameter and examining its impact on **G** with numerous forward model predictions. The ultimate goal of such analysis is to resolve a set of nonlinear parameters that minimizes $e_w{}^T e_w$ [10]. There are many different sampling approaches, including classical grid search and probabilistic random search [73]. The former conducts the parameter search over a predefined grid to find an optimal solution [63]. However, this strategy is usually biased by the researcher's expectations and achieves a poor solution resolution. On the contrary, the probabilistic type of Monte Carlo sampling randomly perturbs the solutions of a nonlinear parameter with more sophisticated and dynamic sampling strategies [10]. These regimes require re-computation of **G** upon each suite of sampled parameters, and hence is more applicable to the earthquake models with few (e.g., less than 10) calibration parameters. Directed stochastic sampling methods such as Monte Carlo Markov Chain (MCMC) and Monte Carlo Simulated Annealing (MCSA), combine the effectiveness of gradient methods and adaptive random sampling to calibrate nonlinear fault parameters of earthquake sources [10].

In the 2016 M6.2 Amatrice, Italy earthquake (AE), Tung et al. [10] used the MCSA method to calibrate a few thousands of nonlinear parameters in FEM-based models of seismic deformation (**Figure 4**). In particular, both a planar and listric dislocation are examined through a series of nonlinear analysis to invert the InSAR data obtained by ESA Sentinel-1 A/B and JAXA ALOS-2 satellite, assuming a uniform slip distribution. On one hand, seven nonlinear parameters, namely, fault dip, δ, strike, ϕ, length, L, width, W, and fault-center location, $[x_c, y_c, z_c]$ are used to designate the geometry and location of a planar fault (**Figure 4**) [10]. On the other hand, the listric fault geometry is constrained by a set of 6 parameters, namely, listric parameter, $[a, b]$, locking depth, Dm, fault horizontal width, H_x, fault length, L and fault location, x_c [10].

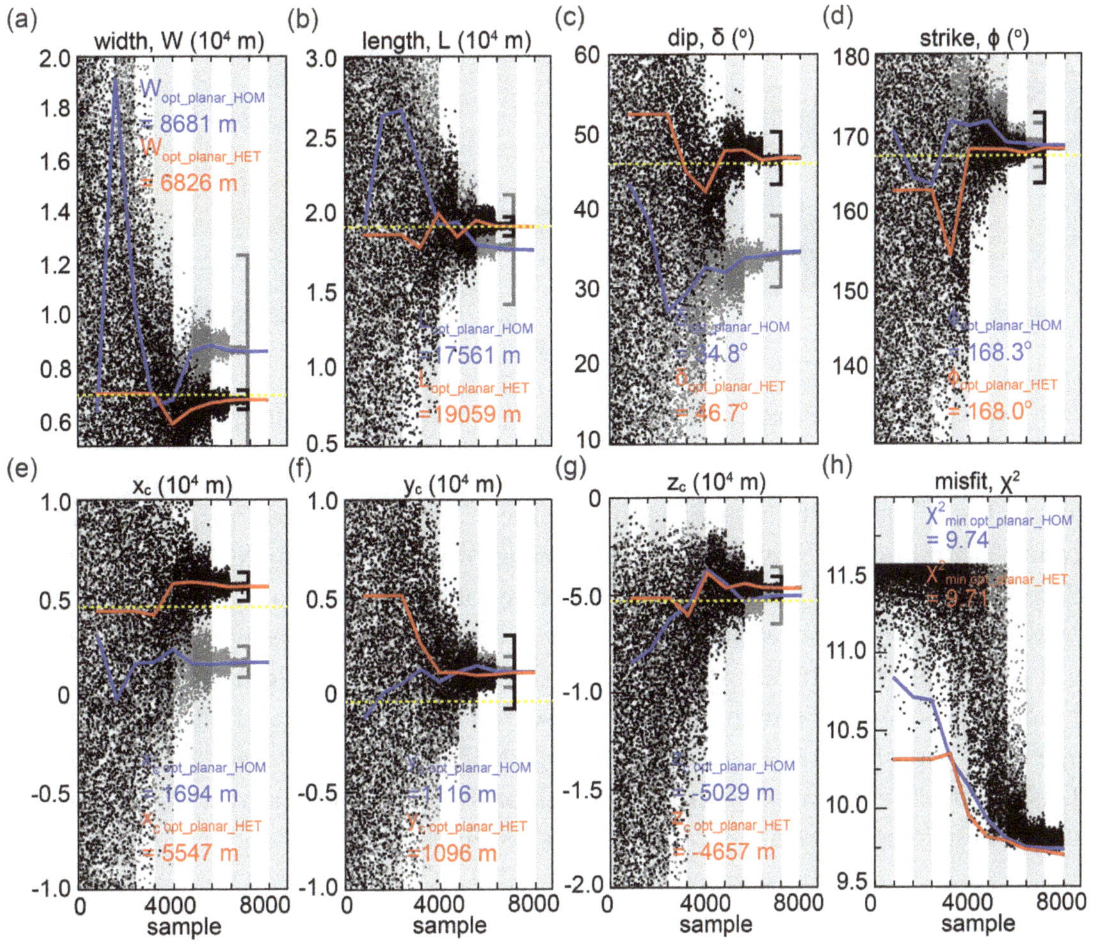

Figure 4. Solution convergence of resolving a planar source geometry in a HOM and HET domain for the 2016 M6.2 Amatrice, Italy earthquake [22]. The configurations of a planar dislocation specified by (a) width, (b) length, (c) dip, (d) strike and (e, f and g) the location of fault center [x_c, y_c, z_c] are optimized by minimizing (h) the model misfit, $\chi 2$ between the predicted and observed LOS displacements. The samples over HOM and HET are denoted as gray and black dots respectively, while the solution convergence is denoted by blue and red lines respectively. The brackets denote the 1-sigma parameter range. The HET solution yielding smaller misfit is more compatible with the fault model (yellow dashed line) suggested by [5].

Once these geometric parameters are fixed, the planar and listric models of distributed coseismic slip can then be derived linearly (**Figure 5**). The nonlinear analysis searches through a few thousands of uniformly-slipping fault models to minimize the weighted error misfit $\boldsymbol{e_w}^T\boldsymbol{e_w}$. This MCSA method combines simulated annealing [74] and nested Monte Carlo method [73] to search for a set of fault parameters that minimize $\boldsymbol{e_w}^T\boldsymbol{e_w}$. The cooling schedule of the MCSA algorithm is described by

$$T_i = T_o\left(\frac{N-i}{N}\right)^k + T_o\omega_{min} \tag{13}$$

$$v_{pi} = \left[\left(\frac{T_i}{T_o}\right)^{k_p} + v_{pmin}\right] v_{bounded\_range} \tag{14}$$

where T_o and T_i refer to the temperature of initial step and i^{th} step (0,1,2, … N-1) respectively; ω_{min} denotes the scaling factor of minimum temperature, $T_{min} \approx \omega_{min} T_o$, N is number of iteration (step);

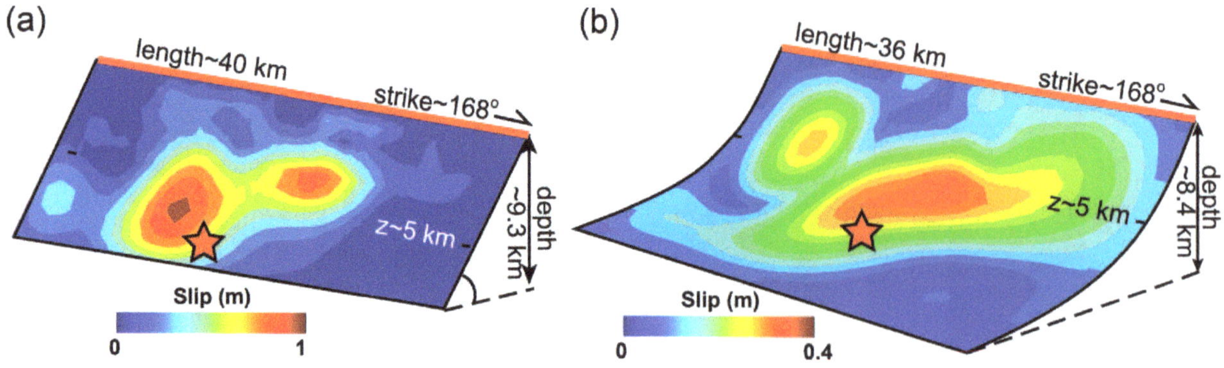

Figure 5. 3D representation of slip distributions resolved for the 2016 M6.2 Amatrice, Italy earthquake over a (a) planar and (b) listric fault [22].

V_{pi} refers to the pth parameter's search range in the ith step; V_{pmin} and $V_{bounded\_range}$ are the last-step scaling factor and the bounded parameter range of the pth parameter respectively; k and k_p respectively indicate the cooling/decaying coefficient of temperature and the search range of the pth parameter. The cooling schedule T_i is an array of decreasing number/temperature guiding the search of parameter at each step as well as the solution precision (**Figure 4**). Sharply changing (larger k and k_p) schedules narrows down the search range earlier, and allow the solutions to converge more rapidly. The unique advantage of MCSA over other Bayesian approaches like MCMC is its effectiveness of searching for the global optimal solution over the local minima of the model misfit. Its resistance to local minima is reinforced by acceptance criteria:

$$\chi_i^2 < \chi_{i-1}^2 \tag{15}$$

$$e^{-\left(\chi_i^2-\chi_{i-1}^2\right)\tau/T_i} > r \tag{16}$$

where χ_i^2 and χ_{i-1}^2 refer to the mean error of the ith step and *i-1th* step respectively; r denotes a random parameter between 0 and 1; τ is a normalization factor. The criterion described in Eq. (15) always accepts a better solution of the subsequent step, while the criterion in described Eq. (16) accepts a worse solution only when the misfit discrepancy, $\chi_i^2 - \chi_{i-1}^2$ between the subsequent steps is small. The cooling schedule T_i is initially designed with a MCMC analysis that does not include the acceptance criteria described in Eqs. (15) and (16). We then retrospectively fine-tune the cooling schedule and the acceptance criteria in a manner until a consistent solution is obtained. A general principle of MCSA is fix the values of T_i and τ so that the initial few steps are able to escape the local minima and arrive at a desired solution precision after all calculation steps (**Figure 4**).

5. Conclusions

The innovative modeling protocols of FEMs are developed to satisfy the need of simulating realistic elastic earthquake systems. By taking advantage of the increasingly data availability of seismic and tomographic studies, complex fault geometry and distributed rock materials are revealed especially within the upper crust. The customary half-space models of fault deformation, which assume a homogeneous domain and rectangular dislocations, cannot fully account for such shallow-crust complexity and hence induce prediction uncertainties when imaging earthquake sources with geodetic observations. New generations of fault model are fashioned in the framework of finite elements such that arbitrary lithological and structural heterogeneity can be accommodated when modeling seismic ruptures, which is particularly essential for earthquake locations of drastically changing lithology such as subduction margins. The modeling results of FEMs are found significantly more accurate than those of the conventional analytical solutions in nonlinear fault-geometry analyses and linear inversion for detailed slip distributions. This chapter, for the first time, describes the basic principles of constructing a sophisticated FEM for modeling elastic dislocation and elaborate how other auxiliary geophysical and geodetic data can be fed into the numerical domain and associated inverse analyses respectively. The resolution of governing equations and the corresponding validations are also discussed to ensure the reliability of the proposed FEM method. The modeling capacities of FEMs can further be extended beyond to simulate earthquake-induced poroelastic [75, 76] and viscoelastic [77] coupling processes which render physical mechanisms of triggering aftershocks and post-seismic surface deformation, summarizing the exceptional advantages of using FEMs for a wide range of earthquake research.

Acknowledgements

This work is supported jointly by a NASA grant NNX17AD96G, NSF grant 1316082, NSF grant OCE-1636653-subaward-4(GG013106-1) and NASA JPL subcontract 1468758. We would also like to acknowledge support from JAXA research program (0414001PI#3357).

Author details

Sui Tung[1]*, Timothy Masterlark[1] and Daniel Sai Huen Lo[2]

*Address all correspondence to: sui.tung@sdsmt.edu

1 Department of Geology and Geological Engineering, South Dakota School of Mines and Technology, Rapid City, SD, USA

2 Department of Civil Engineering, The University of Hong Kong, Hong Kong SAR

References

[1] Graymer R, Ponce D, Jachens R, Simpson R, Phelps G, Wentworth C. Three-dimensional geologic map of the Hayward fault, northern California: Correlation of rock units with variations in seismicity, creep rate, and fault dip. Geology. 2005;**33**:521-524

[2] Bartlow NM, Si M, Bradley AM, Segall P. Space-time correlation of slip and tremor during the 2009 Cascadia slow slip event. Geophysical Research Letters. 2011:38

[3] Tung S, Masterlark T. Coseismic slip distribution of the 2015 Mw7.8 Gorkha, Nepal, earthquake from joint inversion of GPS and InSAR data for slip within a 3-D heterogeneous domain. Journal of Geophysical Research: Solid Earth. 2016;**121**:3479-3503. DOI: 10.1002/2015JB012497

[4] Marone CJ, Scholtz C, Bilham R. On the mechanics of earthquake afterslip. Journal of Geophysical Research: Solid Earth. 1991;**96**:8441-8452

[5] Okada Y. Surface deformation due to shear and tensile faults in a half-space. Bulletin of the Seismological Society of America. 1985;**75**:1135

[6] King GC, Stein RS, Lin J. Static stress changes and the triggering of earthquakes. Bulletin of the Seismological Society of America. 1994;**84**:935-953

[7] Stein RS. The role of stress transfer in earthquake occurrence. Nature. 1999;**402**:605-609

[8] Bassin C, Laske G, Masters G. The current limits of resolution for surface wave tomography in North America. Eos Transactions American Geophysical Union. 2000;81

[9] Plesch A, Shaw JH, Benson C, Bryant WA, Carena S, Cooke M, Dolan J, Fuis G, Gath E, Grant L. Community fault model (CFM) for southern California. Bulletin of the Seismological Society of America. 2007;**97**:1793-1802

[10] Tung S, Masterlark T. Resolving source geometry of the august 24, 2016 Amatrice, Central Italy earthquake from InSAR data and 3D finite element modeling. Bulletin of the Seismological Society of America. 2018;**108**:553-572

[11] Hearn EH, Bürgmann R. The effect of elastic layering on inversions of GPS data for coseismic slip and resulting stress changes: Strike-slip earthquakes. Bulletin of the Seismological Society of America. 2005;**95**:1637-1653

[12] Williams CA, Wallace LM. Effects of material property variations on slip estimates for subduction interface slow-slip events. Geophysical Research Letters. 2015;**42**:1113-1121

[13] Trasatti E, Kyriakopoulos C, Chini M. Finite element inversion of DInSAR data from the mw 6.3 L'Aquila earthquake, 2009 (Italy). Geophysical Research Letters. 2011;**38**

[14] Eleonora R, Walter M, Maurizio B. The edge dislocation problem in a layered elastic medium. Geophysical Journal International. 2002;**149**:508-523

[15] Fernandez J, Yu TT, Rundle JB. Deformation produced by a rectangular dipping fault in a viscoelastic-gravitational layered earth model. 1. Thrust fault - FLTGRV and FLTGRH FORTRAN programs. Computers & Geosciences. 1996;**22**:735-750

[16] He WW, Yao Z. Static deformation due to shear and tensile faults in a layered half-space. Bulletin of the Seismological Society of America. 2003;**93**:2253-2263

[17] Hughes KLH, Masterlark T, Mooney WD. Poroelastic stress-triggering of the 2005 M8. 7 Nias earthquake by the 2004 M9. 2 Sumatra–Andaman earthquake. Earth and Planetary Science Letters. 2010;**293**:289-299

[18] Jovanovich DB, Husseini MI, Chinnery MA. Elastic dislocations in a layered half-space—I. Basic theory and numerical methods. Geophysical Journal International. 1974;**39**:205-217

[19] Kyriakopoulos C, Masterlark T, Stramondo S, Chini M, Bignami C. Coseismic slip distribution for the mw 9 2011 Tohoku-Oki earthquake derived from 3-D FE modeling. Journal of Geophysical Research: Solid Earth. 2013;**118**:3837-3847

[20] Masterlark T, Hughes KLH. Next generation of deformation models for the 2004 M9 Sumatra-Andaman earthquake. Geophysical Research Letters. 2008;**35**:19310

[21] Pan E. Green's functions in layered poroelastic half-spaces. International Journal for Numerical and Analytical Methods in Geomechanics. 1999;**23**:1631-1653

[22] Sato R. Crustal deformation due to dislocation in a multi-layered medium. Journal of Physics of the Earth. 1971;**19**:31-46

[23] Savage J. Effect of crustal layering upon dislocation modeling. Journal of Geophysical Research: Solid Earth (1978–2012). 1987;**92**:10595-10600

[24] Savage J. Displacement field for an edge dislocation in a layered half-space. Journal of Geophysical Research: Solid Earth (1978–2012). 1998;**103**:2439-2446

[25] Wang RJ, Lorenzo-Martin F, Roth F. PSGRN/PSCMP - a new code for calculating co- and post-seismic deformation, geoid and gravity changes based on the viscoelastic-gravitational dislocation theory. Computers & Geosciences. 2006;**32**:527-541

[26] Reinoso J, Paggi M, Linder C. Phase field modeling of brittle fracture for enhanced assumed strain shells at large deformations: Formulation and finite element implementation. Computational Mechanics. 2017;**59**:981-1001

[27] Korelc J, Šolinc U, Wriggers P. An improved EAS brick element for finite deformation. Computational mechanics. 2010;**46**:641-659

[28] Bouchart V, Brieu M, Kondo D, Abdelaziz MN. Implementation and numerical verification of a non-linear homogenization method applied to hyperelastic composites. Computational Materials Science. 2008;**43**:670-680

[29] Klinkel S, Gruttmann F, Wagner W. A robust non-linear solid shell element based on a mixed variational formulation. Computer methods in applied mechanics and engineering. 2006;**195**:179-201

[30] Reese S. On a physically stabilized one point finite element formulation for three-dimensional finite elasto-plasticity. Computer Methods in Applied Mechanics and Engineering. 2005;**194**:4685-4715

[31] Areias P, César de Sá J, António C, Fernandes A. Analysis of 3D problems using a new enhanced strain hexahedral element. International Journal for Numerical Methods in Engineering. 2003;**58**:1637-1682

[32] Sze K, Lo S, Yao LQ. Hybrid-stress solid elements for shell structures based upon a modified variational functional. International Journal for Numerical Methods in Engineering. 2002;**53**:2617-2642

[33] Sze K. Efficient formulation of robust hybrid elements using orthogonal stress/strain interpolants and admissible matrix formulation. International Journal for Numerical Methods in Engineering. 1992;**35**:1-20

[34] Simo JC, Rifai M. A class of mixed assumed strain methods and the method of incompatible modes. International Journal for Numerical Methods in Engineering. 1990;**29**:1595-1638

[35] Pian TH. Finite elements based on consistently assumed stresses and displacements. Finite Elements in Analysis and Design. 1985;**1**:131-140

[36] Lee C, Lo S. Automatic adaptive 3-d finite element refinement using different-order tetrahedral elements. International Journal for Numerical Methods in Engineering. 1997; **40**:2195-2226

[37] Lee C, Lo S. Automatic adaptive refinement finite element procedure for 3D stress analysis. Finite Elements in Analysis and Design. 1997;**25**:135-166

[38] Maas SA, Ellis BJ, Rawlins DS, Weiss JA. Finite element simulation of articular contact mechanics with quadratic tetrahedral elements. Journal of Biomechanics. 2016;**49**:659-667

[39] Regueiro R, Ebrahimi D. Implicit dynamic three-dimensional finite element analysis of an inelastic biphasic mixture at finite strain: Part 1: Application to a simple geomaterial. Computer Methods in Applied Mechanics and Engineering. 2010;**199**:2024-2049

[40] Dwivedi S, Hayashi D. Modeling the contemporary stress field and deformation pattern of eastern Mediterranean. Journal of Earth Science. 2010;**21**:365-381

[41] Lu Y, Yang S, Chen L, Lei J. Mechanism of the spatial distribution and migration of the strong earthquakes in China inferred from numerical simulation. Journal of Asian Earth Sciences. 2011;**40**:990-1001

[42] Mitsakaki C, Sakellariou MG, Tsinas D. A study of the crust stress field for the Aegean region (Greece). Tectonophysics. 2013;**597**:50-72

[43] Steffen R, Wu P, Steffen H, Eaton DW. On the implementation of faults in finite-element glacial isostatic adjustment models. Computers & Geosciences. 2014;**62**:150-159

[44] Aitken A, Altinay C, Gross L. Australia's lithospheric density field, and its isostatic equilibration. Geophysical supplements to the. Monthly Notices of the Royal Astronomical Society. 2015;**203**:1961-1976

[45] Turpeinen H, Maniatis G, Hampel A. Slip on normal faults induced by surface processes after the cessation of regional extension—Insights from three-dimensional numerical modelling. Geomorphology. 2015;**237**:79-87

[46] Medvedev S. Understanding lithospheric stresses: Systematic analysis of controlling mechanisms with applications to the African plate. Geophysical Journal International. 2016;**207**:393-413

[47] Bagge M, Hampel A. Postseismic coulomb stress changes on intra-continental dip-slip faults due to viscoelastic relaxation in the lower crust and lithospheric mantle: Insights from 3D finite-element modelling. International Journal of Earth Sciences. 2017;**106**: 2895-2914

[48] Ronchin E, Masterlark T, Molist JM, Saunders S, Tao W. Solid Modeling Techniques to Build 3D Finite Element Models of Volcanic Systems: An Example from the Rabaul Caldera System. Papua New Guinea: Computers & Geosciences; 2012

[49] Masterlark T, Tung S. Finite Element Models of Elastic Volcano Deformation. London, United Kingdom: IntechOpen; 2018

[50] Lo DS. Finite Element Mesh Generation. Boca Raton, Florida, USA: CRC Press; 2014

[51] Wu D, Sze K-Y, Lo S-H. Two-and three-dimensional transition element families for adaptive refinement analysis of elasticity problems. International Journal for Numerical Methods in Engineering. 2009;**78**:587-630

[52] Wu D, Lo S, Sheng N, Sze K. Universal three-dimensional connection hexahedral elements based on hybrid-stress theory for solid structures. International Journal for Numerical Methods in Engineering. 2010;**81**:307-334

[53] Fichtner A, Trampert J, Cupillard P, Saygin E, Taymaz T, Capdeville Y, Villasenor A. Multiscale full waveform inversion. Geophysical Journal International. 2013;**194**:534-556

[54] Restrepo D, Bielak J. Virtual topography: A fictitious domain approach for analyzing free-surface irregularities in large-scale earthquake ground motion simulation. International Journal for Numerical Methods in Engineering. 2014;**100**:504-533

[55] Carson JP, Kuprat AP, Jiao X, Dyedov V, Del Pin F, Guccione JM, Ratcliffe MB, Einstein DR. Adaptive generation of multimaterial grids from imaging data for biomedical Lagrangian fluid–structure simulations. Biomechanics and modeling in mechanobiology. 2010;**9**:187-201

[56] Spear AD, Hochhalter JD, Cerrone AR, Li SF, Lind JF, Suter RM, Ingraffea AR. A method to generate conformal finite-element meshes from 3D measurements of microstructurally small fatigue-crack propagation. Fatigue & Fracture of Engineering Materials & Structures. 2016;**39**:737-751

[57] Simmons G and Brace W. Comparison of static and dynamic measurements of compressibility of rocks. Journal of Geophysical Research. 1965;**70**:5649-5656

[58] Turcotte D, Schubert G. Geodynamics. Cambridge, UK: Cambridge University Press; 2014

[59] Dragert H, Hyndman R, Rogers G, Wang K. Current deformation and the width of the seismogenic zone of the northern Cascadia subduction thrust. Journal of Geophysical Research: Solid Earth (1978–2012). 1994;**99**:653-668

[60] Barnhart W, Murray J, Yun SH, Svarc J, Samsonov S, Fielding E, Brooks B, Milillo P. Geodetic constraints on the 2014 M 6.0 South Napa earthquake. Seismological Research Letters. 2015;**86**:335-343

[61] Massonnet D, Rossi M, Carmona C, Adragna F, Peltzer G, Feigl K, Rabaute T. The displacement field of the Landers earthquake mapped by radar interferometry. Nature. 1993; **364**:138-142

[62] Masterlark T, Feigl KL, Haney M, Stone J, Thurber C, Ronchin E. Nonlinear estimation of geometric parameters in FEMs of volcano deformation: Integrating tomography models and geodetic data for Okmok volcano, Alaska. Journal of Geophysical Research: Solid Earth (1978–2012). 2012:117

[63] Huang MH, Fielding EJ, Liang C, Milillo P, Bekaert D, Dreger D, Salzer J. Coseismic deformation and triggered landslides of the 2016 mw 6.2 Amatrice earthquake in Italy. Geophysical Research Letters. 2017

[64] Wang, Liu L. An analysis of 2004 Sumatra earthquake fault plane mechanisms and Indian Ocean tsunami. Journal of Hydraulic Research. 2006;**44**:147-154

[65] Matsuyama M, Walsh J, Yeh H. The effect of bathymetry on tsunami characteristics at Sisano lagoon, Papua New Guinea. Geophysical Research Letters. 1999;**26**:3513-3516

[66] Tung S, Masterlark T. Sensitivities of nearfield tsunami forecasts to megathrust deformation predictions. Journal of Geophysical Research. 2018;**123**:1711-1735

[67] Masterlark T, Donovan T, Feigl KL, Haney M, Thurber CH, Tung S. Volcano deformation source parameters estimated from InSAR: Sensitivities to uncertainties in seismic tomography. Journal of Geophysical Research: Solid Earth. 2016;**121**:3002-3016

[68] Hartsuijker C, Welleman J. Engineering mechanics: Volume 1: Equilibrium. In: Berlin, Germany: Springer; Vol. 1. 2007

[69] Sadd MH. Elasticity: Theory, Applications, and Numerics. Cambridge, Massachusetts, US: Academic Press; 2009

[70] Masterlark T. Finite element model predictions of static deformation from dislocation sources in a subduction zone: Sensitivities to homogeneous, isotropic, Poisson-solid, and half-space assumptions. Journal of Geophysical Research. 2003;**108**:1029-2510

[71] Menke W. Geophysical Data Analysis: Discrete Inverse Theory. Cambridge, Massachusetts, US: Academic press; 2012. DOI: 10.1016/B978-0-12-397160-9.00001-1

[72] Simons M, Minson SE, Sladen A, Ortega F, Jiang J, Owen SE, Meng L, Ampuero J-P, Wei S, Chu R. The 2011 magnitude 9.0 Tohoku-Oki earthquake: Mosaicking the megathrust from seconds to centuries. Science. 2011;**332**:1421-1425

[73] Press WH. Numerical recipes. In: The Art of Scientific Computing. 3rd ed. Cambridge, UK: Cambridge University Press; 2007

[74] Jónsson S, Zebker H, Segall P, Amelung F. Fault slip distribution of the 1999 mw 7.1 Hector mine, California, earthquake, estimated from satellite radar and GPS measurements. Bulletin of the Seismological Society of America. 2002;**92**:1377-1389

[75] Tung S, Masterlark T. Transient poroelastic stress coupling between the 2015 M7.8 Gorkha, Nepal earthquake and its M7.3 aftershock. Tectonophysics. 2018;**733**:119-131

[76] Tung S, Masterlark T. Delayed poroelastic triggering of the 2016 October Visso earthquake by the august Amatrice earthquake, Italy. Geophysical Research Letters. 2018;**45**:2221-2229

[77] Bürgmann R, Dresen G. Rheology of the lower crust and upper mantle: Evidence from rock mechanics, geodesy, and field observations. Annual Review of Earth and Planetary Sciences. 2008;**36**:531-567

7

Structural Seismic Input Model on the Condition of Slope Site

Li Jianbo and Liu Weihong

Abstract

The viscoelastic artificial boundary model is widely used in the field of earthquake resistance analysis of water conservancy projects and nuclear power stations. However, for the analysis of soil-structure dynamic interaction on the sloping site, some problems will arise while using that method. The large size difference of side facades on the outer boundary will cause inconsistent horizontal seismic forces, which may lead to the divergent results or the drift of displacement. In this chapter, based on the basic formula of dynamic interaction and seismic input mechanism, a virtual symmetrical substructure system is built to solve those problems, which not only satisfies the consistence of the whole seismic input on the outer boundary but also simulates the seismic wave propagation. Finally, the accuracy and the stability of the new method are verified through numerical examples on the sloping site.

Keywords: slope site, structure-foundation interaction, viscoelastic boundary, symmetrical substructure, virtual symmetric substructure

1. Introduction

With the development of numerical methods, the advanced dynamic model of infinite foundation [1, 2] is widely used in the field of earthquake resistance analysis of water conservancy projects and nuclear power stations. Now, the viscous boundary [3], viscoelastic boundary [4] and transmitting boundary [5, 6] are more prevalent in engineering. Among them, the viscous boundary is the earliest applied, but it only reflects the energy absorbing effect of boundary damping; under the static or slow dynamic loading, the overall structure drift easily occurs. The transmitting boundary [7, 8] is more accurate, but it cannot reflect the elastic supporting effect of the far-field foundation, and similar to the viscous boundary, the numerical result drifts easily under the combination of static and dynamic forces; moreover, the problem of high-frequency instability is obvious which limits its application in engineering to a certain extent. In contrast, the viscoelastic boundary [9, 10] can not only effectively simulate the elastic recovery properties and the radiation damaging effect of the medium outside the artificial boundary [11] which ensures the numerical stability and precision that fulfills the requirements of engineering but implement in large software and achieve computational efficiency that endures broad application prospects.

1.1. The research of the viscoelastic artificial boundary

As a stress topical artificial boundary, the viscoelastic artificial boundary gave rise to the artificial boundary condition by Deeks [12] based on the assumption that the two-dimensional scattered wave is cylindrical wave. Liu Jingbo [13] introduced the conversion of the ground motion input into the equivalent load form on the artificial boundary which can deal with the problem of oblique incidence. Later, Liu Jingbo deduced the three-dimensional viscoelastic artificial boundary condition based on the assumption of the scattering wave as a spherical wave and established a three-dimensional viscoelastic static and dynamic unified artificial boundary [10] for static and dynamic combination analysis. Du Xiuli et al. applied the viscoelastic artificial boundary to the analysis of the seismic response of the camber and compared the analysis results of the transmitting boundary method. Lu Huaxi et al. [14] used the viscoelastic artificial boundary to study the interaction of pile foundation structures considering the dynamic nonlinear properties of soil. Gao Feng et al. [15] studied the method of static and dynamic artificial boundary transformation of underground structures and recommended a reasonable method. The viscoelastic artificial boundary is studied and developed for more complex and practical engineering applications.

1.2. The viscoelastic artificial boundary model under the condition of the slope site

Earthquake is an important threat to the safety of engineering structures, which is a natural phenomenon with serious consequences. The analysis of soil-structure dynamic interaction is an important basic method to evaluate reasonable and reliable structures such as engineering structures. The viscoelastic artificial boundary foundation model with good engineering applicability and computational robustness is a widely used model of interaction analysis in the engineering field. How to define the input ground vibration of the viscoelastic artificial boundary foundation model [16] is the basic calculation condition, which is not only related to the seismic event but also involves the dynamic analysis of the ground motion field of the free field. This chapter puts forward the reasonable solution to promote the viscoelastic artificial boundary model applied in the field of large engineering structures for more site conditions of slope soil-structure interaction input conditions and analysis. This slope site condition is often encountered in the seismic analysis of large seawalls, slope protection and water diversion structures. Due to limitations on space, the sketch is described in the example.

In the actual engineering seismic response analysis, the viscoelastic boundary is generally divided into five steps: a system composed of intercepting straight artificial boundaries on the foundation and structure, finite element discretization, applying artificial boundaries, equivalent load generation and input and dynamic time history analysis. For the structure built on a flat field, the height of each lateral face of the foundation model intercepted is generally equal, and the generation and input of equivalent loads are issued [7]. However, there are often slope sites in engineering, such as in the northeast, the water intake structures of a nuclear power plant are built in the gentle slope of the coastal zone where the terrain is in a slope overall. After checking the overall stability of the slope site structure, the viscoelastic artificial boundary is adopted to analyze seismic responses of structures. Intercepting the straight boundary that leads to a matrix as shown in **Figure 1**, this model is characterized by setting different heights of the lateral face. This difference causes two obvious problems for equivalent load generation and direct application: the traveling wave effect with different time delay between the incident and reflected wave causes different displacement time history and stresses of wave propagation in lateral faces with different heights and the resultant force of seismic load relates to the area, area difference of the boundary face which causes different resultants with each side making the force on its side illogical. In fact, the limited model causes problems. The actual source is far away from the surface, so the fluctuation of the surface of the ground does not make a significant difference in the input of the seismic waves. There is no discussion on the method of seismic wave input for slope site in present literature.

In this chapter, based on the basic formula of dynamic interaction and seismic input mechanism, we overcome this problem that the seismic load input is consistent between lateral faces with different heights. A method existing nominally of only an equal-sized slope foundation composing of slope symmetry system with the original slope model is adopted. A virtual symmetrical substructure system is built to solve those problems, which not only satisfies the consistency of the whole seismic input on the outer boundary but also simulates the seismic wave propagation. Finally, the accuracy and the stability of the new method are verified through numerical examples on the sloping site.

With the application of the finite element analysis technique, the classical viscoelastic artificial boundary model has been developed and improved, and its accuracy and applicability have been verified by Liu Jingbo [9, 10, 17], Du Xiuli et al. [7]. The model proposed in this chapter is to solve the input conditions under the conditions of the slope site by defining the virtual interaction analysis model and then obtain the dynamic response of the structure. In this chapter, the theses of Liu Jingbo [9, 10, 17] and Du Xiuli [7] are quoted in this chapter, which

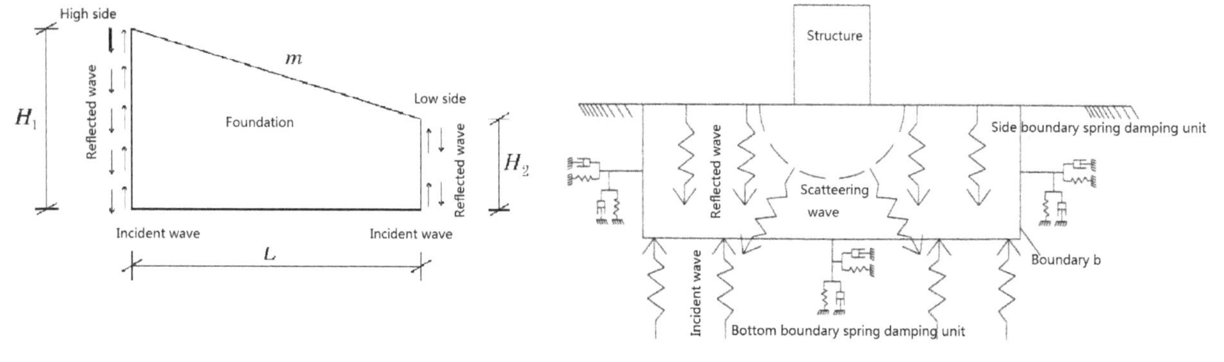

Figure 1. (1) Abstract sketch of slope site model. (2) The sketch of structure-foundation interaction.

mainly refer to the model of input ground vibration in the slope field of this chapter, which conforms to the basic assumption of the viscoelastic artificial boundary model. On this basis, the new model extends the applicability of the viscoelastic artificial boundary model and does not affect its accuracy.

2. The theory of viscoelastic artificial boundary

As shown in **Figure 1 (2)**, the spring damping element is applied to the outer boundary of the topical foundation that is intercepted; the wave from the bottom into the structure foundation system arrives at a time later at the structure, causing structural vibration, and the vibration will also be sent back to the foundation in the form of waves, which is called secondary scattering field fluctuation [18, 19]. So, the vibration of a point on the foundation contains the superposition of three forms of waves: the free field incident wave, the free field reflection wave and the scattered wave of the secondary scattering field. It is assumed that the scattered field wave is divided into cylindrical spherical and waves, respectively, that obtain two- and three-dimensional spring damper coefficients. The specific derivation process can be referred to in the literature [9, 10]. An example of a two-dimensional compressional wave (P-wave) is given here to illustrate the derivation process.

2.1. Formula derivation

Assuming the plane strain condition, the wave front is the cylindrical wave by the example of P-wave; the wave equation of radial displacement is:

$$\frac{\partial^2 u}{\partial t^2} = \frac{2G + \lambda}{\rho}\left(\frac{\partial^2 u}{\partial r^2} + \frac{1}{r}\frac{\partial u}{\partial r} - \frac{u}{r^2}\right) \tag{1}$$

$$\lambda = \frac{\mu E}{(1 + \mu)(1 - 2\mu)} \tag{2}$$

where u is the radial displacement, G is the shear modulus of material at the boundary, λ is *Lamb* constant, E is Young's modulus and μ is Poisson's ratio.

According to the physical and geometric equation, we can obtain:

$$\sigma_\theta = \lambda\varepsilon_r + (2G + \lambda)\varepsilon_\theta \tag{3}$$

$$\sigma_r = \lambda\varepsilon_\theta + (2G + \lambda)\varepsilon_r \tag{4}$$

$$\varepsilon_\theta = \frac{u}{r} \tag{5}$$

$$\varepsilon_r = \frac{\partial u}{\partial r} \tag{6}$$

Use the displacement potential function to represent radial displacement and put it into Eq. (1), we can obtain:

$$\frac{\partial}{\partial r}\frac{\partial^2 \Phi}{\partial t^2} = \frac{2G+\lambda}{\rho}\frac{\partial}{\partial r}\left(\frac{\partial^2 \Phi}{\partial r^2} + \frac{1}{r}\frac{\partial \Phi}{\partial r}\right) \tag{7}$$

Take the squeezing wave velocity as the compressional wave speed $c_p = \sqrt{\frac{2G+\lambda}{\rho}}$ and put it into Eq. (7) and integrating can obtain:

$$\frac{\partial^2 \Phi}{\partial t^2} = c_p^2\left(\frac{\partial^2 \Phi}{\partial r^2} + \frac{1}{r}\frac{\partial \Phi}{\partial r}\right) \tag{8}$$

The approximate general solution for Eq. (8) is:

$$\Phi(r,t) = \frac{1}{\sqrt{r}}f\left(\frac{r}{c_p} - t\right) \tag{9}$$

$$u(r,t) = \frac{\partial \Phi}{\partial t} = -\frac{1}{2}r^{-\frac{3}{2}}f + \frac{1}{c_p}r^{-\frac{1}{2}}f' \tag{10}$$

$$\frac{\partial u}{\partial t}(r,t) = \frac{1}{2}r^{-\frac{3}{2}}f' - \frac{1}{c_p}r^{-\frac{1}{2}}f'' \tag{11}$$

$$\frac{\partial^2 u}{\partial t^2}(r,t) = -\frac{1}{2}r^{-\frac{3}{2}}f'' + \frac{1}{c_p}r^{-\frac{1}{2}}f''' \tag{12}$$

$$\varepsilon_r = \frac{3}{4}r^{-\frac{5}{2}}f - \frac{1}{c_p}r^{-\frac{3}{2}}f' + \frac{1}{c_p^2}r^{-\frac{1}{2}}f'' \tag{13}$$

$$\varepsilon_\theta = -\frac{1}{2}r^{-\frac{5}{2}}f + \frac{1}{c_p}r^{-\frac{3}{2}}f' \tag{14}$$

$$\varepsilon_r + \varepsilon_\theta = \frac{\partial^2 \Phi}{\partial r^2} + \frac{1}{r}\frac{\partial \Phi}{\partial r} = \frac{1}{c_p^2}\frac{\partial^2 \Phi}{\partial t^2} \tag{15}$$

Combining Eq. (4), the radial positive stress obtained is:

$$\sigma_r = (2G+\lambda)r^{-\frac{1}{2}}\frac{1}{c_p^2}f'' - 2G\left(\frac{1}{c_p}r^{-\frac{3}{2}}f' - \frac{1}{2}r^{-\frac{5}{2}}f\right) \tag{16}$$

Comparing with Eq. (10), we can obtain:

$$\sigma_r = (2G+\lambda)r^{-\frac{1}{2}}\frac{1}{c_p^2}f'' - \frac{2G}{r}u \tag{17}$$

$$\frac{\partial \sigma_r}{\partial t} = -(2G+\lambda)r^{-\frac{1}{2}}\frac{1}{c_p^2}f''' - \frac{2G}{r}\frac{\partial u}{\partial t} \tag{18}$$

Synthesizing the Eq. (12, 17, 18), we obtain

$$\sigma_r + \frac{2r}{c_p}\frac{\partial \sigma_r}{\partial t} = -\frac{2G}{r}u - \frac{4G}{c_p}\frac{\partial u}{\partial t} - 2r\rho\frac{\partial^2 u}{\partial t^2} \tag{19}$$

We simulate Eq. (19) with spring, damper and mass unit system, as shown in **Figure 3**.

Dynamic equilibrium differential equations are listed in the mechanical model as shown in **Figure 2**:

$$ku_1 + c(\dot{u}_1 - \dot{u}_2) = -f(t) \tag{20}$$

$$m\ddot{u}_2 + c(\dot{u}_2 - \dot{u}_1) = 0 \tag{21}$$

where u_1、u_2 represent two degrees of freedom of the displacement and $f(t)$ represents the force applied to the boundary point.

Replace the Eq. (20) into (21) to get:

$$f + \frac{m}{c}f' = -ku_1 - \frac{mk}{c}\dot{u}_1 - m\ddot{u}_1 \tag{22}$$

In contrast, the factor of the element is obtained by analogy:

$$m = 2\rho r, \quad c = \rho c_p, \quad k = \frac{2G}{r} \tag{23}$$

For the mechanical model in **Figure 2**, neglecting the quality and fixing on the end of the mass-connected dampers which can simplify the calculation and improve the accuracy of the calculation results form the spring damping element in **Figure 3**, which can be easily combined with large general-purpose finite element software.

2.2. The two- and three-dimensional spring damping coefficient

The derivation is based on the spring damping coefficient obtained by the P-wave. Similarly, there is a similar coefficient form for the shear wave.

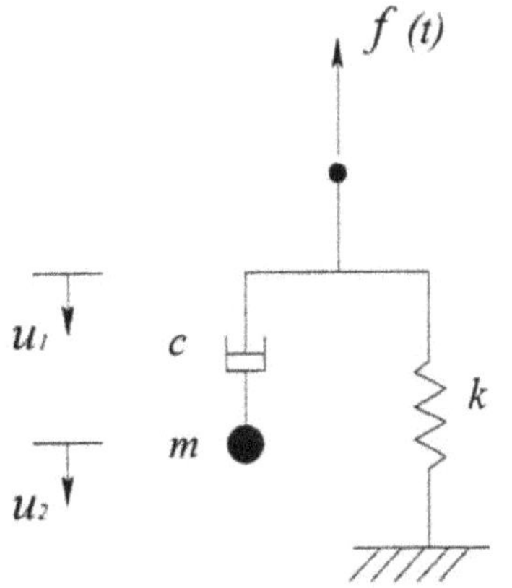

Figure 2. Viso-spring model.

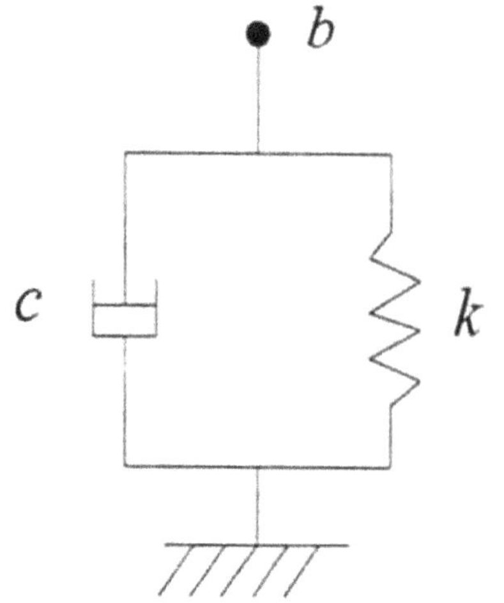

Figure 3. Viso-spring component.

The damping coefficients of the two- and three-dimensional viscoelastic artificial boundaries are summed up.

Two dimension:

Compressional wave (P-wave):

$$k = \frac{2G}{r}, \quad c = \rho c_p \tag{24}$$

Shear wave (SV, SH wave):

$$k = \frac{2G}{r}, \quad c = \rho c_s \tag{25}$$

Three dimension:

Compressional wave (P-wave):

$$k = \frac{4G}{r}, \quad c = \rho c_p \tag{26}$$

Shear wave (SV, SH wave):

$$k = \frac{2G}{r}, \quad c = \rho c_s \tag{27}$$

Based on the Eqs. (24)–(27) the damping coefficient of spring is summed up as follows:

$$K = \alpha \frac{G}{r} \sum_{i=1}^{n} A_i, \quad C = \rho c \tag{28}$$

where k c are the spring damping coefficients at the boundary; r is the distance from the scattering wave to the boundary; ρ and G are the medium density and the shear modulus,

Type	Direction	α
Two-dimensional viscoelastic artificial boundary	Normal direction in plane	2.0
	Tangential direction in plane	1.5
	Tangential direction out of plane	0.5
Three-dimensional viscoelastic artificial boundary	Normal direction	4.0
	Tangential direction	2.0

Table 1. The visco-elastic artificial boundary parameter α.

respectively, c is the wave velocity in the medium where the P-wave velocity c_p and the S-wave velocity c_s are the P-wave velocity and the S-wave velocity in the normal and the tangential direction, respectively. $\sum_{i=1}^{n} A_i$ is the area represented by a boundary node obtained by *arnode* function in ANSYS. The value of the parameter α can be found in the **Table 1**.

3. Input method of equivalent load fluctuation

Liu Jingbo introduced a method for seismic waves that are converted to the equivalent load seismic waves on the viscoelastic artificial boundary to conduct wave inputs. The method calculating structural seismic responses by the combination of viscoelastic artificial boundary and large general-purpose finite element program can be summed up in five steps in the earlier flowchart [19].

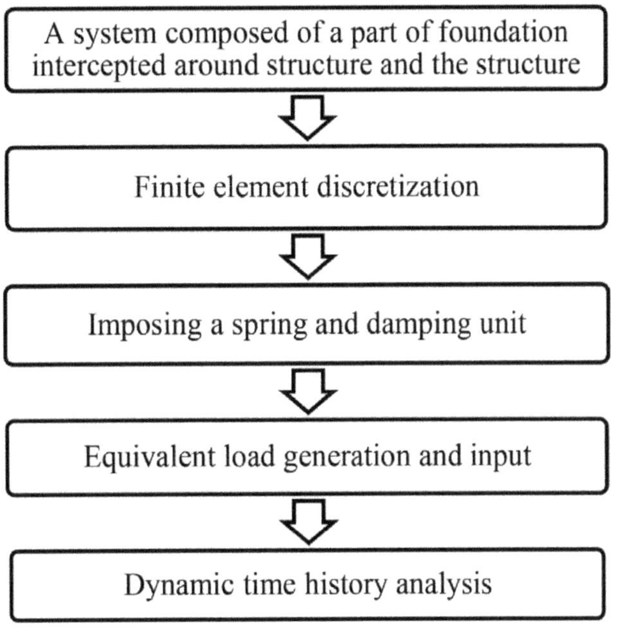

3.1. Wave field separation

The total field displacement U_b^t is decomposed into the incident field U_b^f (including incidence and reflection) and scattering field U_b^s, as shown in **Figure 1**.

$$\begin{bmatrix} M_s & \\ & M_b \end{bmatrix}\begin{bmatrix} \ddot{U}_s \\ \ddot{U}_b^t \end{bmatrix} + \begin{bmatrix} C_{ss} & C_{sb} \\ C_{bs} & C_{bb}+C \end{bmatrix}\begin{bmatrix} \dot{U}_s \\ \dot{U}_b^t \end{bmatrix} + \begin{bmatrix} K_{ss} & K_{sb} \\ K & K_{bb}+K \end{bmatrix}\begin{bmatrix} U_s \\ U_b^t \end{bmatrix} = \begin{bmatrix} 0 \\ \tau_0 \end{bmatrix} \quad (29)$$

$$\begin{bmatrix} M_s & \\ & M_b \end{bmatrix}\begin{bmatrix} \ddot{U}_s \\ \ddot{U}_b^t \end{bmatrix} + \begin{bmatrix} C_{ss} & C_{sb} \\ C_{bs} & C_{bb} \end{bmatrix}\begin{bmatrix} \dot{U}_s \\ \dot{U}_b^t \end{bmatrix} + \begin{bmatrix} K_{ss} & K_{sb} \\ K & K_{bb} \end{bmatrix}\begin{bmatrix} U_s \\ U_b^t \end{bmatrix} = \begin{bmatrix} 0 \\ \tau_0 - C\dot{U}_b^t - KU_b^t \end{bmatrix} \quad (30)$$

Equivalent load is $F_e = \tau_0 - C\dot{U}_b^t - KU_b^t$, in this, $U_b^t = U_b^f + U$, and if we take one derivative of this we get $\dot{U}_b^t = \dot{U}_b^f + \dot{U}$. Equivalent load becomes $F_e = \tau_0 - C\dot{U}_b^f - KU_b^f - C\dot{U}_b^s - KU_b^s$; the scattering field fluctuation will be absorbed by the yellow damping unit and the general expression of the equivalent load will be obtained, and it is $F_e = \tau_0 - C\dot{U}_b^f - KU_b^f$.

3.2. Equivalent load input

Equivalent load expression is $F_e = KU + C\dot{U} + \tau_0$; note $KU + C\dot{U}$ as the velocity displacement term and τ_0 as the stress term.

As shown in **Figure 4**, through the stress analysis of the microelement on the boundary, we obtained in the first stress state, the Poisson effect of the cell is considered, and the shear stress in the second stress state is equal to each other. Therefore, the lateral boundary stress τ_0 caused by seismic waves can be discussed on two situations of P-wave and S-wave (wave vertical incidence).

The stress when P-wave enters the lower boundary: $\tau_0 = \frac{\nu\rho C_p \dot{U}_b^f}{1-\nu.}$

The stress when S-wave enters the lower boundary: $\tau_0 = \rho C_s \dot{U}_b^f$

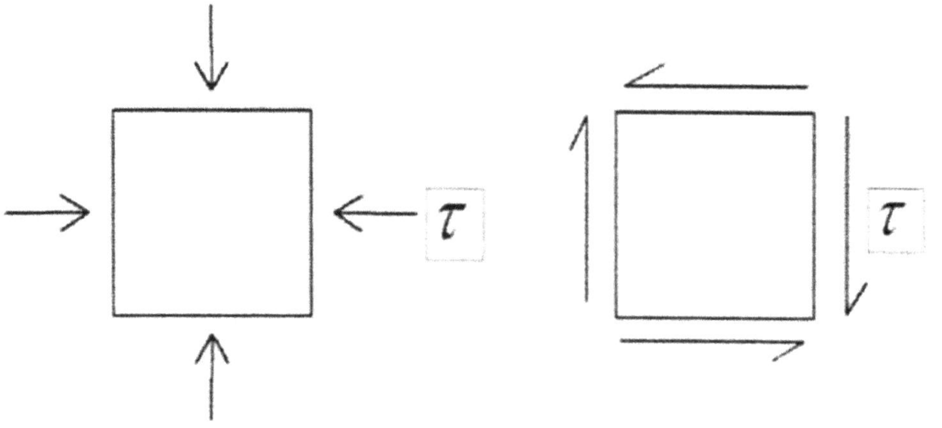

Figure 4. Soil element at the boundary.

3.2.1. *Two-dimensional equivalent load input*

The value of the spring damping coefficient is based on "a direct method of the analysis of the dynamic interaction of the structural foundation" [6, 9], and the specific values are:

Surface normal: $k_N = \frac{G}{2r}\sum A, \quad C_N = \rho C_p \sum A$

Surface tangential: $k_T = \frac{G}{2r}\sum A, \quad C_T = \rho C_s \sum A$

where k_N is the coefficient of the normal spring coefficient, C_N is the coefficient of the normal damper, k_T represents the tangential spring coefficient and C_T is the coefficient of the tangential damper.

The spring damping coefficients of each boundary and the direction of the velocity displacement and stress in the equivalent load of each boundary line when seismic waves are inputs from two directions are given as shown in **Figure 5**.

The boundary line 1:

The spring damper coefficient:

Surface normal (external normal Y, minus): $k_N = \frac{G}{2r}\sum A, \quad C_N = \rho C_p \sum A$

Surface tangential(X): $k_T = \frac{G}{2r}\sum A, \quad C_T = \rho C_s \sum A$

Equivalent load in all directions (two-way input):

X dimension: Velocity displacement term (X+); stress term (X+)

Y dimension: Velocity displacement term (Y+); stress term (Y+)

The boundary line 2:

The spring damper coefficient:

Surface normal (external normal X, plus): $k_N = \frac{G}{2r}\sum A, \quad C_N = \rho C_p \sum A$

Surface tangential(Y): $k_T = \frac{G}{2r}\sum A, \quad C_T = \rho C_s \sum A$

Equivalent load in all directions (two-way input):

X dimension: Velocity displacement term (X+); stress term (X−)

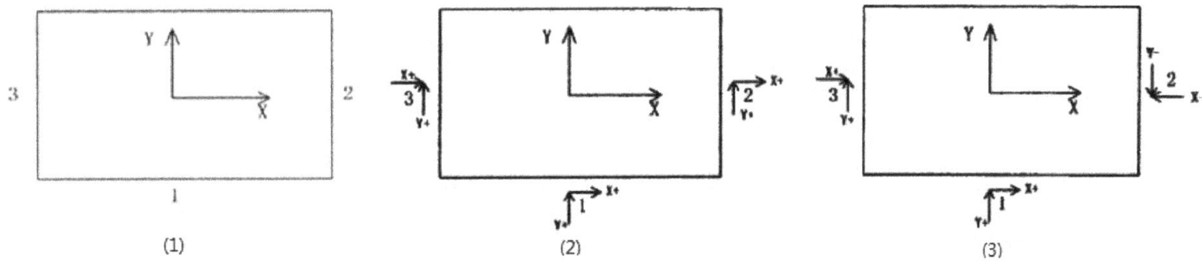

Figure 5. The input of equivalent load in two dimensions. (1) Two-dimensional numbering schematic. (2) Input of equivalent load term. (3) Input of stress term.

Y dimension: Velocity displacement term (Y+); stress term (Y−)

The boundary line 3:

The spring damper coefficient:

Surface normal (external normal X, minus): $k_N = \frac{G}{2r}\sum A, \quad C_N = \rho C_p \sum A$

Surface tangential(Y): $k_T = \frac{G}{2r}\sum A, \quad C_T = \rho C_s \sum A$

Equivalent load in all directions (two-way input):

X dimension: Velocity displacement term (X+); stress term (X−)

Y dimension: Velocity displacement term (Y+); stress term (Y−)

3.2.2. Three-dimensional equivalent load input

As shown in the three-dimensional model of **Figure 6**, the coordinate system is in conformity with the laws of the corkscrew, the "1" is the bottom, the "2" is the front interface, the "3" is the right side of the boundary surface, the "4" is the back interface and the "5"is the left side of the boundary surface, and the faces from 2 to 5 are numbered counter-clockwise.

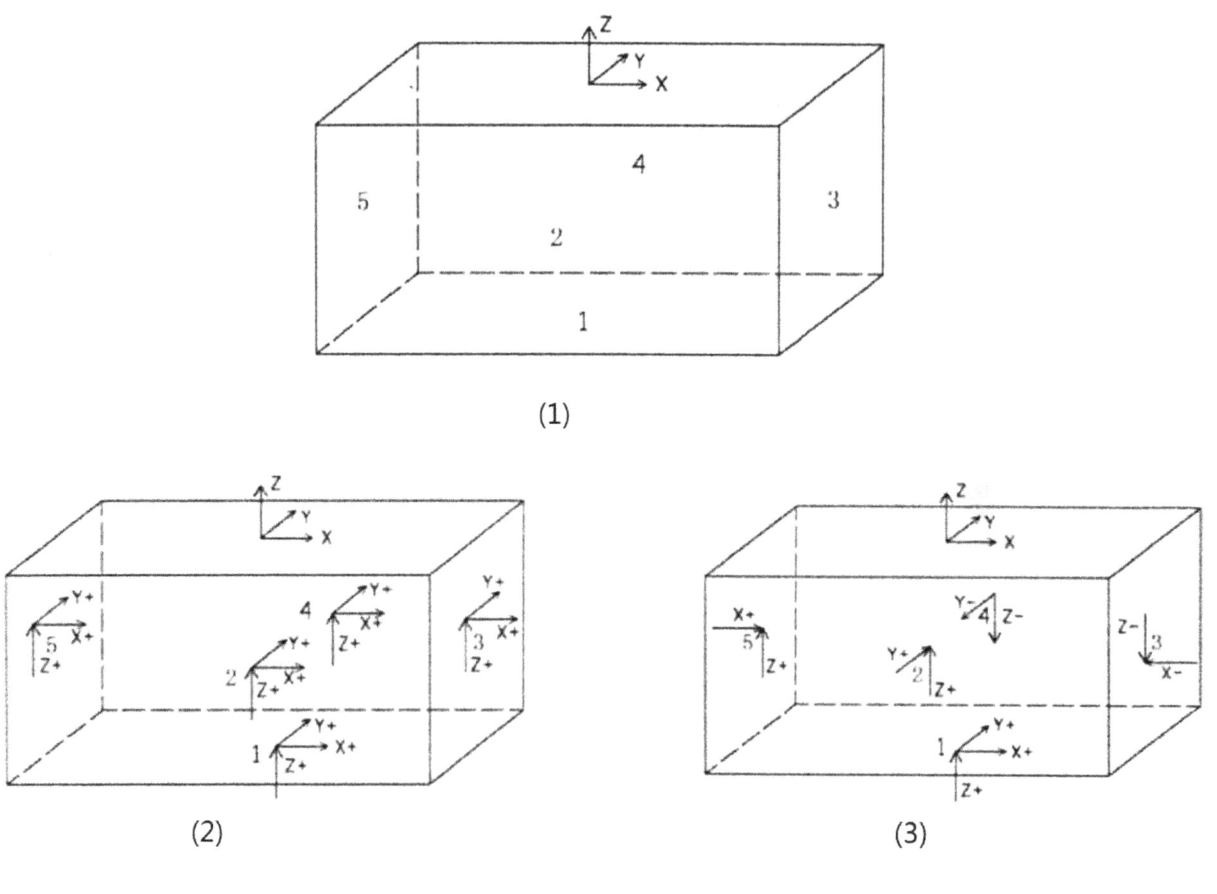

Figure 6. The input of equivalent load in three dimensions. (1) Three-dimensional numbering diagram. (2) Input of equivalent load term. (3) Input of stress term.

The value of the spring damping coefficient is based on "three-dimensional time-domain viscoelastic artificial boundary in wave problem" [7, 10], and the specific values are:

Face normal: $k_N = \frac{4G}{r}\sum A, \quad C_N = \rho c_p \sum A$

Surface tangential: $k_T = \frac{2G}{r}\sum A, C_T = \rho c_s \sum A$

where k_N is the coefficient of the normal spring coefficient, C_N is the coefficient of the normal damper coefficient, k_T represents the tangential spring coefficient and C_T is the coefficient of the tangential damper.

The spring damping coefficients of each boundary and the direction of the velocity displacement and stress in the equivalent load of each boundary line when seismic waves are inputs from three directions are given as shown in **Figure 6**.

Face 1:

The spring damper coefficient:

Surface normal (external normal Z, minus): $k_N = \frac{4G}{r}\sum A, c_N = \rho c_p \sum A$

Surface tangential (XY): $k_T = \frac{2G}{r}\sum A,\ c_T = \rho c_s \sum A.$

Equivalent load in all directions (three-way input):

X dimension: Velocity displacement term (X+); stress term (X+)

Y dimension: Velocity displacement term (Y+); stress term (Y+)

Z dimension: Velocity displacement term (Z+); stress term (Z+)

Face 2:

The spring damper coefficient:

Surface normal (external normal Y, minus): $k_N = \frac{4G}{r}\sum A, c_N = \rho c_p \sum A$

Surface tangential (XZ): $k_T = \frac{2G}{r}\sum A,\ c_T = \rho c_s \sum A.$

Equivalent load in all directions (three-way input):

X dimension: Velocity displacement term (X+); stress term (nothing)

Y dimension: Velocity displacement term (Y+); stress term (Y+)

Z dimension: Velocity displacement term (Z+); stress term (Z+)

Face 3:

The spring damper coefficient:

Surface normal (external normal X, plus): $k_N = \frac{4G}{r}\sum A, c_N = \rho c_p \sum A$

Surface tangential (YZ): $k_T = \frac{2G}{r}\sum A, \quad c_T = \rho c_s \sum A.$

Equivalent load in all directions (three-way input):

X dimension: Velocity displacement term (X+); stress term (X-)

Y dimension: Velocity displacement term (Y+); stress term (nothing)

Z dimension: Velocity displacement term (Z+); stress term (Z-)

Face 4:

The spring damper coefficient:

Surface normal (external normal Y, plus): $k_N = \frac{4G}{r}\sum A, c_N = \rho c_p \sum A$

Surface tangential (XZ): $k_T = \frac{2G}{r}\sum A, \quad c_T = \rho c_s \sum A.$

Equivalent load in all directions (three-way input):

X dimension: Velocity displacement term (X+); stress term (nothing)

Y dimension: Velocity displacement term (Y+); stress term (Y-)

Z dimension: Velocity displacement term (Z+); stress term (Z-)

Face 5:

The spring damper coefficient:

Surface normal (external normal X, minus): $k_N = \frac{4G}{r}\sum A, c_N = \rho c_p \sum A$

Surface tangential (XZ): $k_T = \frac{2G}{r}\sum A, \quad c_T = \rho c_s \sum A.$

Equivalent load in all directions (three-way input):

X dimension: Velocity displacement term (X+); stress term (X+)

Y dimension: Velocity displacement term (Y+); stress term (nothing)

Z dimension: Velocity displacement term (Z+); stress term (Z+)

3.2.3. The free field motion considering the traveling wave effect

$$u(k,t) = \begin{cases} u_{inp}(k,t) & t \leq \dfrac{d}{c} \\ u_{inp}(k,t) + u_{ref}\left(k, t - \dfrac{2H-d}{c}\right) & \dfrac{d}{c} < t < t_{end} \\ u_{ref}\left(k, t - \dfrac{2H-d}{c}\right) & t \geq t_{end} \end{cases} \tag{31}$$

where k represents a point at the boundary point, d represents the vertical distance from the bottom boundary point k, c represents the velocity of wave propagation, u_{inp} and u_{ref} represent the displacement wave and the displacement wave displacement, respectively.

In the process of wave propagation in elastic medium, deformation, stress and particle will spread at the same speed in the same way. The boundary point K at a certain time is the superposition of the incident wave displacement and the reflected wave displacement. This is similar for velocity.

4. Seismic input model of slope site constructed by virtual symmetry substructure

4.1. Equivalent load of node in slope site

As in **Figure 7**, the normal direction of the two sides with different heights is fixed in the direction of X, considering the traveling wave effect that the expression of the equivalent load on the two sides of the slope is:

$$\sigma_x = K\left(u_{inp}(k,t) + u_{ref}\left(k, t - \frac{2H-d}{c}\right)\right) + C\left(u_{inp}(k,t) + u_{ref}\left(k, t - \frac{2H-d}{c}\right)\right) \tag{32}$$

$$\sigma_{x\text{高}} = K\left(u_{inp}(k,t) + u_{ref}\left(k, t - \frac{2H_1-d}{c}\right)\right) + C\left(u_{inp}(k,t) + u_{ref}\left(k, t - \frac{2H_1-d}{c}\right)\right) \tag{33}$$

$$\sigma_{x\text{低}} = K\left(u_{inp}(k,t) + u_{ref}\left(k, t - \frac{2H_2-d}{c}\right)\right) + C\left(u_{inp}(k,t) + u_{ref}\left(k, t - \frac{2H_2-d}{c}\right)\right) \tag{34}$$

For (7), take $\Delta t = \frac{2H_1 - 2H_2}{c}$, so

$$\sigma_{x\text{低}} = K\left(u_{inp}(k,t) + u_{ref}\left(k, t - \frac{2H_1-d}{c} + \Delta t\right)\right) + C\left(u_{inp}(k,t) + u_{ref}\left(k, t - \frac{2H_1-d}{c} + \Delta t\right)\right) \tag{35}$$

where K and C are the coefficient of spring dampers, and the other symbolic meanings are the same as (s4).

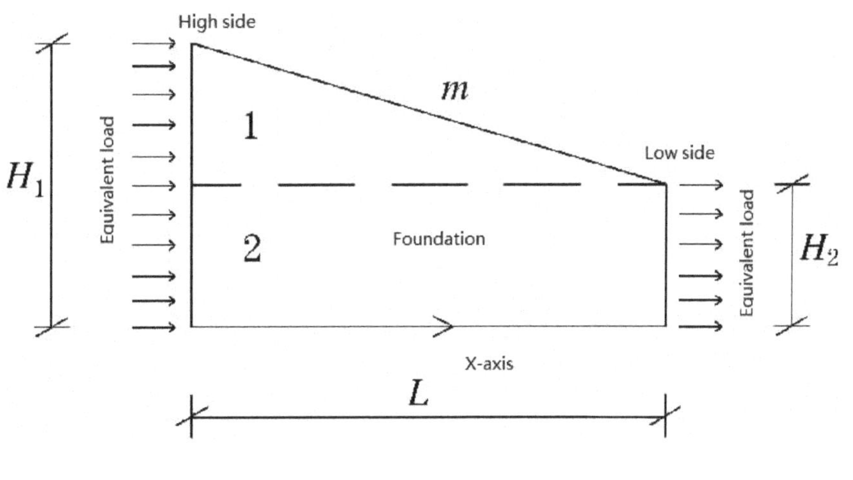

Figure 7. Equivalent load on the slope site.

Comparing Eqs. (33) and (35), the difference is Δt between the displacement and velocity time history which causes the difference in the size of the two side nodes with different heights. The ground is artificially divided into one and two parts in **Figure 3**, and the reasons for the inconsistency of the equivalent load can be seen directly. Part 1 has only boundaries without the corresponding equivalent load. The force is equal to the product of the stress and the area, and steeper the slope, the greater the difference between the area of two sides; the difference of the force will be more obvious.

4.2. The solving steps of the virtual symmetric substructure system

The analysis of the expression of the equivalent load shows that the difference in height and area can cause the inconsistency of the equivalent load of the two sides. By analyzing the mechanical properties, we can come to a conclusion that the cause of displacement divergence or response eccentricity is the inconsistency of the equivalent load resultant force in the two sides with different heights. Therefore, a virtual symmetric substructure is used to construct a viscoelastic artificial boundary seismic input model suitable for the slope site; the specific steps are as follows:

1. As shown in **Figure 8**, the boundary of the symmetry system composed by the original slope model and an existing nominally only equal-sized slope foundation about line 2 is line 1 and 3 based on two dimensions. Based on the high consistency of the boundary line, the equivalent load generation and input method of the flat site can be directly applied to calculate the seismic response in this case, and the node displacement and velocity on line 2 of the symmetry axis can be extracted.

2. Calculate the equivalent load of the node of line 2; only the actual slope is calculated that the virtual foundation is canceled where the boundary line is changed into line 1 and 2. The displacement and velocity of nodes extracted from step (1) are the total displacement and total velocity at any time that we do not need to consider the traveling wave effect, so the formula for calculating the equivalent load is: $F_e = \dot{C}U_{tb} + KU_b^t$. Take the calculated equivalent load of the line 2 node and the equivalent load of the original line 1 into the slope model to calculate the seismic response of the structure.

The steps for solving the three-dimensional slope model are the same as earlier. In theory, the equivalent load of this structure is equal on the boundary line, and it can effectively simulate the wave propagation.

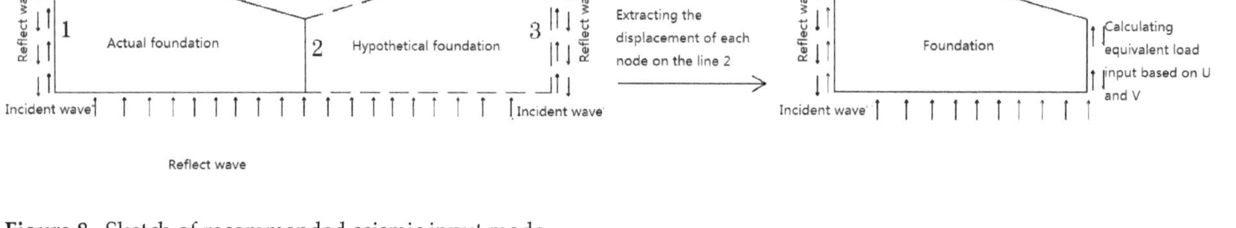

Figure 8. Sketch of recommended seismic input mode.

5. Example

5.1. Input seismic waves to the free field from three directions

Firstly, the correctness of the three-dimensional viscoelastic boundary procedure under the conditions of free field is verified, namely, the surface displacement analytical solution of the homogeneous free field is two times than the incident displacement time history when considering the delay due to the traveling wave effect. The seismic input from 3 directions all is simple harmonic sine wave. Input two cycles, and view the stability of the result.

The model is a cuboid, the length and width of it are 800 m, the height of it is 400 m and the size is shown in **Figure 9**. The shear modulus is 5.292×10^9 Pa, the Poisson ratio is 0.25, the density is 2700 kg/m^3 and the shear wave velocity is 1400 m/s. The type of the solid unit is solid45, size of the unit is 40 m and three-dimensional spring damping unit Combin14 on the bottom boundary and four side boundaries is applied. The coordinate origin is selected at the center point of the top surface and is taken as the observation point, marked O.

As shown in **Figure 10**, P wave's velocity is greater than S wave's and reaches the top after 0.180 s; it has simple harmonic vibration with the size as twice as the bottom input, and at the same time the response of the S wave is affected; S wave shows slight fluctuations between 0.180 and 0.285 s and then is followed by a twofold unit sinusoidal vibration. The response of the top of the three directions is two times than the bottom input, and the displacement responses of the x direction and y direction are basically the same. We verify the correctness of the seismic input method and the procedure and then study the seismic response under the slope.

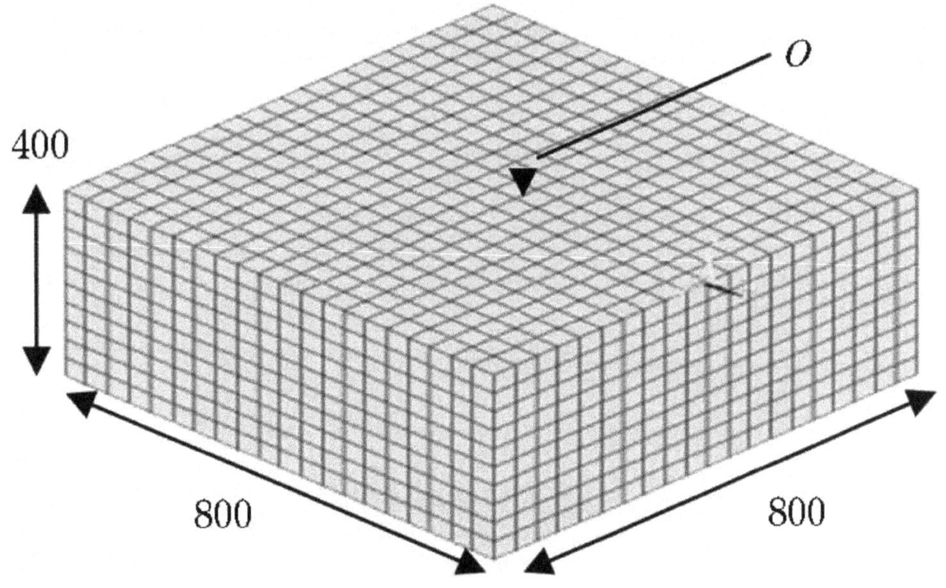

Figure 9. Three dimensional finite element model of free-field.

5.2. Input the one-way ground motion in slope field

In order to make a comparison with the method of the symmetrical substructure system, the equivalent load is generated in the form of a flat free field, and the slope model is directly taken as the input to calculate it is seismic response. The slope model (**Figure 11**) is selected and there

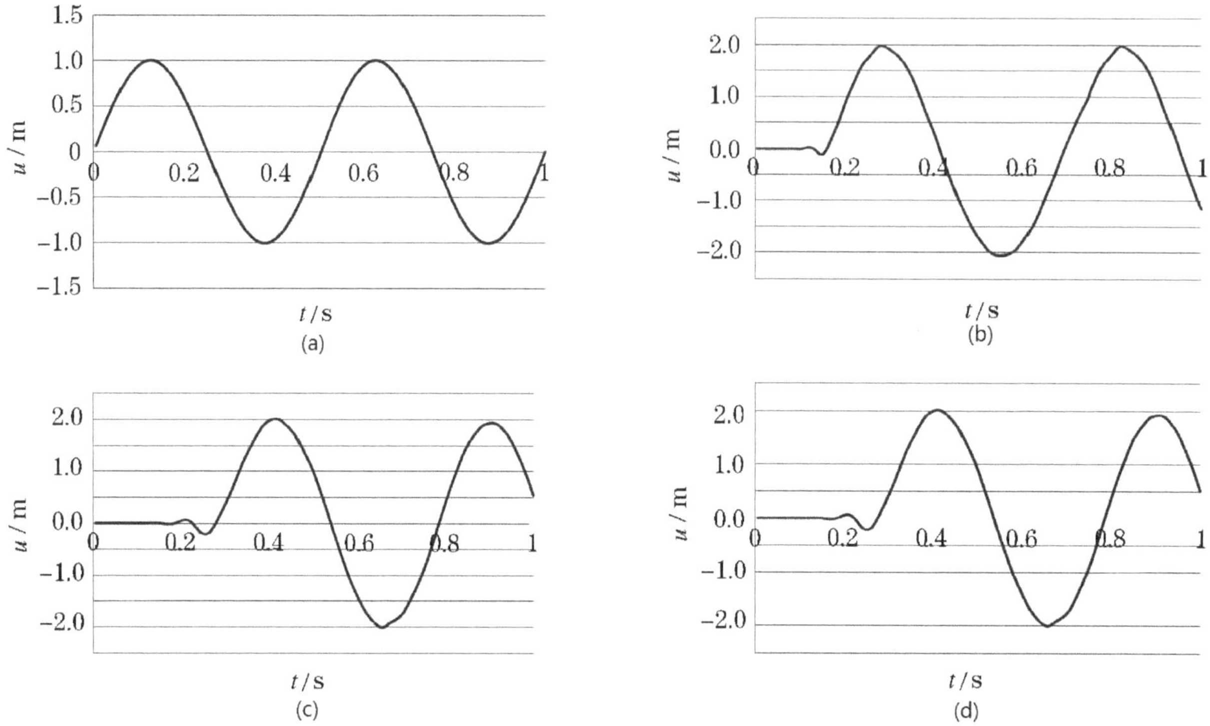

Figure 10. Dynamic responses of free-field. (a) The unit sine wave entered at the bottom. (b) The z-directional displacement time history of the observation point O (P-wave, α-angle). (c) The x-directional displacement time history of the observation point O (S-wave, α-angle). (d) The y-directional displacement time history of the observation point O (S-wave, α-angle).

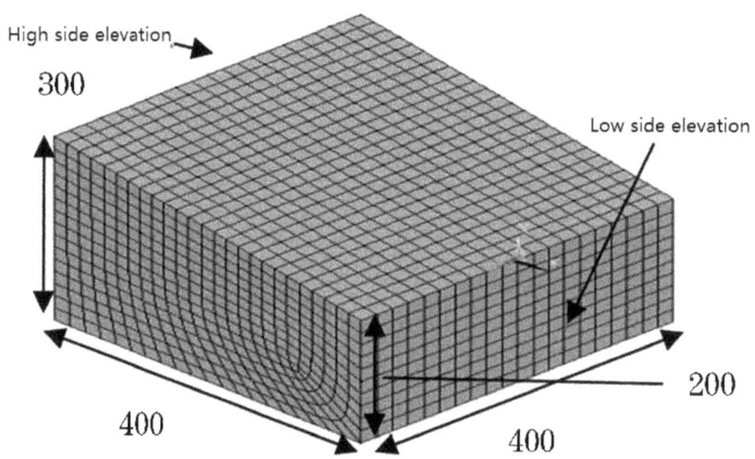

Figure 11. The map of finite element grid division of slope.

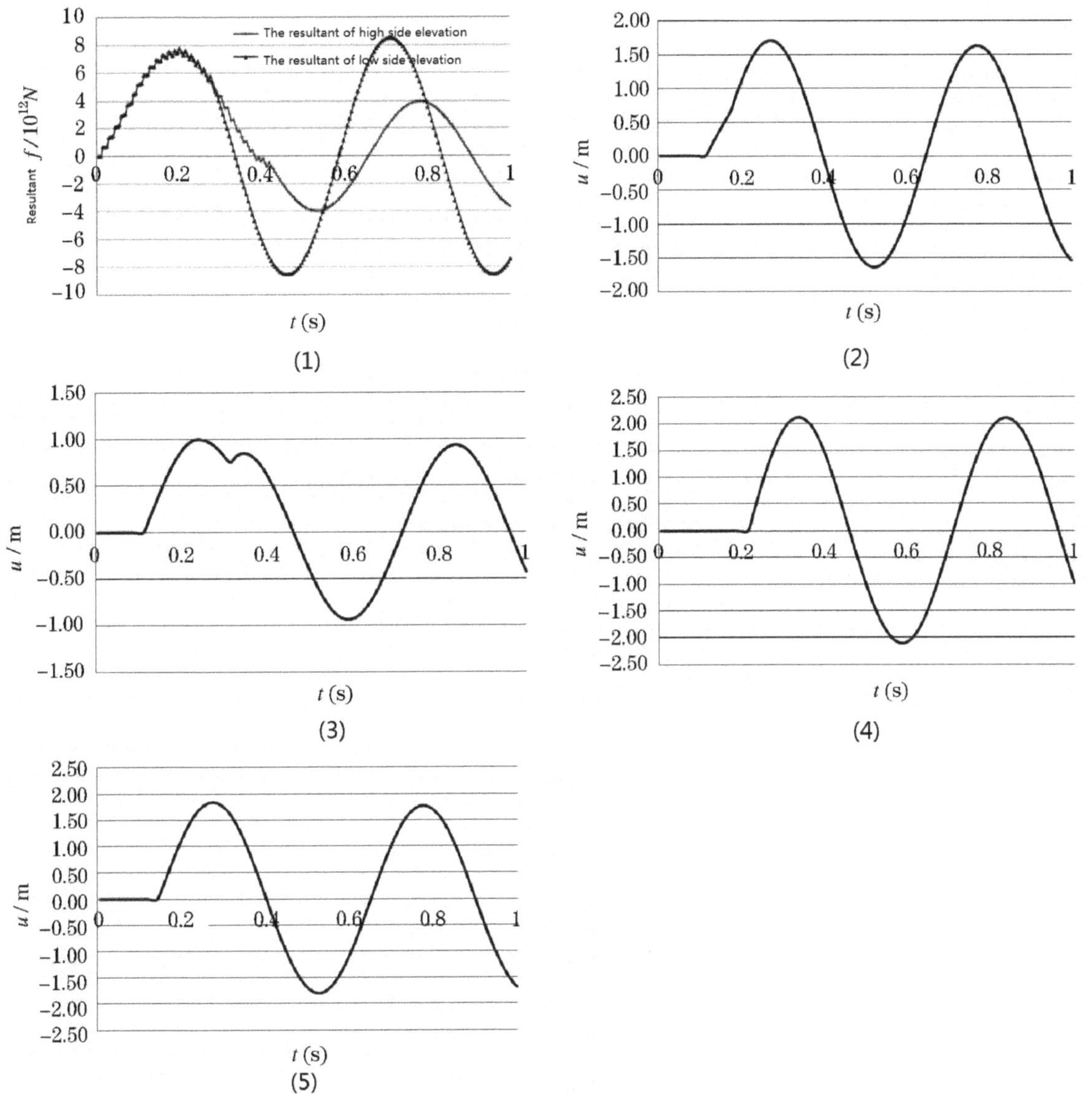

Figure 12. Dynamic responses of slope site. (1) The comparison of the resultant force on the side elevation. (2) The displacement time history of lower side elevation at the height of 160 m. (3) The displacement time history of higher side elevation at the height of 160 m. (4) Vertex displacement time history higher side elevation. (5) Vertex displacement time history higher side elevation.

is a high difference in the side elevation of one direction. The base is square and the length and width is 400 m. Take the normal direction on both sides of the plane as the X-axis, and so, the height of both sides in X is 300 and 200 m, respectively, and gradient of the slope is 0.25; the detailed dimensions are shown in the figure. The material parameters are the same as the free field. The entity unit type is solid45, the size of the element is 20 m in the direction of the boundary line; the mapping subdivision is performed within the surface. We apply a three-dimensional spring damping unit Combin14 on the bottom boundary and four side boundaries. In order to clearly see the change of equivalent load force, only input the unit sinusoidal shear wave in X; the waveform is the same as the input of the three-dimensional free-field calculation example.

As **Figure 12** shows:

1. In **Figure 12 (1)**, after 0.25 s, the equivalent load in x of the low-side elevation is far greater than that of the high-side elevation, and the resultant force is unbalanced.
2. Compared with the two points with elevation of 160 m in two sides, the distance time history between the two points is completely different, as shown in **Figure 12 (2)** and **(3)**. The displacement of **Figure 12 (2)** presents obviously the change of the cycle which is led by the superposition, the amplitude is 1, and **Figure 12 (3)** shows the complete sinusoidal waveform with the amplitude of 1.72. Furthermore, the displacement time history of the vertex on both sides of the elevation is further compared to see the accuracy and stability of the results. In **Figure 12 (4)**, the displacement amplitude of the first period is 2.16, the displacement amplitude of the second period is 2.19, showing an increasing trend and faring from the relationship of two times. In **Figure 12 (5)**, the displacement amplitude of the first period is 1.76, the displacement amplitude of the second period is 1.73 and we see a decrease.
3. So you can see that the model is biased towards one side; in this chapter, it is considered that the imbalance of the overall resultant force of the model leads to this eccentricity. Presumably, this eccentricity is more apparent when the slope is larger. Because the seismic wave calculating actually is not a regular periodic wave, so directly taking the equivalent load calculated by flat free field into the slope site model will cause the eccentricity, or even wrong results.

5.3. The slope site model is calculated using a symmetric substructure

In accordance with the method recommended in Section 2 and the slope model, the model size is the same and the material parameters are unchanged. According to step (1), a symmetric system is constructed by setting phonily the isometrical foundation and the finite element mesh of the symmetric model is shown in **Figure 13**. The height of the symmetric face of the model is 200 m, the height of the two sides is 300 m, the base length is 800 m and the width is

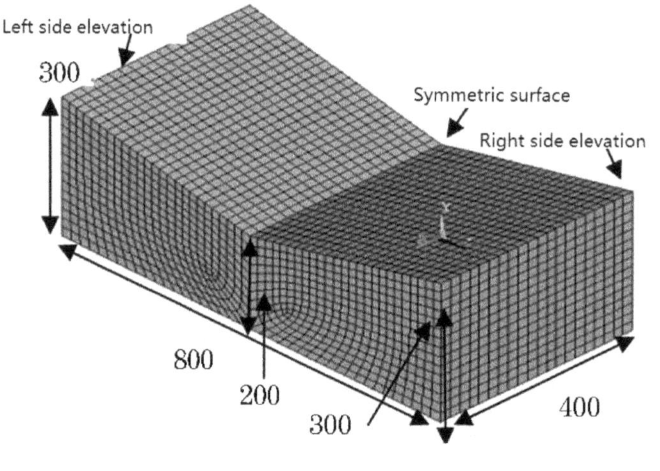

Figure 13. Mesh division of finite element of symmetrical model.

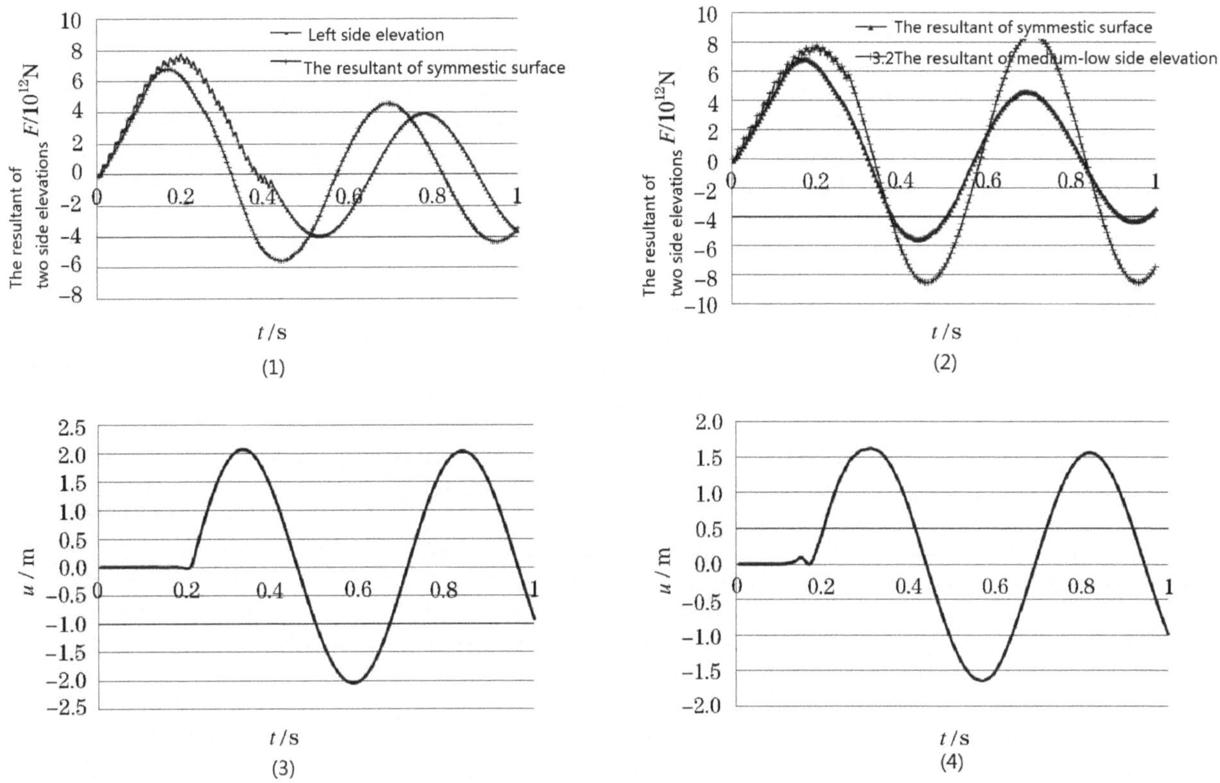

Figure 14. Dynamic response of symmetrical model. (1) Resultant force contrast between symmetry plane and left side elevation. (2) Resultant force contrast between symmetry plane and low side elevation. (3) Displacement time history of the vertex of left side elevation. (4) Displacement time history of the point with the height of 250 m.

400 m. The solid unit of Solid45 is used to divide the grid, and the size of the finite element mesh on the boundary line is 20 m, with a total of 23,876 units. The model in step (2) is the same as **Figure 11**. Since the symmetrical face in step (1) is the same face as the low side elevation of the slope model in step (2), they will be replaced by symmetry plane in the following.

Figure 14 (1) is the contrast diagram of the equivalent load force of every moment of the symmetry plane and the equivalent load of the left side. **Figure 14 (2)** is the contrast diagram of the equivalent load force of the symmetry plane and the equivalent load of the low-side elevation in Section 2.

You can see from **Figure 14**:

1. The equivalent load size of the symmetrical face in **Figure 14 (1)** is close to that of the left elevation, and the change of equivalent load conforms to the theoretical expectation. As shown in **Figure 14 (2)**, the numerical variation of the resultant force of the lower lateral equivalent load of the slope model is significant, indicating the effectiveness of the proposed method.

2. In **Figure 14 (3)** and **(4)**, the displacement time history is numerically stable and the rationality of the seismic input mode of constructing the symmetric substructure is proved from the results.

6. Conclusion

In this chapter, an earthquake input mode for viscoelastic artificial boundary under the condition of the slope is proposed by constructing the virtual symmetric substructure system. The numerical examples verify the stability of the results and the following conclusions are obtained:

1. Based on the numerical expression of the equivalent load at the viscoelastic artificial boundary, the chapter deduces that the height and area difference of the side elevation are the important reasons for the inconsistency of the seismic input load.

2. By constructing a virtual symmetric structure system, we can ensure the symmetry of foundation calculation area outside the boundaries, which is easy to simulate the propagation characteristics of the free field seismic waves and determine the external boundary input load.

3. The method ensures the accordance of resultant force of seismic input load in the condition of the slope field, makes the dynamic response results more reasonable and is easy to implement in a large general finite element software-ANSYS and can be used in the analysis of seismic responses of the nuclear structure under the conditions of two-dimensional and three-dimensional slope field.

Author details

Li Jianbo* and Liu Weihong

*Address all correspondence to: jianboli@dlut.edu.cn

Institute of Earthquake Engineering, Dalian University of Technology, Dalian, China

References

[1] Liao ZP. Introduction to Wave Motion Theories in Engineering. Beijing: Science Press; 2002

[2] Du X. Theories and Methods of Wave Motion for Engineering. Beijing: Science Press; 2009

[3] Lysmer J, Kulemeyer RL. Finite dynamic model for infinite media. Journal of Engineering Mechanics, ASCE. 1969;**95**(4):759-877

[4] Deeks AJ, Randolph MF. Axisymmetric time-domain transmitting boundaries. Journal of Engineering Mechanics, CE. 1994;**20**(1):25-42

[5] Liao ZP, Wong HL. A transmitting boundary for the numerical simulation of elastic wave in stratified media. Computers in Physics. 1992;**101**:386-418

[6] Liao ZP. Guidance of Engineering Wave Theory. 2nd ed. BeiJing: Science press; 2002

[7] Du X, Zhao M. Analysis method for seismic response of arch dams in time domain based on viscous-spring artificial boundary condition. SHUI LI XUE BAO. 2006;**37**(9):1063-1069

[8] Du X, Zhao M. A local time-domain transmitting boundary for simulating cylindrical elastic wave propagation in infinite media. Soil Dynamics and Earthquake Engineering. 2010;**30**:937-946

[9] Liu J, Lv Y. A direct method for the analysis of dynamic interaction of structural foundations. China Civil Engineering Journal. 1998;**31**(3):55-64

[10] Liu J, Wang Z, Du X, et al. Three-dimension visco-elastic artificial boundaries in domain for wave motion problems. Engineering Mechanics. 2005;**22**(6):46-51

[11] Zhi F. Dynamic response and artificial boundary analysis in soil structure interaction system. Chinese Quarterly of Mechanics. 2009;**30**(3):475-480

[12] Deeks AJ, Randolph MF. Axisymmetric time-domain transmitting boundaries. Journal of Engineering Mechanics, ASCE. 1994;**120**(1):25-42

[13] Liu J, Li B. Three-dimensional viscoelastic static power unified artificial boundary. Chinese science E. 2005;**35**(9):996-980

[14] Lu H, Liang Y, Shouping S. Calculation and analysis of dynamic interaction system of pile foundation structure on layered foundation. Journal of Geotechnical Engineering. 2007;**29**(5):701-711

[15] Gao F, Fengbing Z. Study on artificial boundary conversion method for static and dynamic analysis of underground structures. Vibration and Shock. 2011;**30**(11):165-170

[16] Xue Z, Pei Q, Wu S, et al. Input seismic motion based on the viscous-spring boundary: Analytical solution and numerical simulation. International Journal of u-and eE-Service Science and Technology. 2016;**9**(6):59-68

[17] Jingbo L, Yixin D, Qiushi Y. Visco-elastic and the ground motion input implement in the general finite element software. Journal of Disaster Prevention and Mitigation Engineering. 2007;**27**:37-42

[18] Li J, Chen JY, Lin G. Finite element damping extraction method for dynamic interaction time domain analysis of nonhomogeneous unbounded rock. Chinese Journal of Geotechnical Engineering. 2004;**26**(2):263-267

[19] Jiang X, Li J, Lin G. Study on the seismic input mode of viscoelastic artificial boundary under the condition of slope site. World Earthquake Engineering. 2013;**29**(4):126-132

8

An Estimation of "Energy" Magnitude Associated with a Possible Lithosphere-Atmosphere-Ionosphere Electromagnetic Coupling Before the Wenchuan M_S8.0 Earthquake

Mei Li, Wenxin Kong, Chong Yue, Shu Song,
Chen Yu, Tao Xie and Xian Lu

Abstract

A large scale of abnormities from ground-based electromagnetic parameters to ionospheric parameters has been recorded during the Wenchuan M_S8.0 earthquake. All these results present different anomalous periods, but there seems one common climax leading to a lithosphere-atmosphere-ionosphere electromagnetic coupling (LAIEC) right on May 9, 3 days prior to the Wenchuan main shock. Based on the electron-hole theory, this chapter attempts to estimate the "energy source" magnitude driving this obvious coupling with the Wenchuan focus zone parameters considered. The simulation results show that the total surface charges fall in ~10^7–10^8 C, and the related upward electric field is ~10^8–10^9 V/m. These corresponding parameters are up to 10^9 C and 10^{10} V/m when the main rupture happens, and the order of the output current is up to 10^7 A. The electric field increasing in the interface between the Earth's surface and the atmosphere, on one hand, can cause electromagnetic parameter abnormities of ground-based observation, with the range beyond 1000 km. On the other hand, it can accumulate air ionization above pre-earthquake zone and lead to ionospheric anomaly recorded by some spatial seismic monitoring satellites.

Keywords: Wenchuan earthquake, electromagnetic abnormity, lithosphere-atmosphere-ionosphere electromagnetic coupling, electron-hole theory, "energy" magnitude

1. Introduction

A substantial number of effects in the Earth's lithosphere possibly associated with earthquakes (EQs) have been revealed in the past several decades. Electromagnetic observation is one of the geophysical methods which are applied in seismic investigation most early. A large number of electromagnetic emissions related to strong EQs have been reported worldwide.

One considers that direct electromagnetic signals in a wide frequency range originate in the Earth's crust at epicenter depth being more than 10 km (even several hundreds of kilometers) and can be recorded on the Earth's surface. The first example is DC seismic electric signal (SES) [1]. Another example is ultra low frequency (ULF) electromagnetic emissions during the Loma Prieta M_s7.1 EQ on 17 October 1989 [2, 3] and the Spitak M_S6.9 EQ on 7 December 1988 [4]. Anomalous ULF emissions were also observed before 8 August 1993 M_S8.0 Guam EQ [5, 6] and the L'Aquila M_S6.3 EQ on 6 April 2009 [7].

At the same time, a large number of effects in the Earth's atmosphere and the ionosphere possibly associated with earthquakes have also been reported in the past 20 years. The ionosphere and the upper atmosphere phenomena correlated with seismic events have been examined in numerous experimental and theoretical studies. As a result of these researches, it has been shown that there are obvious ionospheric perturbations associated with strong seismic activities especially after the launch of the DEMETER satellite in 2004 [8]. The total electron content (TEC) abnormities with an increase of about 67% have been recorded 4–6 days prior to the L'Aquila 6.2 earthquake [9–11]. Using GPS and CHAMP data, Hasbi et al. [12] investigated the ionospheric variations before some large earthquakes occurring during 2004–2007 in Sumatra, and the result shows positive and negative anomalies in TEC and electron density within a few hours to several days before the events. The results attained by Hayakawa et al. [13] have presented subionospheric very low frequency (VLF) perturbations before the 2007 Niigata Chuetsu-oki 6.8 earthquake and the 2010 Haiti 7.0 earthquake with an abnormal radius up to several kilometers.

Furthermore, in very recent years, more and more evidences have been shown that at the last stage of the long-term preparation of an earthquake, there could be a transfer and interchange of energy among the involved layers of lithosphere, atmosphere, and ionosphere, so as to introduce the concept of a lithosphere-atmosphere-ionosphere coupling (LAIC) of the Earth system (e.g., [8, 14–20]).

However, there is still a primary stage for study on the generation and propagation mechanisms of electromagnetic signals related to seismic activities. Many rock-pressure experiments have been performed in order to understand the mechanism of earthquake-related electromagnetic variations and pre-earthquake abnormous phenomena, for example, "selectivity" of the information, and one commonsense is that electromagnetic signals, as well as strong electrical current, are produced during a rock pressure [21–29].

A strong electrical current is produced when rocks are stressed during laboratory rock experiments, especially at the stage of the main rupture, although no clear explanation has been given to interpret the generation of EM emissions and electrical currents observed either during seismic activity or in the laboratory experiments up to now. So at the present, the study based on a physical model [30–32] and mathematical model [33, 34] is an effective way to investigate electromagnetic generation and propagation mechanism and interpret electromagnetic phenomena recorded before some seismic events.

A seismic activity is a dynamic evolution process, and electromagnetic anomaly emissions are accompanying at every stages of its preparation, development, and occurrence. It is a reasonable explanation that using dynamic-electromagnetic coupling mechanism to explain piezoelectric effect related to seismic electric signals, although Tuck et al. [35] have proposed that the piezoelectric effect of natural rocks is too weak to generate an observable electromagnetic signal for completely random orientation of piezoelectric grains. Not to mention that Sasaoka et al. [36] attained that the percentage of the effective electrokinetic effect during a natural rock experiment is only about 0.1–1%. However, using a model considering the compensation of piezoelectric effect and the dislocation theory of fault, Huang [37] attempted to investigate theoretically the generation of coseismic electric signals. He attained that the induced electric signals would be negligible in the case of the completely random orientation of quartz crystals. While if there are some preferred orientations or piezoelectric fabrics of quartz-bearing rocks in a fault area, it would form transient piezoelectric anomalous fields, which could be large enough to be detected as some seismic EM anomalies. Using a relatively simple model of an underground current source colocated with the earthquake hypocenter, Bortnik et al. [38] proposed that the expected seismotelluric current magnitudes fall in the range ~10–100 kA for an observed 30 nT magnetic pulse at 1 Hz before the Alum Rock M_W5.6 earthquake.

The work of Freund [21–25, 29] has gained a new insight into the production of current and electromagnetic signals in stressed rocks, and the electron-hole theory has been created subsequently. Based on this electron-hole theory, Kuo et al. [39, 40] constructed a theoretical coupling model for the stressed rock-Earth surface charges—atmosphere-ionosphere system—and the stressed rock acts as the dynamo to provide the currents for the coupling system. Their simulation results show that the total surface charges of stressed rock are 1.3×10^6 C when an earthquake fault surface area being 200×30 km and the surface current density 0.5 μA/m^2 are taken into account.

A large EQ of magnitude M_S8.0 hit Wenchuan, Sichuan province, at 14:28:01 CST (China standard time) on 12 May 2008 with an epicenter located at 103.4°N and 31.0°E and a depth of 19 km. This event caused major extensive damage, and 69,000 people lost their lives. At the same time, there are an increasing number of studies to investigate the possible existence of seismic-related electromagnetic precursors. One of the most important phenomena is that the climax of electromagnetic abnormities from ground-based observation to ionospheric observation happened right on 2008 May 9, 3 days before the Wenchuan event, which indicates a lithosphere-atmosphere-ionosphere electromagnetic coupling (LAIEC). One question is that what and how big is the "energy" source which drives this kind of coupling. So in this paper, a probable LAIEC process associated with the Wenchuan M_S8.0 earthquake is given in Section 2 firstly. Then Electron-hole theory is introduced in Section 3. For specified parameters on the Wenchuan earthquake, an "energy" magnitude driving this LAIEC is estimated using electron–hole theoretical model in Section 4. Discussion and conclusions are given in Section 5 and Section 6, respectively.

2. LAIEC before the Wenchuan M_S8.0 EQ

2.1. Possible mechanism of "energy" source driving a LAIEC

2.1.1. Different time-scale variations prior to the Wenchuan EQ

An EQ is a dynamic phenomenon of a long-term slow-strain accumulation and culminating with a sudden rupture and displacement of blocks of rock in the rigid lithosphere [41]. The evolutionary process of EQs is characterized by some complex features from stochastic to chaotic or pseudo-periodic dynamics. These events are so sophisticated that they are full of variations for different parameters at their different development stages from a long term to a short term, even several hours. Scientists acknowledge that any seismic electromagnetic anomaly is a climax of some process which begins a few days before the main event and stays until a few days after it (De Santis [42]).

The intensive compressive movement between the Qinghai-Tibet Plateau and the Sichuan basin has generated many strong EQs. On 12 May 2008, the Wenchuan M_S8.0 EQ ruptured the middle part of the Longmenshan (LMS) thrust belt [43], with a total length of about 400 km along the west edge of the Sichuan basin and the eastern margin of the Tibetan plateau in China. An increasing large number of preseismic phenomena have been reported after the occurrence of the Wenchuan EQ. Long-term variations basically include seismicity anomalies, deformation measurement anomalies, and strain/stress measurement anomalies. Jiang and Wu [44] have shown decade-scale quiescence along the Longmenshan fault zone. Yu et al. [45] also demonstrated a "load-unload response ratio (LURR)" variation of the seismicity along the Longmenshan fault zone up to 3 years before the Wenchuan EQ. Fang et al. [46] presented the "oscillation anomalies" since 2007 by analyzing GPS horizontal time series, and Zhang and Liu [47] also found a month-to-year-scale disturbance in the borehole strain measurement by analyzing four-component borehole strain observations. Fan and Jiao [48] analyzed fault activity characteristics in Sichuan-Yunnan area and found a two-year-scale increase of regional stress. These regional strain/stress changes and redistributions subsequently lead to a gradual variation of the electrical properties in the lithosphere beneath the Earth, conductivity/resistivity, for example. Lu et al. [49] analyzed shallow apparent resistivity data at Pixian, Jiangyou, and Wudu stations on the Longmenshan tectonic zone in western China and found that, at Pixian station 35 km away from the Wenchuan epicenter, a gradual decrease of apparent resistivity of up to 6.7% had been recorded clearly about 2 years prior to the *M*8.0 Wenchuan event, and a small coseismic change of −0.61% occurred during the earthquake.

However, one must keep in mind that electromagnetic emissions should be a candidate mentioned almost in all big events during the climax stage and plays an important role not to be ignored during pursuing the precursors prior to EQs. Ma and Wu [50] performed a work by collecting more than 200 papers published since the Wenchuan EQ, including seismic deformation, strain/stress, structure variations, gravity and broadband seismic recordings, and geomagnetic, geothermal, atmospheric, and ionospheric abnormities. **Figure 1** shows their statistical result according to the time scale of different anomalies collected this time. One must notice that the percentage of the anomalies appearing 2–3 days prior to the Wenchuan

EQ is 29%, the biggest one among 12 kinds of abnormities with different time scales from 30 to 10 years to 6–0 hours (see **Figure 1**). The most important is these anomalies are almost electromagnetic fields recorded by ground-based observation and ionospheric parameters recorded by GPS and DEMETER satellite and the rest are geothermal and thermosphere parameters if we check the context of the paper.

2.1.2. Possible "energy" source driving a LAIEC

The concept of lithosphere-atmosphere-ionosphere (LAI) coupling among the three involved layers of the Earth system has been proposed due to the transfer of energy among these three layers during some last stages of the long-term process of preparation [8, 15–17, 19, 20, 51, 52]. A few hypotheses for this LAI coupling have been proposed [53]: (1) chemical channel based on radon emanation [19], (2) atmospheric oscillation channel [51, 54], and (3) electrostatic channel due to positive hole carriers [55]. The main player of the first channel is the emanation of radon, which induces changes in the atmospheric conductivity and then atmospheric electric field and finally leads to the ionospheric plasma change. The third channel based on the generation of positive holes during the stressed rocks is likely to generate the electrostatic electric field, influencing the ionosphere. The second channel seems to be different from those mechanisms, in which some kinds of changes on the ground (or emanation of charged aerosols) are the possible exciters of atmospheric oscillations (either acoustic or internal gravity waves). Though a lot of evidences in support to this mechanism have been obtained, internal gravity waves tend to propagate obliquely so that it might be difficult to explain the most intense perturbation above the EQ epicenter. As the conclusion, this mechanism is most plausible but is poorly understood at the moment because of the lack in observational facts [52]. So Pulinets and Davidenko [56] have thought it is the smooth transition from acoustic-driven mechanisms to the electromagnetic coupling. The reason of this is very simple: acoustic coupling

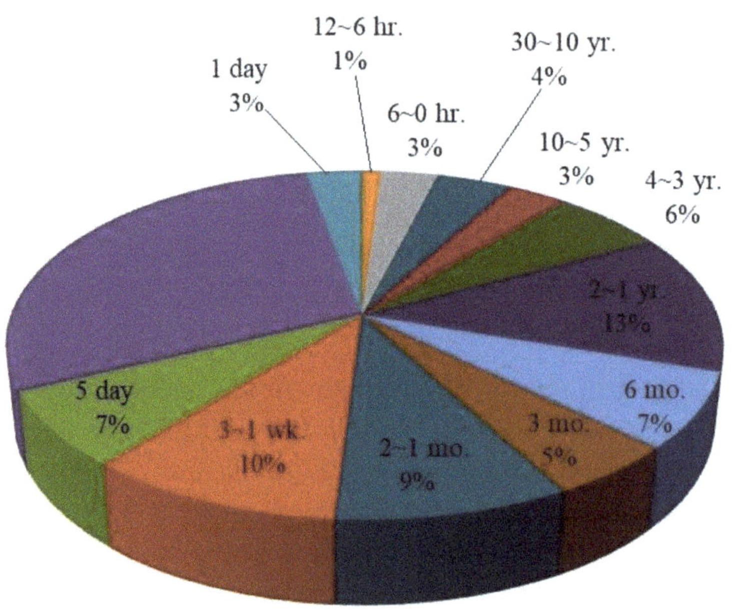

Figure 1. Temporal distribution of the appearance of all the reported anomalies (adapted from [50]).

mechanism has shown recently its very low effectiveness that it is difficult to produce an order of the ionospheric variations which are really observed before EQs.

Except for the electron-hole theory proposed by Freund [24], the results of rock-pressure experiments conducted by Qian et al. [28] indicate that they are rich in self-potential and magnetic signals with obviously different shapes during the early, mid, and late terms of the experiments. However, the large shorter-period magnetic pulses appearing at the last stage of the experiment may be induced by instantaneous electric current of the accumulated charges during the main cracking acceleration. These experiment results would effectively support a large number of electromagnetic observations during strong EQs.

2.2. Electromagnetic variations in the atmosphere

It is inferred that a northward pushing of the India plate and an intensive southeast movement of Qinghai-Tibet Plateau are strongly resisted by the Sichuan basin, from which the dynamic resources of the Wenchuan EQ probably originate. There is a long-time elastic strain accumulation and strengthening in Longmenshan nappe structural belt, and eventually this strain is beyond a threshold of friction strength of the faults, which leads to the Wenchuan M_S8.0 EQ on 12 May 2008 [57].

Li et al. [58] have reported an unprecedented ULF (0.1–10 Hz) electromagnetic emission anomaly appeared at Gaobeidian observing station lying in north China and 1440 km away from the epicenter. **Figure 2a** shows the daily total cumulative amplitude A (black bar) of electromagnetic information recorded at Gaobeidian station from January 2007 to December 2008. This emission is characterized by a 6-month-scale term by appearing weak information at the end of October 2007, on one hand, and, on the other hand, an abrupt increase of total information by appearing larger magnitude and higher frequency signals for both south-north (SN) and east-west (EW) directions at the same time on 9 May 2008, and this climax does not stop until May 17, 5 days after the main event on 12 May 2008 (see **Figure 2a** and **b**).

A "double low-point" phenomenon of vertical Z recorded on 9 May 2008 at most ground-based DC-ULF (0–0.3 Hz) geomagnetic observing stations around the epicenter of the Wenchuan EQ in China has been presented, and the amplitude variation of the Z component in most of the middle and western stations is up to 10–20 nT [59, 60].

2.3. Electromagnetic variations in the ionosphere

The pre-earthquake dynamo releases large number of charges and induces a strong electrical field over the epicenter areas. This field may provide air ionizations, leading to the presence of both positive and negative ions in the air, and enhance the air conductivity. The enhancement of the air conductivity advantageously contributes to a transmission of the electromagnetic fields from near ground surface to the atmosphere. The fields reach the bottom of the ionosphere and cause broadly ionospheric changes. The work of Hayakawa et al. [61] and Li et al. [32] shows lower ionospheric perturbations, as well as weak lithospheric radiation in ULF frequency by analyzing the vertical Z and horizontal H based on ground-based geomagnetic observation.

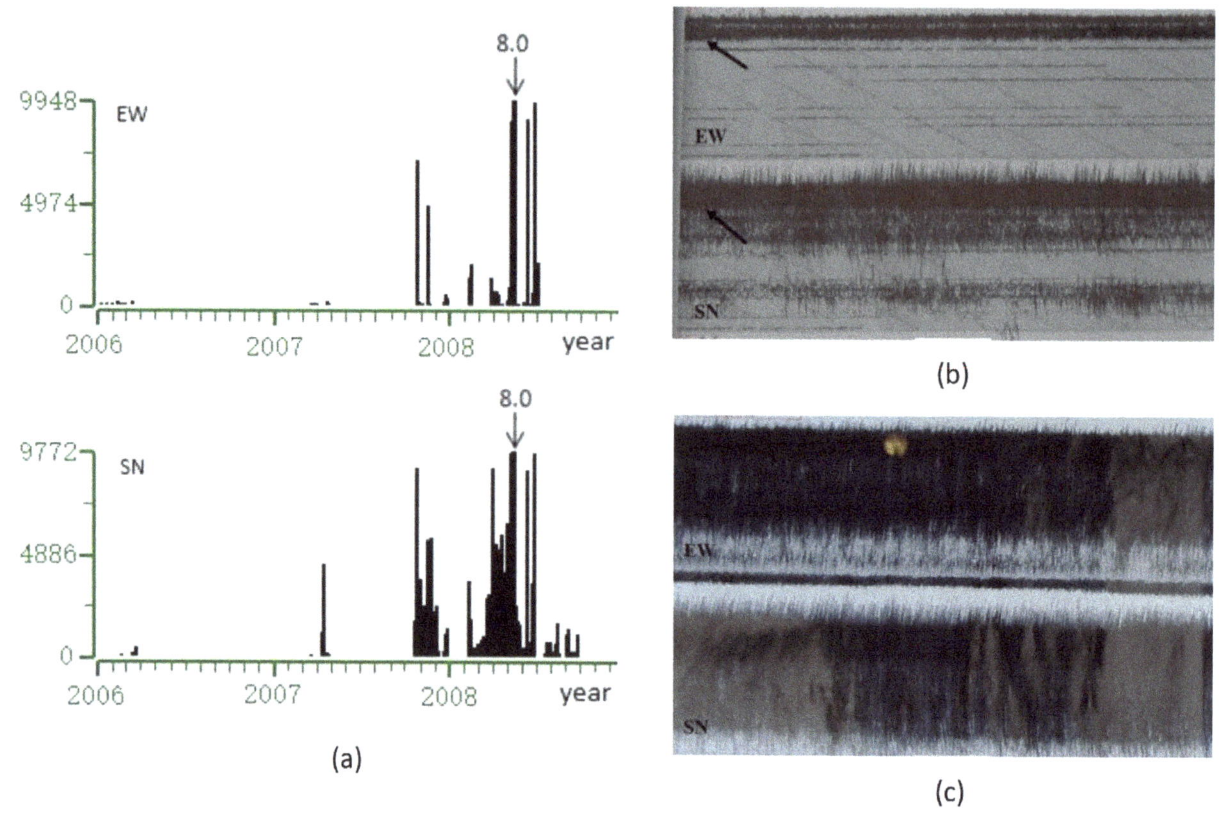

Figure 2. ULF electromagnetic variations during the Wenchuan M_S8.0 EQ (modified from [58]). (a) Timeline of daily total cumulative amplitude A (black bar) of electromagnetic information recorded at Gaobeidian station from January 2007 to December 2008. The Wenchuan EQ is indicated by an arrow. Copy of a part of an original record (b) from 9:00 AM, 9 May 2008, to 9:00 AM, 9 May 2008, and (c) from 9:00 AM, 9 May 2008, to 9:00 AM, 10 May 2008, at Gaobeidian station. Two observation components are shown with EW and SN, respectively. The abrupt increasing time on 9 May 2008 of emissions is denoted by black arrows for these two components in the right top panel.

The ionospheric variations, occurred right on the same day, i.e., 9 May 2008, partly are the followings. An obvious strength of 10–15 TECU for TEC happened at 16–18 LT in the range of ~20° in the southeast of the Wenchuan epicenter, as well as an obvious increase in its magnetically conjugated point area (see **Figure 3a**), and 40–80% increase for f0F2 from noon to sunset on May 9 (see **Figure 3b**). The ionospheric F2 peak electron density, NmF2, and height, hmF2, significantly decrease approximately 40% and descend about 50–80 km, respectively, when the GPS TEC anomalously reduces (e.g. [9, 62–70]).

The observed effects were strongest on May 9 for ionospheric ELF electromagnetic field [71]. By analyzing the data of DEMETER ion density O^+, Zhang et al. [64] found that the value reaches to its minimum of 10,000/cm^3 on May 9, 3 days before the event (see **Figure 3c**). The minimum percent of −2.40% on May 9 on O^+ is also found by Akhoondzadeh et al. [9]. There is an abrupt change for electron density Ne on May 9, and the value reaches its minimum with a 3% decrease during 1 second recorded by DEMETER [72]. All these parameters from ground-based electromagnetic observation to spatial ionospheric observation recorded a climax on 9 May 2008, 3 days before the Wenchuan M_S8.0 EQ, which seems to indicate a clear lithosphere-atmosphere-ionosphere electromagnetic coupling.

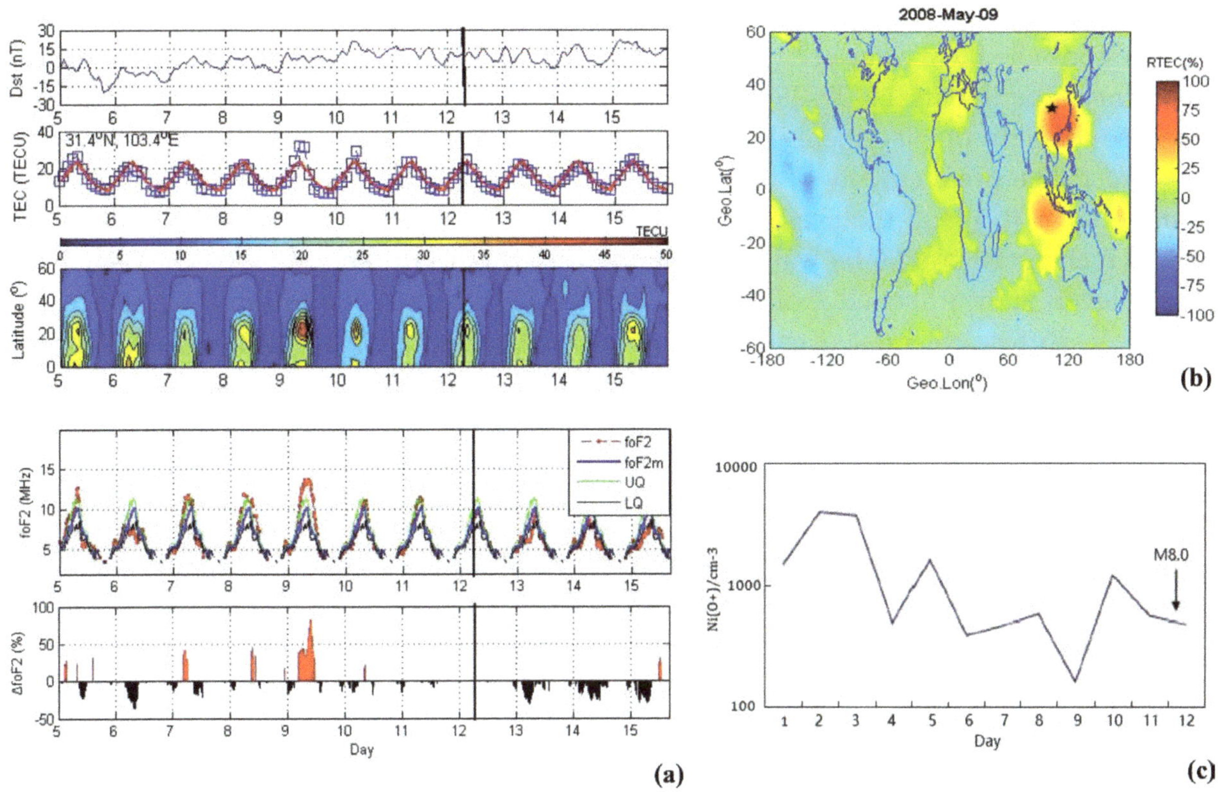

Figure 3. Variations of ionospheric parameters during the Wenchuan M_S8.0 EQ. (a) Variations of Dst, TEC, and f0F2 during 5–15 May 2008 (modified from [62]). (b) Distribution of TEC variation on 9 May 2008 (from [63]). (c) Daily averaged Ni (O+) on local nighttime during 1 to 12 May 2008 (from [64]).

In summary, all these results present different anomalous periods, but there seems one common climax leading to a lithosphere-atmosphere-ionosphere coupling (LAIC) just right on May 9, 3 days prior to the Wenchuan main shock. In this sense, LAIC is induced by a sudden change in ground and underground electrical properties, which is high probably originated from a fast development of the main rupture. At the same time, this evolutionary process is highly coincident with the LAI electromagnetic coupling model established by Kuo et al. [39, 40] that a current originated from the stressed rock in the focal zone propagates along the magnetic lines from the epicenter area of an earthquake, via the ionosphere, and to its magnetically conjugated point, causing electromagnetic disturbances, respectively, on the Earth's surface, in the atmosphere and the ionosphere and its conjugate point. Some magnetically conjugated ionospheric abnormities have also been reported (e.g., [62, 67, 68]; see also **Figure 3b**).

3. The "energy" source magnitude associated with the Wenchuan M_S8.0 earthquake

3.1. Electron-hole theory model

The concept of defect electrons or positive holes had been proposed firstly by Freund and Wengeler [22]. They first identified that peroxy defects consist of pairs of covalently bonded oxygen anions in the valence state 1^- instead of the usual 2^- in MgO single crystals. Then the electron-hole theory has been of a development and an improvement gradually during laboratory experiments [21, 23–25]. It demonstrates that, as rocks come under stress, stresses give rise to slight displacements of mineral grains in the rocks, which then activates peroxy defects that preferentially sit on or across grain boundaries. The peroxy breakup causes positive holes ($h^\bullet$), and the positive holes $h^\bullet$ are able to flow from stressed to unstressed rock, traveling fast and far by way of a phonon-assisted electron hopping mechanism using energy levels at the upper edge of the valence band [29]. At the rock-air interface, they cause (i) positive surface potential, (ii) field ionization of air molecules, and (iii) corona discharges.

Positive surface charges are produced at the surface as an accumulation of positive hole charge carriers $h\bullet$ as shown in **Figure 4**. The electric field at the edge of rock surface may cause air molecules to be ionized, that is, $O^2 \rightarrow O^{2+} + e^-$, producing O^{2+} ions and electrons. There is more $h\bullet$ influx in the bulk rock as electrons escape out to the rock surface. The effect of the field ionization of air molecules can be seen as to move some of the positive surface charges to the air with O^{2+} ions across the air-rock interface [25]. Electric currents along the stress-gradient direction with current density J_{rock} is generated during these processes, as illustrated in **Figure 4**.

3.2. Electron: hole theoretical formulas

We still use some theoretical formulas of the electron-hole theory described by Kuo et al. [39], and only a simple description is given here. For the Earth-air interface, the electric field E_{air} and the current density J_{air} near the air are the following:

$$E_{air} = q_s/\varepsilon_0 \tag{1}$$

$$J_{air} = \sigma_{air} E_{air} \tag{2}$$

Let q_s be the surface charge density. ε_0 is the electric permittivity of free space (i.e., 8.854×10^{-12} Farad/m), and σ_{air} is the air conductivity. The electric field will drive an upward current density J_{air} for the finite air conductivity.

Let A be the surface area of the stressed rock, and $Q = A\,q_s$ will be the total surface charges above the rock. From Eqs. (1) and (2), the total surface charges of stressed rock can be expressed as

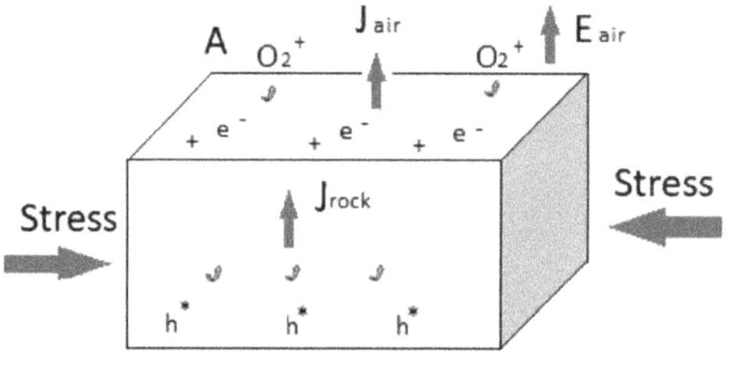

Figure 4. Sketch of electronic charge carriers and electric currents produced as rocks are subjected to a stress.

$$Q = Aq_s = AJ_{air}(\varepsilon_0/\sigma_{air}) \quad (3)$$

The time variation of Q can be written as

$$dQ/dt = AJ_{rock} - AJ_{air} = AJ_{rock} - \sigma_{air} Q/\varepsilon_0 \quad (4)$$

If we ignore the conductivity caused by the stress-induced surface positive charge carriers, the analytic solution for the total surface charges Q in Eq. (4) can be obtained from

$$Q(t) = Q_0(1 - e^{-t/\tau_0}) \quad (5)$$

where $\tau_0 = \varepsilon_0/\sigma_0$,

$$Q_0 = AJ_{rock}\tau_0 = AJ_{rock}(\varepsilon_0/\sigma_0) \quad (6)$$

where Q_0 is the asymptotic value of Q as $t \to \infty$ and $\tau_0 = \varepsilon_0/\sigma_0$(~440 s) near ground. More details can be referred to Kuo et al. [39].

Figure 5 shows Q as a function of time. The total surface charge Q as a function of charging time t, $Q(t) = Q_0(1 - e^{-t/\tau_0})$, where $Q_0 = AJ_{rock}\tau_0$ is the asymptotic value of Q when $t \to \infty$ and $\tau_0 = \varepsilon_0/\sigma_0$

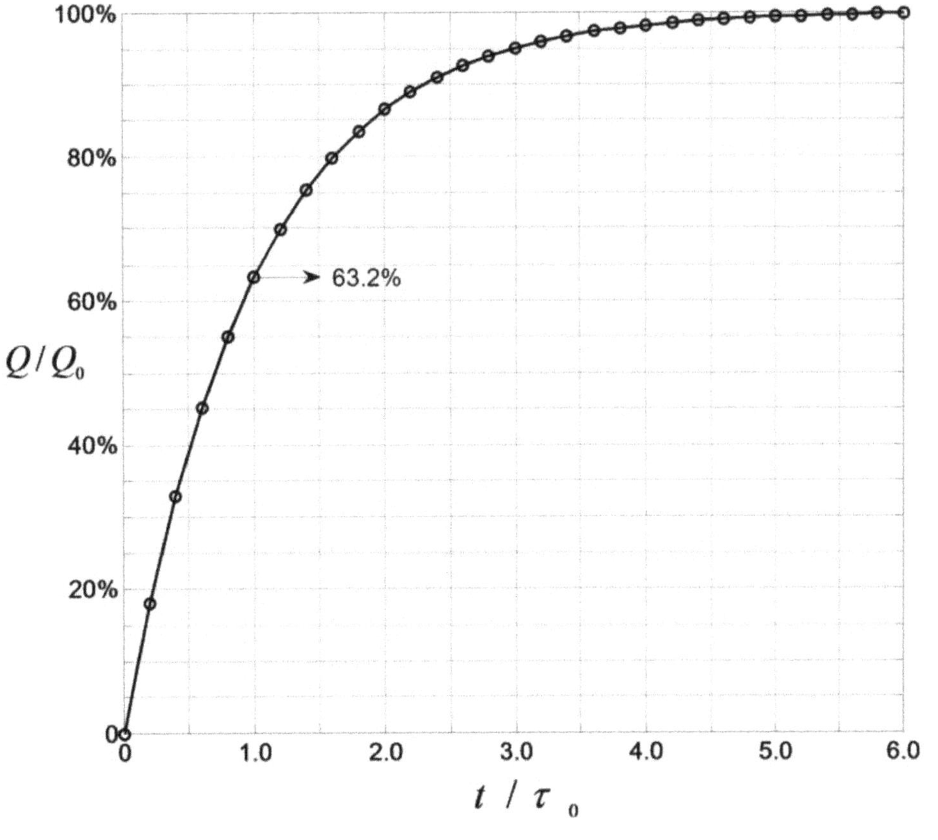

Figure 5. The total surface charge Q as a function of charging time t.

(~440 s) near ground. From **Figure 5**, one can see that Q(t) is quickly close to its asymptotic Q_0 as the charging time t increases. In Eq. (5) and **Figure 5**, τ_0 is the characteristic time for the buildup of surface charges to reach 63.2% of Q_0. When Q reaches its asymptotic value $Q_\infty = Q_0$, the output total current AJ_{air} into atmosphere equals input total current $I_{rock} = AJ_{rock}$ from rock, since dQ/dt = 0.

4. The "energy source" magnitude associated with the Wenchuan LAIEC

In order to attain the "energy" magnitude related to the Wenchuan event, two principal parameters should be confirmed according to input total current $I_{rock} = AJ_{rock}$ from rock into atmosphere based on the electron-hole theory illustrated above. One is the effective stressed area A, and another is the surface current density J_{rock} output into the air associated with the Wenchuan M_S8.0 earthquake.

4.1. Estimation of surface current density J_{rock}

A gabbro sample with a porosity of ~0.3 and <1% total water and with a size of 30 × 15 × 10 cm from Shanxi, China, was used in a test performed by Freund [24]. Before loading, the background ion current was in the low pA range. It remained low at low loads. A positive ion current of 10–25 nA is presented over an area of collector plate 200 cm^2 when the load is between 10,000 and 25,000 lbs. However, a 55 nA current has recorded about 2 s before failure, with the load being at about 30,000 lbs. (~13,600 kg), and the maximum spike reaches 450 nA when the main failure took place. According to this experiment, the surface current density out of stressed rock J_{rock} is of ~0.5–2.75 μA/m^2 before the failure, and it reached up to 22.5 μA/m^2 when the main failure happened.

The intensive compressive movement between the Qinghai-Tibet Plateau and the Sichuan basin results in the Wenchuan M_S8.0 EQ on 12 May 2008, with a total length of 240–350 km and width of 30 km surface rupture belt along the edge of the Sichuan basin and the eastern margin of the Tibetan plateau, in the middle of the north-south seismic belt of China [57, 73, 74]. The micro-epicenter of this event lies in the middle fault of the Longmenshan rupture belt, and there are full of granitic layers with different thickness surrounding its hypocenter [75].

An et al. [76] conducted the ground stress measurement with hydraulic fracturing in nine boreholes along both sides of the LMS fault zone. The results show the measured maximum major horizontal principal stress is 14 MPa (about two times of the stress used by Freund [24]) for the Wenchuan borehole, 400 m away from the Wenchuan epicenter, and the horizontal principal stress increases as the depth. Hao et al. [27] found in their experiment that the self-potential ΔV is proportional to the stress rate Δγ when the rock sample is under a biaxial compression. Therefore, with the depth effect on the horizontal principal stress ignored, the surface charge density J_{rock} would be less than ~5.5 μA/m^2 (2.75 × 2 = 5.5 μA/m^2) in the Wenchuan epicentral areas if its minimum value of the charge density is presumed to be the tested value ~0.5 μA/m^2 above attained by Freund [24]. While the charge density in the

Wenchuan preparation zone will be up to J_{rock} ~45.0 μA/m^2 (22.5 × 2 = 45.0 μA/m^2) during micro-cracks, developing quickly leads to the main failure.

4.2. Estimation of effective stressed plate A

The radius of seismic area where changes can be expected can be estimated using the Dobrovolsky formula $R = 10^{0.43M}$, where R is the radius of the EQ preparation zone and M is the EQ magnitude [77]. It gives that the radius of this Wenchuan M_S8.0 case preparation zone is more than 2,700 km although this event only ruptured a 300 × 30 km ground surface rupture belt. At the same time, Liu et al. [66] have found that GPS TEC extremely enhances in the afternoon of day 3 before the Wenchuan earthquake. The spatial distributions of the anomalous indicate that the earthquake preparation area is about 1650 and 2850 km from the epicenter in the latitudinal and longitudinal directions, respectively.

The investigation results of Li et al. [78] have presented that the 7 day's synthetic surface latent heat flux (SLHF) abnormity area during 6–12 May 2008 covers around Longmenshan Faults and the epicenter of the main shock locates in the center of the abnormity area (see **Figure 6**). From **Figure 6**, one can see that the SLHF basically forms a~6 × 6 degree square anomalous area. The Longmenshan faults are consisted of three fault belts, named Shanqian fault, Middle fault, and Houshan fault. These three faults are paralleled to each other and extend about 500 km from southwest to northeast direction. From the data listed above, we select the effective area bearing the principal stress of the Wenchuan earthquake to be 500 × 500 km, which will be used to calculate in the following section.

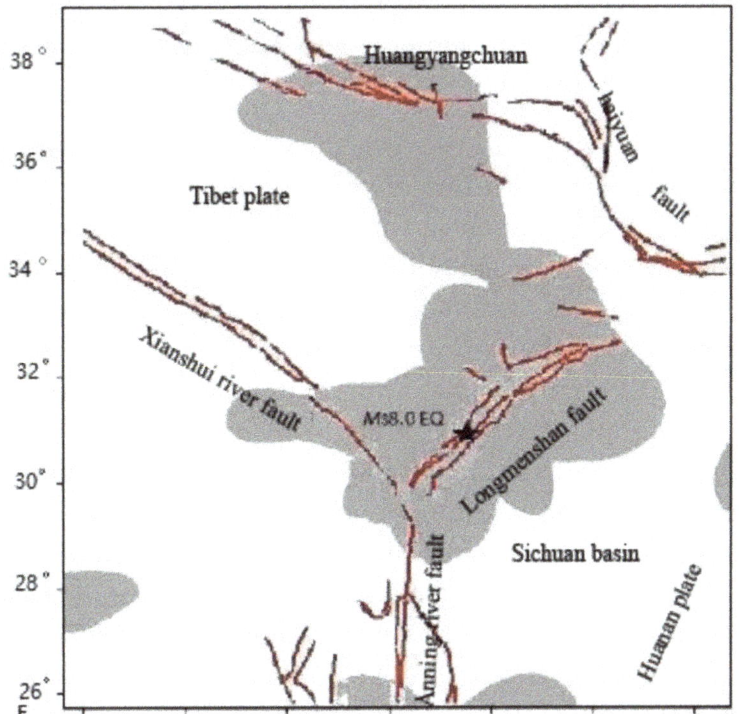

Figure 6. Synthetic surface latent heat flux abnormity during 6–12 May 2008 [78].

4.3. Calculation of "energy source" magnitude

Considering the Wenchuan earthquake plate area of $A = 500 \times 500$ km^2, output surface current density $J_{rock} = 0.5–5.5$ μA/m^2, the air conductivity $\sigma_0 = 2.0 \times 10^{-14}$ S/m, and the electric permittivity of free space $\varepsilon_0 = 8.854 \times 10^{-12}$ Farad/m, we get the total surface charges of the $Q_0 = (5.6–61) \times 10^7$ C and the upward electrical field $E_{air} = (6.9–75) \times 10^8$ V/m based on the Eqs. (1) and (6). $Q_0 = 5.0 \times 10^9$ C and $E_{air} = 6.1 \times 10^{10}$ V/m, respectively, for J_{rock} ~45.0 μA/m^2 when the main rupture occurred.

Presuming the asymptotic value of the total surface charges Q being Q_∞, when the total charges Q reaches its asymptotic Q_∞, then $Q_\infty = Q_0$. At this time, $dQ/dt = 0$ and the total current input air $I_{air} = AJ_{air}$ should be equal to output current from the stress fault $I_{rock} = AJ_{rock}$. The corresponding calculated output current is $I_{rock} = (1.3–14) \times 10^5$ A. The corresponding value is $I_{rock} = 1.1 \times 10^7$ A as the main rupture took place. This is greater than the threshold of required E field for air ionization (~3×10^6 V/m near ground). The presence of these charges and electric field will modify the air conductivity and further cause the air ionizations at the ground-to-air interface.

5. Discussion

All rock-pressure experiments reach one common result that a strong current and geomagnetic climax occurred as the load increases, especially at the main rupture stage, although these experiments present different anomalous periods. Stressed rocks are like a dynamo (battery) to drive currents along the earthquake faults and propagating in the surrounding areas. This process gives rise to obvious variations from near-ground electromagnetic parameters like electrical field, geomagnetic field, and others to ionospheric parameters and leads to an ultima LAIE coupling prior to the Wenchuan event.

The intensive compressive movement between the Qinghai-Tibet Plateau and the Sichuan basin is a long-time elastic strain accumulation and strengthens the Longmenshan nappe structural belt. The strain accumulation can activate hole carriers. The carriers form current and produce additional electrical fields. A large number of holes are produced mainly along the main fault and spread around surrounding areas gradually arousing geomagnetic parameters' changes in near surface of the Earth, and the anomalous area can be beyond 1000 km. A notable example is that ULF electromagnetic information becomes strong at the beginning of April 2008 at Gaobeidian station in north China, 1440 km away from the Wenchuan EQ. On May 9, 3 days before the main shock, the seismic fault comes into its main culture, and strong current induces electrical field propagating in different directions. This process gives rise to higher frequency and big amplitude signals for both SN and EW directions at the same time, and electromagnetic climax occurred subsequently. The corresponding electrical field is up to 1.3 mV/m. In addition, a "double low-point" phenomenon on vertical Z has been recorded on May 9, 2008 at most ground-based DC-ULF (0–0.3 Hz) geomagnetic observing stations around the epicenter of the Wenchuan EQ.

Based on the electron-hole theoretical formulas established, the outflow current and the corresponding field are $I_{rock} = 1.1 \times 10^7$ A and $E_{air} = 6.1 \times 10^{10}$ V/m if the Wenchuan effective area

A = 500 × 500 km^2 and surface current density J_{rock} ~45.0 μA/m^2 being considered during micro-cracks developing quickly lead to the main failure. This current result is near to what have been gained by Li et al. [30]. Using a finite length dipole current source colocated at the Wenchuan hypocenter in a three-layer (Earth-air-ionosphere) physical model and a two-layer (Earth-air) model, they have gotten that the seismotelluric currents with and without ionospheric effect are about 5.0 × 10^7 and 3.7 × 10^8 A, respectively, for an observable ULF signal ~1.3 mV/m recorded at 1440 km Gaobeidian station. While using an infinitesimally short horizontal dipole colocated with the earthquake hypocenter, Bortnik et al. [38] have attained the expected seismotelluric current is of ~10–100 kA for the "Alum Rock" M_W = 5.6 earthquake for an observed 30 nT pulse at 1 Hz and D = 2 km. These results are probably in a reasonable range.

An abrupt strengthening of local electrical field accelerates ionization of surrounding air molecules, leading to a charge accumulation near the ground in the preparation zone of the forthcoming Wenchuan EQ. This makes air molecule to ionize further. Air ionization could strengthen the conductivity of the air. It is advantageous for charges spreading to the height of the ionosphere, leading to a large-scale ionospheric abnormities of ionospheric parameters, such as TEC, foF2, Ni, Ne, and so on, and the magnitude is up to 60% on May 9, 3 days before the Wenchuan shock.

However, in this study, there are still two uncertainty parameters, i.e., the effective area of stressed plate A and outflow surface current density J_{rock}, during the Wenchuan calculations. The result of the ground stress measurement with hydraulic fracturing shows that the measured maximum major horizontal principal stress increases as the depth. The hypocenter depth of the Wenchuan EQ is 19 km. So the major horizontal principal stress should be larger than the number of 14 MPa used during our calculations, and the probable electrical results would be also larger than what we get now. Thus, the larger charges can lead to a larger ionospheric disturbance. However, whether these charges can give rise to a recorded ground-ionospheric electromagnetic variation or not needs a further test.

6. Conclusions

In this paper, a double pressing effect of the Wenchuan fault between the Qinghai-Tibet Plateau and Huanan plate is equivalent to a rock-pressure mode in order to simulate and interpret the reason that there are large-scale accompanying electromagnetic abnormities 3 days before the Wenchuan M_S8.0 earthquake.

According to the electron-hole theory, strong stress activates hole carriers, and the carriers accumulate rapidly in the surface of the stressed Longmenshan belt. Then a current pulse and an additional electrical field are generated and flow mainly along the main faults, especially at the stage of the main rupture. With the Wenchuan earthquake plate area of A = 500 × 500 km^2 and output surface current density J_{rock} = 0.5–5.5 μA/m^2 considered, we get the total surface charges of the Q_0 = (5.6–61) × 10^7 C and the upward electrical field E_{air} = (6.9–75) × 10^8 V/m. We also attain that the output current is I_{rock} = 1.1 × 10^7 A and the corresponding upward electrical field E_{air} = 6.1 × 10^{10} V/m for J_{rock} ~45.0 μA/m^2 when the main rupture took place.

Therefore, substantial charge carriers are produced along the main fault when the Wenchuan fault is under the bilateral effect. Stressed rocks are like a dynamo (battery) that drive currents along the earthquake faults and propagate in the surrounding areas. This process gives rise to, on one hand, obvious variations of near ground electromagnetic parameters like electrical field, geomagnetic field, and others. On the other hand, an abrupt strengthening of local electrical field accelerates ionization of surrounding air molecules. Air ionization could strengthen the conductivity of the air. It is advantageous for charges spreading to the height of the ionosphere, leading to a large-scale ionospheric abnormities of ionospheric parameters recorded by satellites. These processes could contribute to a LAIE coupling on 9 May 2008, 3 days before the Wenchuan M_S8.0 earthquake.

Acknowledgements

The work has been funded from the National Natural Science Foundation of China (NSFC) under grant agreements n°41774084 and n°41704062.

Author details

Mei Li[1]*, Wenxin Kong[2], Chong Yue[1], Shu Song[2], Chen Yu[1], Tao Xie[1] and Xian Lu[1]

*Address all correspondence to: mei_seis@163.com

1 China Earthquake Networks Center, China Earthquake Administration, Beijing, China

2 China University of Geosciences, Beijing, China

References

[1] Varotsos PA, Sarlis NV, Tanaka HK, Skordas ES. Similarity of fluctuations in correlated systems: The case of seismicity. Physical Review E. 2005;**72**(4):041103

[2] Bernardi A, Fraser-Smith AC, McGill PR, Villard Jr OG. Magnetic field measurements near the epicenter of the M_s7.1 Loma Prieta earthquake. Physics of the Earth and Planetary Interiors. 1991;**68**:45-63

[3] Fraser-Smith AC, Bernardi AE, McGill PR, Ladd ME, Helliwell RA, Villard OG Jr. Low-frequency magnetic measurements near the epicenter of the M_s7.1 Loma Prieta earthquake. Geophysical Research Letters. 1990;**17**:1465-1468. DOI: 10.1029/GL017i009p01465

[4] Kopytenko YA, Matiashvili TG, Voronov PM, Kopytenko EA, Molchanov OA. Detection of ultra-low frequency emissions connected with the Spitak earthquake and its aftershock activity, based on geomagnetic pulsations data at Dusheti and Vardzia observatories. Physics of the Earth and Planetary Interiors. 1993;**77**:85-95. DOI: 10.1016/0031-9201(93)90035-8

[5] Hayakawa M, Kawate R, Molchanov OA, Yumoto K. Results of ultra-low-frequency magnetic field measurements during the Guam earthquake of 8 august 1993. Geophysical Research Letters. 1996;**23**:241-244. DOI: 10.1029/95GL02863

[6] Kawate R, Molchanov OA, Hayakawa M. Ultra-low frequency magnetic fields during the Guam earthquake of 8 august 1993 and their interpretation. Physics of the Earth and Planetary Interiors. 1998;**105**:229-238. DOI: 10.1016/S0031-9201(97)00094-0

[7] Prattes G, Schwingenschuh K, Eichelberger HU, Magnes W, Boudjada M, Stachel M, Vellante M, Villante U, Wesztergom V, Nenovski P. Ultra low frequency (ULF) European multi station magnetic field analysis before and during the 2009 earthquake at L'Aquila regarding regional geotechnical information. Natural Hazards and Earth System Sciences. 2011;**11**:1959-1968. DOI: 10.5194/nhess-11-1959-2011

[8] Hayakawa M, Molchanov OA, editors. Seismo-Electromagnetics: Lithosphere-Atmosphere-Ionosphere Coupling. Tokyo: TERRAPUB; 2002

[9] Akhoondzadeh M, Parrot M, Saradjian MR. Electron and ion density variations before strong earthquakes (M>6.0) using DEMETER and GPS data. Natural Hazards and Earth System Sciences. 2010;**10**:7-18

[10] Dabas RS, Das RM, Sharma K, Pilai KGM. Ionospheric precursors observed over low latitudes during some of the recent major earthquakes. Journal of Atmospheric and Solar: Terrestrial Physics. 2007;**69**:1813-1824

[11] Stangl G, Boudjada MY. Investigation of TEC and VLF space measurements associated to L'Aquila (Italy) earthquakes. Natural Hazards and Earth System Sciences. 2011;**11**:1019-1024

[12] Hasbi AM, Mohd Ali MA, Misran N. Ionospheric variations before some large earthquakes over Sumatra. Natural Hazards and Earth System Sciences. 2011;**11**:597-611. DOI: 10.5194/nhess-11-597-2011

[13] Hayakawa M, Horie T, Yoshida M, Kasahara Y, Muto F, Ohta K, Nakamura T. On the ionospheric perturbation associated with the 2007 Niigata Chuetsu–oki earthquake, as seen from subionospheric VLF/LF network observations. Natural Hazards and Earth System Sciences. 2008;**8**:573-576

[14] Kamogawa M. Pre-seismic lithosphere-atmosphere-ionosphere coupling. Eos, Transactions American Geophysical Union. 2006;**87**(40)

[15] Molchanov OA, Fedorov E, Schekotov A, Gordeev E, Chebrov V, Surkov V, Rozhnoi A, Andreevsky S, Iudin D, Yunga S, Lutikov A, Hayakawa M, Biagi PF. Lithosphere-atmosphere-ionosphere coupling as governing mechanism for preseismic short-term events in atmosphere and ionosphere. Natural Hazards and Earth System Sciences. 2004; **4**(5/6):757-767

[16] Pulinets SA, Legen'ka AD, Alekseev VA. Pre-earthquakes effects and their possible mechanisms. In: Dusty and Dirty Plasmas, Noise and Chaos in Space and in the Laboratory. New York: Plenum Publishing; 1994. pp. 545-557

[17] Pulinets SA, Alekseev VA, Legen'ka AD, Khegai VV. Radon and metallic aerosols emanation before strong earthquakes and their role in atmosphere and ionosphere modification. Advances in Space Research. 1997;**20**:2173-2176

[18] Pulinets SA, Khegai VV, Boyarchuk KA, Lomonosov AM. Atmospheric electric field as a source of ionospheric variability. Physics-Uspekhi. 1998;**41**:515-522

[19] Pulinets SA, Boyarchuk KA, Hegai VV, Kim VP, Lomonosov AM. Quasielectrostatic model of atmosphere-thermosphere-ionosphere coupling. Advances in Space Research. 2000;**26**:1209-1218

[20] Pulinets SA, Ouzounov D. Lithosphere-atmosphere-ionosphere coupling (LAIC) model, an unified concept for earthquake precursors validation. Journal of Asian Earth Sciences. 2011;**41**:371-382

[21] Freund F, Sornette D. Electro-magnetic earthquake bursts and critical rupture of peroxy bond networks in rocks. Tectonophysics. 2007;**431**:33-47

[22] Freund F, Wengeler H. The infrared spectrum of OH^-compensated defect sites in C-doped MgO and CaO single crystals. Journal of Physics and Chemistry of Solids. 1982; **43**:129-145

[23] Freund F. Charge generation and propagation in igneous rocks. Journal of Geodynamics. 2002;**33**:543-570

[24] Freund F. Stress-activated positive hole charge carriers in rocks and the generation of pre-earthquake signals. In: Hayakawa M, editor. Electromagnetic Phenomena Associated with Earthquakes. Trivandrum, India: Transworld research network, Chapter 3; 2009. pp. 41-96

[25] Freund F. Toward a unified solid state theory for pre-earthquake signals. Acta Geophysica. 2010;**58**(5):719-766

[26] Hao JQ, Qian SQ, Gao JT, Zhou JG, Zhu T. ULF electric and magnetic anomalies accompanying the cracking of rock sample. Acta Seismologica Sinica. 2003;**25**(1):102-111 (in Chinese with English abstract)

[27] Hao JQ, Liu LQ, Long HL, Ma SL, Guo ZQ, Qian SQ, Zhou JG. New result of the experiment on self-potential change of rocks under biaxial compression. Chinese Journal of Geophysics. 2004;**47**(3):475-482 (in Chinese with English abstract)

[28] Qian SQ, Ren KX, Lv Z. Experimental study on VLF, MF, HF and VHF electromagnetic radiation characteristics with the rock breaking. Acta Seismologica Sinica. 1996;**18**(3):346-351 (in Chinese with English abstract)

[29] Scoville J, Sornette J, Freund F. Paradox of peroxy defects and positive holes in rocks part II: Outflow of electric currents from stressed rocks. Journal of Asian Earth Sciences. 2015;**114**:338-351

[30] Li M, Tan H, Cao M. Ionospheric influence on the seismo-telluric current related to electromagnetic signals observed before the Wenchuan M_S 8.0 earthquake. Solid Earth. 2016;**7**:1405-1415. DOI: 10.5194/se-7-1405-2016

[31] Li M, Tan H, Wang Z, Zhang X, Cao M. Using the electron-hole theory to estimate the "energy" magnitude related with the electromagnetic abnormities before the Wenchuan M_s8.0 earthquake. Acta Seismologic Sinica. 2015a;**37**(5):842-852 (in Chinese with English abstract)

[32] Li Q, Schekotov A, Asano T, Hayakawa M. On the anomalies in ULF magnetic field variations prior to the 2008 Sichuan earthquake. Open Journal of Earthquake Research. 2015b;**4**:55-64. DOI: 10.4236/ojer.2015.42005

[33] Huang QH, Lin YF. Selectivity of seismic electric signal (SES) of the 2000 Izu earthquake swarm: A 3D FEM numerical simulation model. Proceedings of the Japan Academy, Ser. B. 2010;**86**(3):257-264

[34] Ren HX, Chen XF, Huang QH. Numerical simulation of coseismic electromagnetic fields associated with seismic waves due to finite faulting in porous media. Geophysical Journal International;**188**(3):925-9442012

[35] Tuck GJ, Stacy FD, Starkey J. A search for the piezoelectric effect in quartz-bearing rocks. Tectonophysics. 1977;**39**(4):T7-T11

[36] Sasaoka H, Yamanaka C, Ikeya M. Measurements of electric potential variation by piezo-electricity of granite. Geophysical Research Letters. 1998;**25**(12):2225-2228

[37] Huang QH. One possible generation mechanism of co-seismic electric signals. Proceedings of the Japan Academy. 2002;**78**(7):173-178

[38] Bortnik J, Bleier TE, Dunson C, Freund F. Estimating the seismotelluric current required for observable electromagnetic ground signals. Annales de Geophysique. 2010;**28**(8): 1615-1624

[39] Kuo CL, Huba JD, Joyce G, Lee LC. Ionosphere plasma bubbles and density variations induced by pre-earthquake rock currents and associated surface charges. Journal of Geophysical Research. 2011;**116**(A10):A10317. DOI: 10.1029/2011JA016628

[40] Kuo CL, Lee LC, Huba JD. An improved coupling model for the lithosphere-atmo-sphere-ionosphere system. Journal of Geophysical Research: Space Physics. 2014;**119**(4): 3189-3205

[41] McCaffrey R. Earthquakes and crustal deformation. In: Harsh KG, editor. Encyclopedia of Solid Earth Geophysics. 218-226, 2011, doi: 10.1007/978-90-481-8702-7

[42] De Santis A, Franceschi GD, Spogli L, Perrone L, Alfonsi L, Qamili E, Cianchini G, Giovambattista RD, Salvi S, Filippi E, Pavon-Carrasco FJ, Monna S, Piscini A, Battiston R, Vitale V, Picozza PG, Conti L, Parrot M, Pincon JL, Balasis G, Tavani M, Argan A, Piano G, Rainone ML, Liu W, Tao D. Geospace perturbations induced by the earth: The state of the art and future trends. Physics and Chemistry of the Earth. 2015;**85-86**:17-33

[43] Burchfiel BC, Royden LH, Van der Hilst RD, Hager BH, Chen Z, King RW, Li C, Lu J, Yao H, Kirby E. A geological and geophysical context for the Wenchuan earthquake of 12 May 2008. Sichuan, People's Republic of China, GSA Today. 2008;**18**(7):4-11. DOI: 10.1130/GSATG18A.1

[44] Jiang CS, Wu ZL. Seismic moment release before the May 12, 2008, Wenchuan earthquake in Sichuan of Southwest China. Concurrency Computation Practice and Experience. 2010;**22**(12):1784-1795

[45] Yu H, Cheng J, Wan Y. Load/unload response ratio and stress accumulation model before large earthquakes. Acta Seismologica Sinica. 2010;**32**(5):517-528. (in Chinese with English abstract)

[46] Fang Y, Jiang Z, Gu G. Oscillation analysis of GPS horizontal time series before the Wenchuan earthquake. Journal of Seismological Research. 2010;**33**(2):125-130 (in Chinese with English abstract)

[47] Zhang J, Liu Q. Processing and analysis of four component borehole strain observations. Journal of Geodesy and Geodynamics. 2010;**30**(6):6-9 (in Chinese with English abstract)

[48] Fan G, Jiao Q. Analysis of fault activity characteristics in Sichuan-Yunnan area before Wenchuan M_S8.0 earthquake. Journal of Geodesy and Geodynamics. 2008;**28**(6):27-30 (in Chinese with English abstract)

[49] Lu J, Xie T, Li M, Wang Y, Ren Y, Gao S, Wang L, Zhao J. Monitoring shallow resistivity changes prior to the 12 May 2008 *M*8.0 Wenchuan earthquake on the Longmen Shan tectonic zone, China. Tectonophysics. 2016;**675**:244-257. DOI: 10.1016/j.tecto.2016.03.006

[50] Ma T, Wu Z. Precursor-like anomalies prior to the 2008 Wenchuan earthquake: A critical—but—constructive review. International Journal of Geophysics. 2012;**1687-885X**:55-64. DOI: 10.1155/2012/583097

[51] Molchanov OA, Hayakawa M. Seismo Electromagnetics and Related Phenomena: History and Latest Results. Tokyo: TERRAPUB; 2008. p. 189. DOI: 10.1016/j.pce.2006.05.001

[52] Sorokin V, Chemyrev V, Hayakawa M. Electrodynamic Coupling of Lithosphere-Atmosphere-Ionosphere of the Earth. Nova Science Publisheres Inc; 2015. 326p

[53] Hayakawa M, Molchanov OA, NASDA/UEC team. Summary report of NASDA's earthquake remote sensing frontier project. Physics and Chemistry of the Earth. 2004;**29**:617-625

[54] Hayakawa M. Lower ionospheric perturbations associated with earthquakes, as detected by subionospheric VLF/LF radio waves. In: Hayakawa M, editor. Electromagnetic Phenomena Associated with Earthquakes. Trivandrum, India: Transworld research network, chapter 6. 2009; pp. 137-185

[55] Freund F. Earthquake forewarning—a multidisciplinary challenge from the ground up to space. Acta Geophysica. 2013;**61**(4):775-807. DOI: 10.2478/s11600-013-0130-4

[56] Pulinets SA, Davidenko D. Ionospheric precursors of earthquakes and global electric circuit. Advances in Space Research. 2014;**53**(5):709-723

[57] Xu XW. Album of 5.12 Wenchuan 8.0 Earthquake Surface Ruptures. Seismological press; 2009 (in Chinese with English abstract)

[58] Li M, Lu J, Parrot M, Tan H, Chang Y, Zhang X, Wang Y. Review of unprecedented ULF electromagnetic anomalous emissions possibly related to the Wenchuan M_S=8.0 earthquake, on 12 May 2008. Natural Hazards and Earth System Sciences. 2013;**13**(2):279-286. DOI: 10. 5194/nhess-13-279-2013

[59] Hu JC, Liu W, Guo MR, Zheng H. The "Double low-points" anomaly of daily vertical component variation of geomagnetic field before the *M*8.0 Wenchuan earthquake. Acta Seismologica Sinica. 2009 (in Chinese with English abstract);**31**(5):589-593. DOI: 10.3321/j.issn:0253-3782.2009.05.012

[60] Wang W, Ding J, Yu S, Zhang Y. Short-term geomagnetic abnormity before Wenchuan *M*s8.0 earthquake and strong earthquake prediction explore. Earthquake Science. 2009;**31**(2):172-179 (in Chinese with English abstract)

[61] Hayakawa M, Schekotov A, Potirakis SM, Eftaxias K, Li Q, Asano T. An integrated study of ULF magnetic field variations in association with the 2008 Sichuan earthquake, on the basis of statistical and critical analyses. Open Journal of Earthquake Research. 2015;**4**:85-93. DOI: 10.4236/ojer.2015.43008

[62] Yu T, Mao T, Wang YG, Wang JS. Study of the ionospheric anomaly before the Wenchuan earthquake. Chinese Science Bulletin. 2009;**54**(6):1086-1092 (in Chinese with English abstract)

[63] Zhao B, Yu T, Wang M, Wan W, Lei J, Liu L, Ning B. Is an unusual large enhancement of ionospheric electron density linked with the 2008 great Wenchuan earthquake? Journal of Geophysical Research. 2008;**113**:A11304. DOI: 10.1029/2008JA013613

[64] Zhang X, Shen X, Liu J, Ouyang X, Qian J, Zhao S. Analysis of ionospheric plasma perturbations before Wenchuan earthquake. Natural Hazards and Earth System Sciences. 2009;**9**:1259-1266

[65] Liu J, Huang J, Zhang X. Ionospheric perturbations in plasma parameters before global strong earthquakes. Advances in Space Research. 2014;**53**(5):776-787. DOI: 10.1016/j.asr.2013.12.029

[66] Liu JY, Chen YI, Chen CH, Liu CY, Chen CY, Nishihashi M, Li JZ, Xia YQ, Oyama KI, Hattori K, Lin CH. Seismoionospheric GPS total electron content anomalies observed before the 12 May 2008 M_W7.9 Wenchuan earthquake. Journal of Geophysical Research. 2009;**114**:A04320. DOI: 10.1029/2008JA013698

[67] Xu T, Hu YL, Wu J, Wu ZS, Suo YC, Feng J. Giant disturbance in the ionospheric F2 region prior to the *M*8.0 Wenchuan earthquake on 12 May 2008. Annales Geophysicae. 2010a;**28**:1533-1538. DOI: 10.5194/angeo-28-1533-2010

[68] Xu T, Hu YL, Wu J, Wu ZS, Suo YC, Feng J, Huang CJ. Abnormal perturbations in the ionospheric F2 region before Wenchuan earthquake on 12 May 2008. Science in China Series D: Earth Sciences. 2010b;**53**(11):1671-1674. DOI: 10. 1007/s11430-010-4046-4

[69] Yan XX, Shan XJ, Cao JB, Tang J, Wang FF. Seismoionospheric anomalies observed before Wenchuan earthquake using GPS and DEMETER data. Seismology and Geology. 2012;**34**(1):160-171. DOI: 10. 3969/J ISSN. 0253-4967. 2012. 01. 015

[70] Zhu FY, Wu Y, Lin J, Zhou YY, Xiong J, Yang J. Anomalous response of ionospheric VTEC before the Wenchuan earthquake. Acta Seismologica Sinica. 2009;**31**(2):180-187 (in Chinese with English abstract)

[71] Blecki J, Parrot M, Wronowski R. Studies of the electromagnetic field variations in ELF frequency range registered by DEMETER over the Sichuan region prior to the 12 May 2008 earthquake. International Journal of Remote Sensing. 2010;**31**:3615-3629. DOI: 10.1080/01431161003727754

[72] Zeng ZC, Zhang B, Fang GY, Wang DF, Yin HJ. The analysis of ionospheric variations before Wenchuan earthquake with DEMETER data. Chinese Journal of Geophysics. 2009;**52**(1):11-19 (in Chinese with English abstract)

[73] Ding ZF, Wu Y, Wang H, Zhou XF, Li GY. Variations of shear wave splitting in the 2008 Wenchuan earthquake. China Science Series D: Earth Sciences. 2008;**51**(12):1712-1716 (in Chinese with English abstract)

[74] Xie ZD, Zhu YQ, Lei XL, Yu HY, Hu XL. Pattern of stress change and its effect on seismicity rate caused by M_S8.0 Wenchuan earthquake. China Science Series D: Earth Sciences. 2010;**53**(9):1260-1270 (in Chinese with English abstract)

[75] Zhu YT, Wang XB, Yu N, Gao SQ, Li K, Shi YJ. Deep structure of magnetotelluric profile on Longmen Mts. and its relation to the M_s8.0 Wenchuan earthquake. Acta Geologica Sinica. 2008;**82**(12):1769-1777 (in Chinese with English abstract)

[76] An QM, Ding LF, Wang HZ, Zhao SG. Research of property and activity of Longmen mountain fault zone. Journal of Geodesy and Geodynamics. 2004;**24**(2):115-119 (in Chinese with English abstract)

[77] Dobrovolsky IP, Zubkov SI, Miachkin VI. Estimation of the size of earthquake preparation zones. Pure and Applied Geophysics. 1979;**117**:1025-1044. DOI: 10.1007/BF00876083

[78] Li M, Kang C, Li Z, Jing F, Xue Y, Yan W. Abnormal surface latent heat flux prior to the Wenchuan M_S8.0 earthquake. Earthquake. 2010;**30**(3):64-71 (in Chinese with English abstract)

9

Frequency-Magnitude Distribution of Earthquakes

Zuhair Hasan El-Isa

Abstract

Frequency magnitude distribution of all types of earthquakes has received considerable attention in the last few decades. Their linear logarithmic relationship remains the most accepted. The a and b constants of this equation, their values and variations have been studied in detail. It is largely agreed that for the seismicity of the whole Earth, its hemispheres, quadrants and large epicentral regions are a=10.0 and b=1.0. The b-variations have long been investigated and reported to occur in different forms and values. Long- and short-term b-variations occur in a continuous cyclic manner and may exceed ± (0.6 – 0.7) of its absolute value. These are observed to occur not only yearly or monthly but also daily. The *b*-value always attains maxima and minima before and after the occurrence of all large earthquakes. Many days before the occurrence of large earthquakes, *b*-values start increasing at variable gradients that are affected by foreshocks. It attains a maximum value shortly before each large earthquake and a minimum on its occurrence. Many factors have been proposed to explain the b-variations including prevailing stress, crustal heterogeneity, focal depth, pore pressure, geothermal gradient, tectonic setting, and other factors. Considering the b-variations of the whole Earth, its hemispheres and quadrants and considering that most proposed factors are directly or indirectly related to the stress, it is concluded that this remains the major factor. It is found that for large earthquakes with Mw ≥ 7, an increase of about 0.20 in the *b*-value implies a stress increase that will result in an earthquake with a magnitude one unit higher.

Keywords: earthquakes, b-value, spatiotemporal, seismicity, frequency-magnitude

1. Introduction

During the last few decades, different frequency-magnitude relationships have been proposed to explain the earthquake distribution in any region. These include the double power law and the exponential and Weibull distributions. The linear empirical relationship of Eq. (1) as originally suggested by [1–3] remains the most used among seismologists worldwide:

$$\text{Log}\, N = a - b\, M \tag{1}$$

where "N" is the total number of earthquakes with magnitudes ≥ M. The variable "a" represents the total seismicity rate and the level of seismic activity of a region, while the slope "b" describes the relative size distribution of its earthquakes. Both parameters are of vital importance in seismicity, seismotectonics, and other related studies. The variable "b" has received much more attention and was subjected to many statistical, analytical, and evaluation techniques within the last few decades. To illustrate the long-term global average of both a and b values, [4] compiled the seismicity data for all earthquakes that occurred globally during the period January 1990–December 2012 with magnitudes M ≥ 4.5 from the National Earthquake Information Centre (NEIC) of the US Geological Survey, the Incorporated Research Institutions for Seismology (IRIS), and the International Seismological Centre (ISC)

	Region		NEIC	IRIS	ISC	Average *b*-values
Jan. 1, 2000–Dec. 31, 2010	Eastern Hemisphere	No. of earthqs	45,775	48,318	81,121	1.01
		b-Values	1.03	0.99	1.01	
	Western Hemisphere	No. of earthqs	22,089	22,117	112,861	1.02
		b-Values	1.08	0.96	1.01	
	Whole Earth	No. of earthqs	67,864	70,435	193,983	1.03
		b-Values	1.07	1.02	1.01	
	Overall *b*-Average					1.02 ± 0.02
Jan. 1, 1990–Dec. 31, 2012	Eastern Hemisphere	No. of earthqs	85,954	46,147	146,939	1.02
		b-Values	1.04	1.06	0.96	
	Western Hemisphere	No. of earthqs	39,809	102,337	204,099	1.05
		b-Values	1.12	1.08	0.96	
	Whole Earth	No. of earthqs	125,763	148,484	351,038	1.03
		b-Values	1.07	1.07	0.96	
	Overall *b*-Average					1.03 ± 0.03

Table 1. The a and b values as calculated by [4] for the whole Earth and its Eastern and Western Hemispheres utilizing compiled seismicity data from NEIC, IRIS, and ISC catalogs for the period Jan. 1, 1990–Dec. 31, 2012.

catalogs and subjected the data to least-square regression analysis for the calculation of both a and b. Their results, **Table 1**, indicate that the a-value is mostly in the range 9.5–10.1 for the whole Earth and its hemispheres. With slight variations between the seismicity data of the three sources and the major regions of the Earth, the a-value averages at about 10.0. These values are in agreement with most previous published results, for example, see [5–13].

Both parameters show a relatively wide range of spatial and temporal variations, particularly the b-value, which displays continuous local, regional, and global fluctuations on various timescales of up to 0.5–0.8 of its absolute value or more [4, 14]. On a local scale, it shows a relatively wide range of variation (0.3 to 2.5) or more, for example, see [1, 4, 7, 11, 12, 14–36] and references therein]. Such b-variations are known to accompany microearthquakes as reported in a few laboratory experiments [17], to induced earthquakes associated with mining, large water reservoirs, and other man-made activities [18, 38–40], to large tectonic earthquakes [4, 14, 41–44, 46] . A few studies, however, contradict with this and claim that such temporal and spatial b-variations are apparent and interpreted them as artifact rather than natural, for example, see [6, 8, 47, 48]. Some studies claim that temporal variations are of second order compared to the spatial variations and are generally more difficult to observe [7, 30, 33]. Details on the regional and global spatiotemporal b-variations, their characteristics, their classifications, and the main factors affecting them are presented and discussed in this chapter.

2. Classification and causes of b-variations

In their review, [4] reported the following classification as compiled from published literature on the b-variations:

1. An increase in the b-value before the occurrence of major earthquakes during variable periods of time that extend many years in some cases and an immediate drop soon after their occurrence. Case studies cover different parts of the world, for example, Venezuela [20], China [22], and New Zealand [24].
2. A decrease in the b-value before the earthquake for variable periods of time that may be many years in some cases and an increase thereafter [5, 7, 17, 18, 29, 31, 33, 35, 39, 43, 44, 49, 50].
3. An increase in the b-value long before the main earthquake followed by a short-term drop shortly before its occurrence [27, 28, 51].
4. Continuous cyclic b-variations that occur over the years, months, and days both locally and regionally. The short-term variations are much larger and may exceed 0.7–0.8 of its absolute value. A few weeks before major earthquakes, the b-value starts increasing gradually with variable gradients that are affected by relatively large foreshocks. Soon before the earthquake, the b obtains a maximum and drops down to a minimum on its occurrence [4, 14].
5. A limited number of studies claim that b-variations are apparent and interpreted to be artifact rather than natural and are thus considered insignificant, for example, see [6, 8, 47, 48].

Such studies are limited compared to the many studies that report and confirm the spatial and temporal b-variations.

The factors that affect all reported spatiotemporal b-variations have been compiled and discussed by [4, 14] who claim that such variations are caused by one or more of the following factors:

1. An increase in the applied stress or effective stress is believed by many researchers to result in a b-decrease and vice versa, e.g., [7, 30–33, 35, 41–44]. This is evidenced by some limited laboratory experiments [15, 17]. Other studies however report that the b-value increases before the major earthquake, i.e., with stress and decreases thereafter [22, 24]. From their calculations to the b-variations of the whole Earth, its two hemispheres and four quadrants [4, 14] concluded that the stress remains the major factor that affects b-variations. The b increases with stress and shows gradual increase to maxima soon before large earthquakes and minima on their occurrence.

2. Variations in crustal heterogeneity, either primary or secondary, the second being brought about by one or both of the following causes: (a) rock deformations due to their folding, faulting, fracturing, cracking, and micro-cracking and (b) liquid and other magmatic materials invading the subsurface rocks. Fluid and pore pressure increase is believed by some to increase the b-value, for example, see [7, 19, 30]. In general, the b-value is reported to increase with the crustal heterogeneity [52–54].

3. Latent heat as introduced to the rocks by intrusions is believed to increase the b-values. Limited laboratory experiments are claimed to support this idea [25, 55]; see also [30, 33, 35, 43, 44].

4. Tectonic characteristics and focal mechanisms have long been suggested to cause b-variations [15, 16, 23]. Normal faulting is believed to be associated with relatively high b-values, strike-slip with intermediate, while thrust faulting is claimed to have the lowest b-values [1, 41, 56–58]. Creeping segments of faults appear to display high b-values, while asperities are reported to be characterized by anomalously low b-values [7, 33]. Ref. [26] have reported low b-values in the upper boundary of the subducting plates and high in the lower plane; see also [29, 56]. Contrary to this, other researchers suggest that there is little variation of b between different tectonic regions, for example, see [6, 8].

5. Petrological, environmental, depth, and some geophysical characteristics of rocks are reported in many cases to display b-variations. From his laboratory measurements, [17] reported that frictional sliding and deformation of a ductile rock were found to have higher b-value than that observed in brittle rocks. Variations in the b-values with depth are reported in different localities, e.g., California, where it is reported to decrease with depth [7, 30, 53]. In some other regions, high and low b-values are related to low and high P-wave velocities, respectively [26, 59]. On land, low Bouguer anomalies are reported to correlate with low b-values, while the opposite is suggested for offshore anomalies [25]. In one case, temporal b-variations are suggested to be caused by seismicity migration [21].

6. In each b-calculation for any region, the magnitude scale for all earthquakes should be the same. Using different scales in the one set of calculation or different calculations will

result in different b-values. [31] deduced a b-value = 1.5 for Mexican earthquakes utilizing the duration magnitude (Md) and a b-value = 0.85 for the same set of earthquakes when magnitudes were converted to Ms.

7. It was reported by [4, 14] that the *b*-variations are also affected by the selected calculating technique and how it is applied. The most used methods in the b-calculations are the regression and the maximum likelihood [60–62]. Both methods are sensitive to small numbers of earthquakes in the time window, but a reasonable correlation is observed between the results of both methods, as long as the numbers of earthquakes in the analyzed windows are large [14]. The calculated *b*-values are also affected, albeit slightly, by the length of the time window. They also show some slight variations between the use of moving and expanding time windows utilizing the same technique, though corelatable results are mostly obtained [4, 14].

3. Spatiotemporal b-variations

A detailed review by [4] briefs details on these variations, their classification, and their causes and includes most related references. The seismicity of the whole Earth, its two hemispheres and four quadrants, and the epicentral regions of some selected recent earthquakes has been evaluated by [4, 14] to conclude that b-variations occur continuously not only yearly but monthly and daily and to show that the applied stress remains the main factor that affects these variations. These were classified as follows [4, 14]:

3.1. Long-term b-variations

Detailed b-calculations on the seismicity of the Earth and its Eastern and Western Hemispheres that occurred during the period 1990–2010 were made by [14] for different periods of 1, 2, 11, and 21 years; see **Table 2**. Results show that the overall average *b*-value for the whole Earth

Region	Jan. 1, 1990, to Dec. 31, 2010 No. of Earthqs. and magn. range *b*-value	Jan. 1, 2000, to Dec. 31, 2010 No. of Earthqs. and magn. range *b*-value	Jan. 1–Dec. 31, 2004 No. of Earthqs. and magn. range *b*-value	Jan. 1–Dec. 31, 2005 No. of Earthqs. and magn. range *b*-value	Jan. 1, 2004–Dec. 31, 2005 No. of Earthqs. and magn. range *b*-value	Jan. 1–Dec. 31, 2010 No. of Earthqs. and magn. range *b*-value
Whole Earth	111,297 1.109 4.5–9.1 ± 0.001	68,043 1.056 4.5–9.1 ± 0.001	6187 0.939 4.5–9.1 ± 0.058	7883 1.034 4.5–8.6 ± 0.018	14,070 0.972 4.5–9.1 ± 0.021	7365 1.064 4.5–8.8 ± 0.004
Eastern Hemisphere	75,451 1.084 4.5–9.1 ± 0.002	45,989 1.029 4.5–9.1 ± 0.002	4215 0.886 4.5–9.1 ± 0.056	6037 0.993 4.5–8.6 ± 0.018	10,252 0.923 4.5–9.1 ± 0.027	4228 0.961 4.5–7.8 ± 0.007
Western Hemisphere	35,846 1.105 4.5–8.8 ± 0.002	22,054 1.074 4.5–8.8 ± 0.002	1972 1.115 4.5–7.2 ± 0.006	1846 1.031 4.5–7.8 ± 0.009	3818 1.107 4.5–7.8 ± 0.002	3137 1.077 4.5–8.8 ± 0.005

Table 2. The *b*-values for the whole Earth and its Eastern and Western Hemispheres as calculated for different long periods of time utilizing the NEIC 1990–2010 seismicity data (M ≥ 4.5), slightly modified from [14].

and its two hemispheres is 1.06 ± 0.02 with limited variations of about (0.09) between the three studied regions. The maximum differences in *b*-values (Δb) for the different periods of time are 0.17, 0.198, and 0.084 for the whole Earth and its Eastern and Western Hemispheres, respectively. Such results confirm that the derived *b*-values show clear temporal and spatial variations but with rather limited values [4, 14].

3.2. Short-term (monthly) b-variations

Seismicity data of the Eastern Hemisphere as reported in the catalogs of the three NEIC, IRIS, and ISC sources during the years 2009–2011 were utilized by [14] to study short-term *b*-variations using fixed time windows 1 month long, **Figure 1** and **Table 3**. Continuous cyclic *b*-variations are observed on the three curves of **Figure 1**. The monthly *b*-differences (Δb) as obtained from the NEIC data are 0.38, 0.33, and 0.62 for the months of the 3 consecutive years in respective order. The difference between the lowest and highest *b*-values during the 3 years together is 0.83. The *b*-results of the IRIS and ISC sets of data correlate with the first and show large systematic cyclic variations and differences that vary mostly in the range 0.40–0.70; see **Table 3**. Further b-calculations were made by [14] on the NEIC seismicity data that occurred within Quadrant 1 during the 2 years 2004 and 2005 using time windows 1 month and 2 months. More calculations are made in this study for the same data using time windows 3 months long. Results are plotted in **Figure 2** and **Figure 3**.

Continuous cyclic monthly *b*-variations have occurred through the months of the 2 years in the range 0.68–1.43. During January and February 2004, it increased from about 0.93 to 1.3–1.4. During the following 2 months of the same year, it decreased to about 0.97–0.98. During the period May–July, it increased to about 1.1 and to about 1.34 in August. A general decrease is observed to have occurred during the period September–December 2004 to about 0.675. Further *b*-variations are observed to have taken place during the year 2005 when the

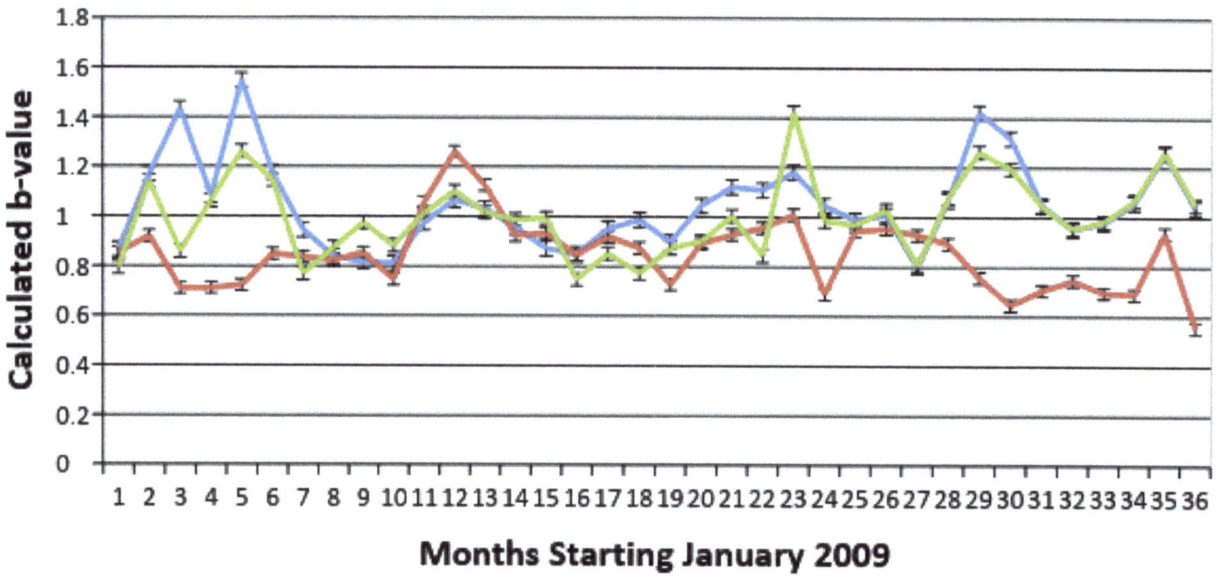

Figure 1. Continuous monthly *b*-variations as calculated by [14] from the seismicity data of NEIC (blue), ISC (red), and IRIS (green) that occurred during the period 2009–2011 with M ≥ 4.5. Bars represent the calculated standard errors.

		2009	2010	2011	2009–2011
NEIC	Earthq. no./month	236–479	228–698	342–2220	0.83
	Δ*b*/year	0.38	0.33	0.62	
	Δ*b*/3 years				
IRIS	Earthq. no./month	183–438	232–767	321–2253	0.68
	Δ*b*/year	0.49	0.68	0.46	
	Δ*b*/3 years				
ISC	Earthq. no./month	437–840	524–1145	530–3058	0.70
	Δ*b*/year	0.55	0.28	0.40	
	Δ*b*/3 years				

Table 3. Ranges of the monthly number of earthquakes and the calculated ***b***-values as calculated by [14] for the Eastern Hemisphere of the Earth utilizing the seismicity data (M ≥ 4.5) of NEIC, IRIS, and ISC for the period 2009–2011.

maximum observed value (1.43) occurred in September. Some 22 earthquakes including all with magnitudes ≥7.0 and some M6.0–M6.9 earthquakes are also marked on the graph according to their times of occurrence. Further *b*-calculations were made to (22 × 2) time windows that include the seismicity data of each of the selected 22 earthquakes on its own; 1 window ends a day before and the other on the day of its occurrence. The time lengths of all windows varied mostly in the range 25 days–35 days. Results are plotted in red in **Figure 2** and **Figure 3**. These show that for all 22 earthquakes, each *b*-value achieved a maximum 1 day before and a minimum soon after its occurrence. Soon after that, the *b*-values continue to change either positively, e.g., the two largest Sumatra earthquakes, or negatively as seen on some cases of **Figure 2**, e.g., the M7.4 earthquake of September 6, 2004, and the M7.2 earthquake of July 2005. The 2-month window results, **Figure 3A**, display continuous *b*-variations with time in the range 0.71–1.37. A general *b*-decrease is observed during the period April–December 2004, and a general increase is observed thereafter up to the end of June 2005 before starting to decrease again. Such results show good correlation with the 1-month results as far as their time and amplitude are concerned [14]. Similar cyclic variations are also observed on the 3-month data of **Figure 3B**.

The monthly b-results clearly indicate that b-variations are cyclic and continuous and may be as large as 0.7–0.8 of the absolute b-value of the Earth or slightly more. These are much higher than long-period *b*-variations as deduced for a full year, 2 years, or larger periods.

3.3. Short-time (daily) *b*-variations

Monthly b-calculations were made by [14] to the seismicity of the whole Earth, its two hemispheres, and its four quadrants for periods of about 1 month that include the dates of occurrence of the largest three earthquakes of the year 2010, namely, the Mw 8.8 Chile earthquake of February 27, Quadrant 3; a Mw 7.8 Sumatra earthquake that occurred on April 6, Quadrant 1; and a Mw 7.8 earthquake that occurred on October 25 in Quadrant 2. Results are included in **Table 4**. These show that the *b*-value of each quadrant and the whole Earth was maxima

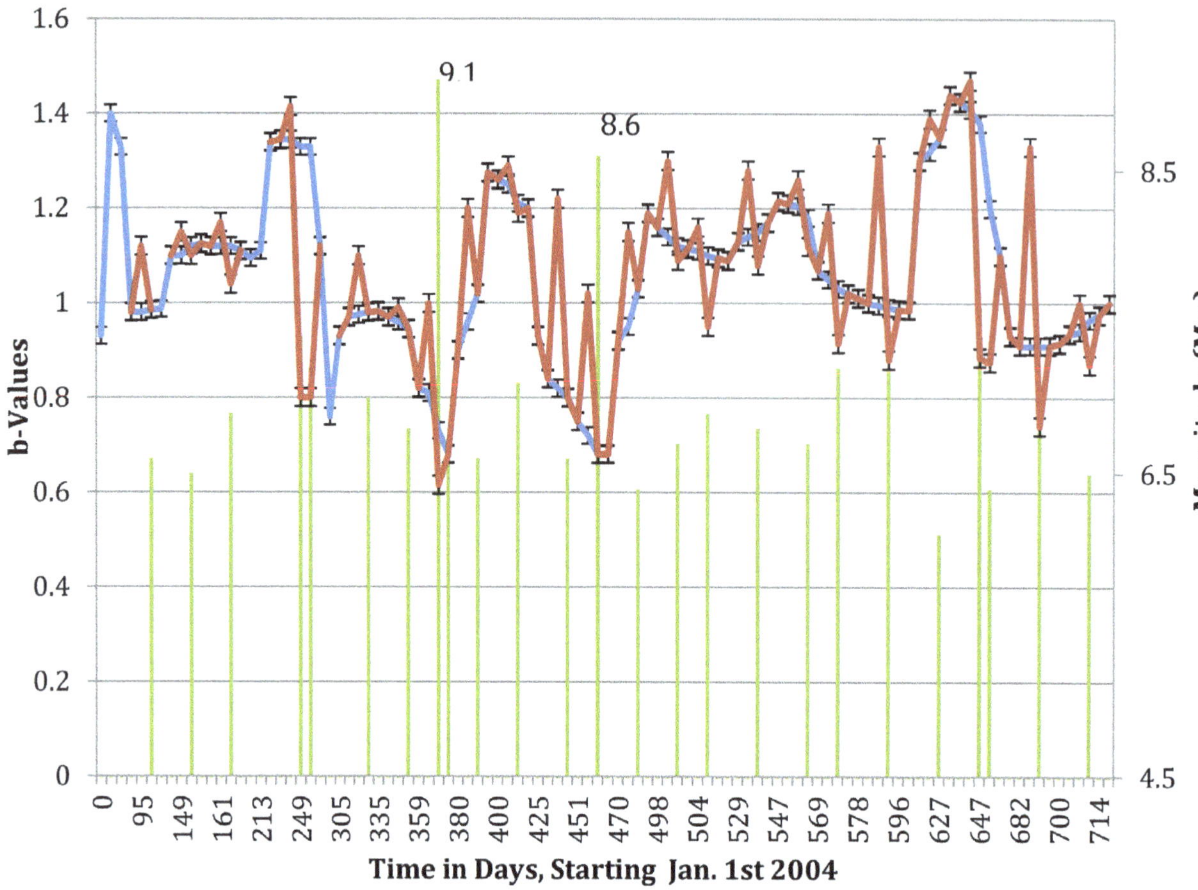

Figure 2. Continuous monthly b-variations as calculated by [14] for Quadrant 1 of the Earth from its 2004–2005 seismicity (blue line) and the calculated b-values 1 day before and soon after the occurrence of 22 selected earthquakes (in red) including the largest two Sumatra earthquakes of Dec. 2004 and March 2005.

soon before the occurrence of its earthquake and got reduced to minima soon after that. The total reduction in the *b*-value of Quadrant 3 where the Chile earthquake occurred is 0.350. The same phenomenon is observed for the whole Earth whose *b*-value decreased in the Chile earthquake case by 0.294. In the Sumatra case, the *b*-value of Quadrant 1 and the whole Earth dropped down by 0.122 and 0.187, respectively. In the third case, the b-value of Quadrant 2 and the Earth got reduced by 0.546 and 0.351 [14].

Further *b*-calculations were made to the seismicity data of Quadrant 3 which occurred a few weeks before the Chile earthquake. Fixed time windows 15 and 20 days long that were moved 1 day at a time and daily expanding time windows were used in the b-calculations. Results are plotted in **Figure 4** which also shows the times and magnitudes of the earthquakes that occurred within the same quadrant during the same period with magnitudes Mw ≥ 5.6. All earthquakes that occurred before the Chili earthquake are not located in its epicentral area. The expanding window results (red color) show that the *b*-value has varied gradually in the range 1.366–0.739 during the 54 days of calculation. The b-gradient has also varied during this period. It decreased with the occurrence of relatively large earthquakes and increased soon afterward; see [14] for details. On the occurrence of this major earthquake and its immediate aftershocks, the calculated *b*-value of Quadrant 3 got reduced to a minimum value, 0.739.

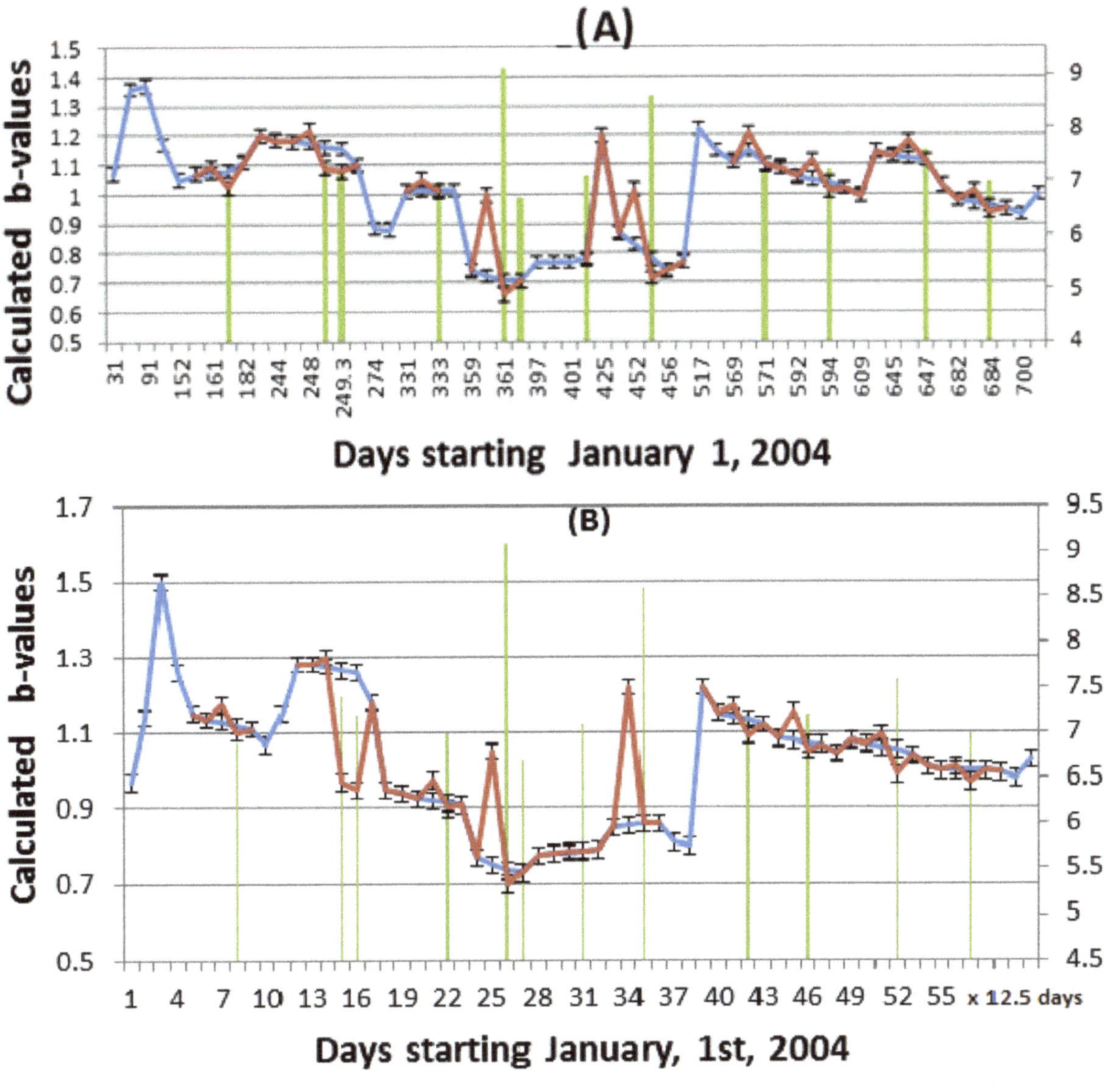

Figure 3. The *b*-values as calculated from the 2004 and 2005 seismicity data of Quadrant 1 utilizing time windows of fixed periods: (A) 2 months [14] and (B) 3 months as shown by the blue lines. The red lines show the calculated *b*-values for some selected large earthquakes utilizing time windows that end 1 day before and after their occurrence.

Continuous b-increase with variable gradients is observed to occur during the following 2 weeks, being largest within the first 1–2 days. All marked earthquakes of this period were epicentered close to the major Chile earthquake and are observed to have resulted in lowering the *b*-gradient; see **Figure 4**. Large b-variations are also noted on the moving-window results (blue and black curves). Differences between the highest and lowest *b*-values are 0.55 and 0.626 for the 15- and 20-day windows, respectively. Both curves show a gradual *b*-increase during the first 12–17 days. The large b-value increase on both curves on January 20 and 25 is simply caused by the removal of the 6.8 earthquake of January 5 from the windows. Maximum b-values are observed on both windows and that of the expanding during the period February 8–9. On the occurrence of the major 8.8 Chile earthquake and its aftershocks, the *b*-value of both windows dropped down to about 0.7. Continuous b-increase is observed on both windows during the following 14 days; for more details, see [14].

Region	Feb. 1–26 No. of *b-value* Earthqs. and magn. range	Feb. 1–27 No. of *b*-value Earthqs. and magn. range	Mar. 1–Apr. 5 No. of *b*-value Earthqs. and magn. range	Mar. 1–Apr. 6 No. of *b*-value Earthqs. and magn. range	Oct. 1–24 No. of *b*-value Earthqs. and magn. range	Oct. 1–25 No. of *b*-value Earthqs. and magn. range
Whole Earth	383 1.063 4.5–7.2 ± 0.006	640 0.769 4.5–8.8 ± 0.059	1032 1.133 4.5–7.2 ± 0.022	1046 1.011 4.5–7.2 ± 0.008	374 1.286 4.5–6.5 ± 0.005	405 0.935 4.5–7.8 ± 0.041
Eastern Hemisphere	212 0.928 4.5–7 ± 0.01	222 0.939 4.5–8.8 ± 0.009	313 0.907 4.5–6.5 ± 0.008	321 0.872 4.5–7.8 ± 0.02	220 1.365 4.5–6.2 ± 0.039	247 0.85 4.5–7.8 ± 0.078
Western Hemisphere	171 1.146 4.5–6.1 ± 0.016	418 0.73 4.5–7.2 ± 0.09	719 1.108 4.5–7.2 ± 0.008	725 1.109 4.5–7.2 ± 0.008	154 1.112 4.5–6.5 ± 0.1	158 1.117 4.5–6.5 ± 0.01
Quadrant 1 Lat 0°–90°N Long 0°–180°E	110 0.80 4.5–7.0 ± 0.01	120 0.821 4.5–7.0 ± 0.016	200 0.997 4.5–6.7 ± 0.002	205 0.810 4.5–7.8 ± 0.003	126 1.226 4.5–6.3 ± 0.007	126 1.226 4.5–6.3 ± 0.007
Quadrant 2 Lat 0°–90°S Long 0°–180°E	102 1.043 4.5–6.2 ± 0.022	102 1.043 4.5–6.2 ± 0.022	113 0.806 4.5–6.8 ± 0.002	116 0.811 4.5–6.8 ± 0.003	94 1.274 4.5–5.9 ± 0.005	118 0.728 4.5–7.8 ± 0.008
Quadrant 3 Lat 0°–90°S Long 0°–180°W	127 1.067 4.5–6.1 ± 0.008	373 0.717 4.5–8.8 ± 0.098	640 1.101 4.5–6.9 ± 0.004	644 1.102 4.5–6.9 ± 0.006	92 1.01 4.5–5.9 ± 0.007	129 1.019 4.5–5.9 ± 0.009
Quadrant 4 Lat 0°–90°N Long 0°–180°W	44 1.041 4.5–5.9 ± 0.002	45 0.954 4.5–5.9 ± 0.005	79 0.766 4.5–7.2 ± 0.004	81 0.769 4.5–7.2 ± 0.004	62 0.868 4.5–6.7 ± 0.01	64 0.873 4.5–6.7 ± 0.015

Table 4. The calculated *b*-values of the whole Earth and its Eastern and Western Hemispheres and four quadrants 1 day before and after the occurrence of the largest three earthquakes of 2010 [14].

Utilizing the expanding time window technique, further b-calculations were made by [14] to the seismicity of Quadrant 1 where the two largest Sumatra earthquakes of December 26, 2004 (M9.1), and March 28, 2005 (M8.6), occurred. In the case of the M9.1 earthquake, 29 *b*-values were calculated utilizing the seismicity data of 29 time windows that have the same starting point, December 1; see **Figure 5A**. A general b-increase is very clear with gradient variations that correlate with the marked earthquakes. On December 6, the b-value got reduced to 0.612 due to the occurrence of a M6.8 earthquake. Soon after that it started increasing at a gradient of about 0.05/day up to December 11. At this time a few earthquakes occurred with M ≤ 5.5 which resulted in lowering the gradient to about 0.02/day. On December 18, the b-value and its gradient were reduced to 0.925 and about 0.011/day, respectively, as caused by the occurrence of a M 6.2 earthquake. On December 25, a maximum b-value of 0.999 is calculated. It

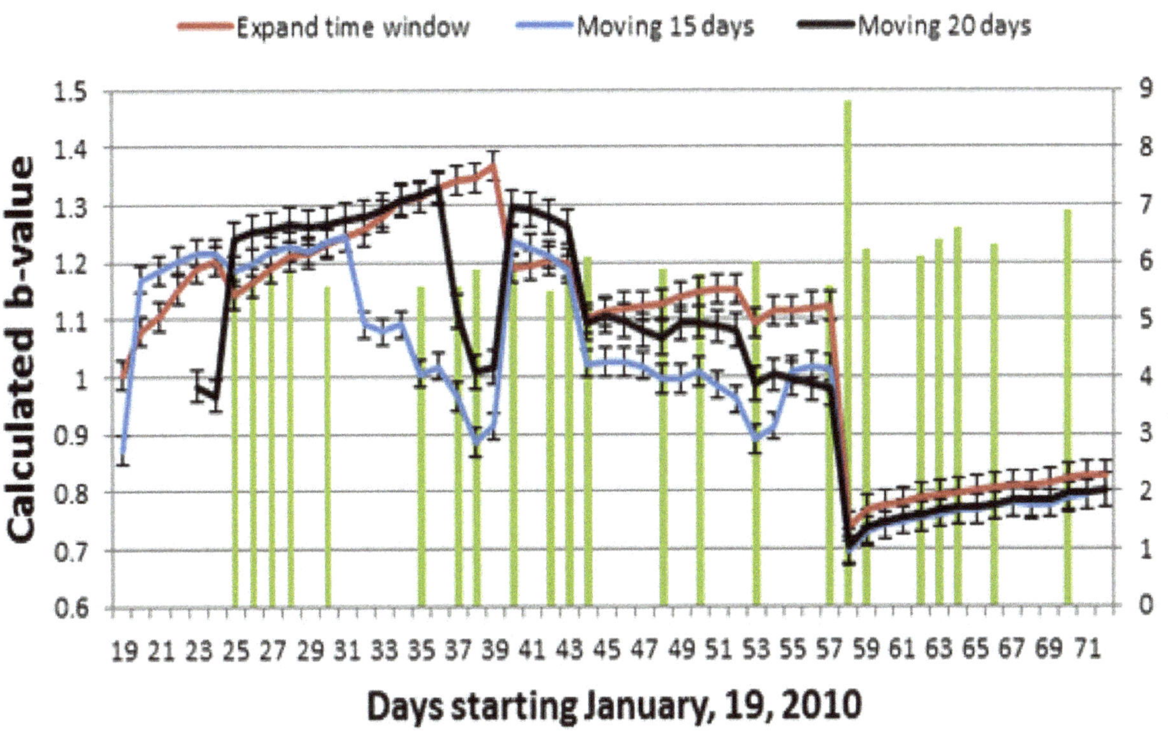

Figure 4. Variation of the *b*-values in Quadrant 3 of the Earth during a period of 39 days before and 14 days after the Chile, 2010, earthquake (Mw = 8.8) as calculated by [14].

dropped down to 0.612 after the occurrence of the major Sumatra earthquake and its aftershocks; see **Figure 5A**. Continuous b-increase is observed afterward with variable gradients. These results clearly demonstrate how the *b*-value of Quadrant 1 of the Earth increased gradually for some time before the Sumatra earthquake of December 26, 2004 (M9.1), to a maximum soon before its occurrence and dropped down to a minimum after that. The occurrence of smaller earthquakes has reduced the *b*-value and its rate of increase [14].

Similar b-evaluations were made by [14] from the seismicity data of Quadrant 1 to study the *b*-variations before and after the occurrence of the second largest Sumatra earthquake of March 28, 2005. **Figure 6B** displays the final results which show a general b-increase up to March 27 with variable gradients, 0.039/day for the first week, 0.017/day during the second week, 0.047/day during the third week, and a smaller gradient of only 0.002/day during the last 2 days before the main earthquake. The largest b-value (1.142) is calculated for March 27. On the occurrence of the main shock, it was reduced to 0.628, i.e., $\Delta b = 0.514$. During the following 6 days, the *b*-values started increasing again at a rate of 0.014/day. Similar to the previous earthquake, the *b*-values and their gradients are reduced with foreshock and aftershock occurrences; see **Figure 6B** [14].

Utilizing the moving and the expanding time windows further, daily b-calculations were made by [14] to the seismicity data of the epicentral area of the Sumatra 2005 earthquake and Quadrant 1 for a period of about 45 days before and 10 days after its occurrence; see **Figure 6**. A good correlation is observed between the two expanding curves (red and black). This correlation leads [14] to conclude that the *b*-values of Quadrant 1 and their variations were mainly controlled by the same factors as those in the Sumatra epicentral area during this period of time. Upon the

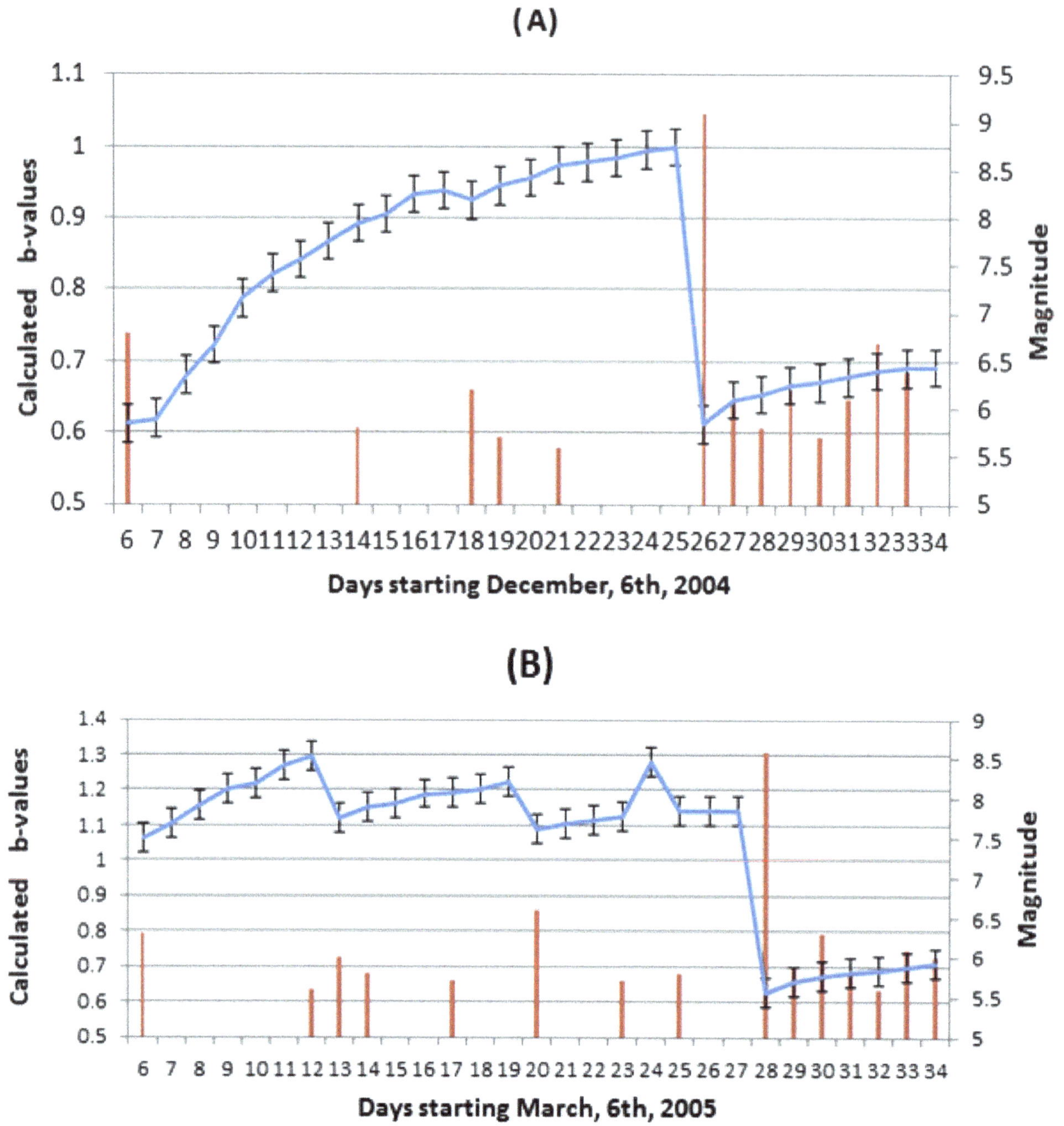

Figure 5. Calculated *b*-values for Quadrant 1 during a period of 29 days before and after the occurrence of (A) the largest Sumatra earthquake of December 26, 2004, and (B) the second largest Sumatra earthquake of March 28, 2005. All earthquakes of the study period that occurred in Quadrant 1 with M ≥ 5.6 are marked with their dates and magnitudes, slightly modified from [14].

occurrence of the major earthquake (M8.6), the *b*-value of the epicentral area got reduced by 0.546 and that of Quadrant 1 by 0.331. The M6.8 foreshock that occurred on February 26 resulted in a reduction of the *b*-value of the epicentral area by 0.260 and that of quadrant 1 by 0.119. Note that during the last 2 days before the 8.6 major earthquake, the rate of *b*-increase of Quadrant 1 got reduced to less than 0.001/day, while that of the epicentral area reduced to about 0.0013/day [14].

The blue and green curves of **Figure 6** represent the *b*-variations as calculated for both epicentral region and Quadrant 1, utilizing a 15-day window that moves daily starting

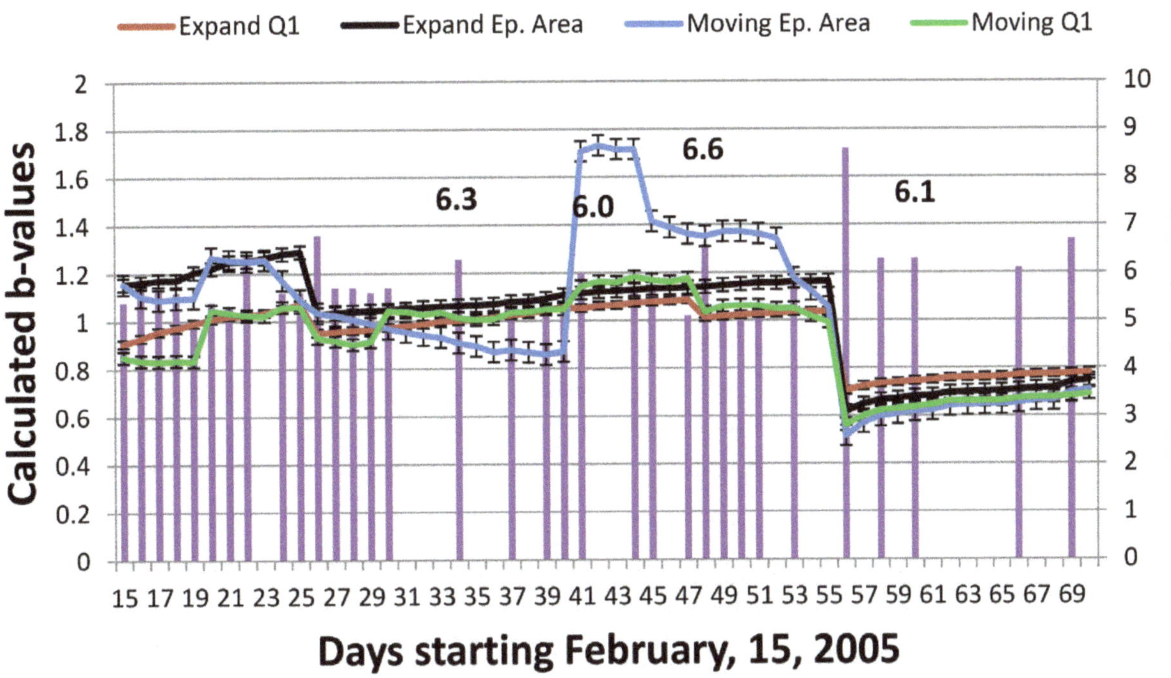

Figure 6. The b-variations as calculated by [14] for Quadrant 1 (red) and the epicentral area (black) of the Sumatra 2005 Mw = 8.6 utilizing daily expanding time windows. The first window expands between Feb. 8 and Feb. 15, while the last expands between Feb. 8 and April 11, 2005. The green and blue curves represent the calculated b-values for both regions utilizing a 15-day window that moves 1 day at a time. All epicentral earthquakes M ≥ 5 are marked. Only four earthquakes (Mw ≥ 6) occurred outside Sumatra epicentral region but within Quadrant 1 are marked with their magnitudes.

February 15, 2005. Both curves show a reasonable correlation and large b-variations. The epicentral values (blue) vary from a maximum of 1.732 on March 13 to a minimum of 0.517 on March 28 soon after the occurrence of the M8.6 earthquake. Between March 13 and 27, the b-value appears to reduce almost continuously to 1.054 1 day before the major earthquake. On March 13, its value jumped from 0.871 the day before, simply because on that day the M6.8 foreshock of February 26 was removed from the moving window. The continuous decrease beyond March 13 appears to correlate with the marked foreshocks, where the M6.6 foreshock of March 20 shows the largest effect. The following 3 days display a b-increase. A sharp decrease is observed soon after, which might be caused by the M5.8 earthquake of March 25 and other smaller magnitude events. The quadrant 1 curve (green) shows the same behavior with the removal of the 6.8 event and the addition of the M6.6, M5.8, and other foreshocks. The b-value got reduced to 0.991 on March 27 and to 0.563 on the occurrence of the M8.6 earthquake. Soon after that day, both curves show gradual increase, good correlation, and some deviations that correlate with the occurrence of aftershocks.

4. Discussion

It has long been documented that the a and b constants of the logarithmic frequency-magnitude distribution of earthquakes are of considerable importance in seismological, seismotectonic, and other related studies. Their regional and global values are 10.0 and 1.0, respectively, but both have long been reported to display considerable spatiotemporal variations [4–14]. The *b*-value is a quantitative measure of the earthquake size distribution within a certain period of time in any region. It has received much more attention, and numerous studies have investigated its value and variations. An increase in the number of small magnitude earthquakes increases its value, while it decreases with the occurrence of only one large earthquake or more. As the occurrence of earthquakes is unpredictable in both time and space, the *b*-value should therefore vary accordingly. It has long been known that the b-variations occur on both local and regional scales and are associated with all types of earthquake activities ranging from micro-tremors to very large tectonic earthquakes. Spatial variations vary from local tectonic areas to large tectonic regions, to a full quadrant of the Earth, to its hemispheres to the whole volume of the Earth [4, 14, 37, 45]. Temporal variations of variable periods are reported to vary from many years to a few years to 1 year to a few months to 1 month and less [1, 4, 12, 14, 34–36, 38, 47, 48, 57].

The most used techniques in the calculation of b-values are the linear regression and the maximum likelihood methods [33, 60, 61, 63–65]. Both methods give corelatable results for the same set of seismicity data but with differences of about 0.1–0.3 of its absolute value or more [4, 14]. The least-square regression method is disproportionately influenced by the largest earthquake magnitudes, while the maximum likelihood method weighs each earthquake equally [65, 66]. The manner in which a method is applied, such as the magnitude range considered, the number of earthquakes per bin, the use of cumulative magnitude statistic, etc. can affect the calculated *b*-values. Other differences are also reported using the same technique depending on how it is used, i.e., using moving time windows of constant or variable lengths or expanding time windows, their lengths, etc.

The fact that the calculated long-period b-variations are much smaller than those of the short period is simply caused by the fact that these variations are of higher frequency and occur continuously through days and months [4, 14]. Most published b-calculations used equal numbers of earthquakes in moving time windows, i.e., unequal times. This results in combining seismicity data over long times, longer than times of variation, and thus, at least some of the temporal b-variations are not captured [4, 14]. The time length of used windows and the total number of earthquakes used in the calculations are delicate factors that must be carefully considered in b-calculations. The use of moving fixed or expanding time windows influences both the number of earthquakes and their size distribution. These b-calculations clearly indicate that an expanding time window that starts at a fixed time appears to be a more effective technique to study the *b*-value and its variations, as all concerned earthquakes and their time distribution are retained within the studied time. In the moving-window technique, adding and removing large foreshocks and aftershocks result in relatively high *b*-variations. It is certain that both fixed and expanding window techniques confirm the b-variations, though their results vary slightly. Therefore combining the results of both techniques is quite useful [4, 14].

The b-calculations made by [4, 14] on the long-time and short-time sets of seismicity data clearly indicate that the *b*-value varies continuously with time in a cyclic manner.

Long-time variations (1 year to many years long) are of relatively low amplitudes, while the short-time sets of data (1 or 2 months or shorter) show that the *b*-variations of the Earth as a whole, its hemispheres, and its quadrants, as well as the epicentral regions of large earthquakes attain continuous cyclic *b*-variations with amplitudes as high as 0.6–0.7 of its absolute value [4, 14]. These variations occur not only yearly or monthly but also daily. Many days before the occurrence of large earthquakes, *b*-values start increasing with variable gradients that are affected by foreshocks. In some cases the gradient is reduced to zero or to a negative value a few days before occurrence of the main earthquake. In general, calculated *b*-values attain maxima 1 day before large earthquakes and minima soon after their occurrence followed by gradual increase thereafter with variable gradients caused by the aftershocks [4, 14].

For large earthquakes, the calculated *b*-variations of [14] show reasonable correlation for those of epicentral regions, the quadrant of the Earth, and the hemisphere they are located in as well as the whole Earth. This led him to conclude that these variations are of local and regional character, i.e., the geodynamic processes that cause the occurrence of such large earthquakes appear to act not only within the epicentral region but have a larger effect within the whole quadrant or more.

Spatiotemporal b-variations are influenced by a number of factors including stress, tectonic environment, crustal heterogeneity, depth, fluid and pore pressure, latent heat and geothermal conditions, and some geophysical and geotectonic characteristics [4, 14]. The *b*-variations of the Earth, its two hemispheres, and its four quadrants clearly indicate that the stress remains the major factor that affects these variations. In this case, the abovementioned regional factors are excluded, and the effects of much of the others are eliminated. It is concluded therefore that stress remains the most important factor that affects the *b* and its variations as, firstly, it causes deformation of rocks including brittle failure, i.e., earthquakes, and, secondly, all other factors are directly related to and influenced by the stress.

5. Conclusions

The following conclusions are made:

1. The logarithmic linear relationship of the frequency magnitude distribution of earthquakes has received much attention during the last decades. The a and b constants are critical parameters that have long been investigated and utilized in seismological, seismotectonic, and seismic hazard studies. Their regional, global, and long-term values are a = 10.0 and b = 1.0.

2. Spatiotemporal variations in the values of these constants, particularly the b, are manifested in numerous studies. The b-variations are expressed in different ways and have long been known to precede and follow the occurrence of large earthquakes. An increase in *b* value for a number of years before an earthquake, followed by a decrease after its

occurrence, or an increase long before the earthquake, followed by a small drop before the earthquake and a major decrease after its occurrence, has long been reported. Yearly, monthly, and daily b-variations are presented from other studies.

3. The *b*-variations are not only influenced by the different calculation techniques but also the way of using the same technique. Both linear regression and maximum likelihood methods give corelatable but variable results but with differences mostly in the range 0.1–0.3. It is found that an expanding time window technique from a fixed starting point is more effective in the study of *b*-variations. Combining the results of both techniques is certainly useful.

4. The calculated long-term and short-term b-variations of this study clearly indicate that b-variations occur in a continuous cyclic manner. The latter are of much higher value that may exceed ± (0.6–0.7) of its absolute value. These variations are observed to occur not only yearly or monthly but also daily, particularly before and after the occurrence of large earthquakes. The *b*-value always attains maxima and minima before and after the occurrence of all large earthquakes irrespective of the used method of calculation and how it is applied.

5. Spatiotemporal b-variations are influenced by a number of factors including stress, tectonic environment, crustal heterogeneity, depth, fluid and pore pressure, latent heat and geothermal conditions, and some geophysical and geotectonic characteristics. The *b*-variations of the Earth, its two hemispheres, and its four quadrants were interpreted by [4, 14] to conclude that the stresst remains the major factor that affects the b and its variations. All other factors remain of secondary effect as all are directly or indirectly related to and influenced by the stress.

6. For large earthquakes, *b*-variations appear to be of both local and regional characters. This may indicate that in the case of large earthquakes, the geodynamic processes acting within an epicentral region appear to influence its concerned quadrant, hemisphere, and the whole Earth. This indicates that the geotectonic factors that cause large earthquakes are of regional and possibly global character.

7. For earthquakes with magnitudes $Mw \geq 7.0$, it is found that an increase of about 0.20 in the *b*-value implies a stress increase that will result in an earthquake with a magnitude one unit higher, i.e., increasing the seismic energy some 28–30 times [14].

Author details

Zuhair Hasan El-Isa

Address all correspondence to: elisaz@go.com.jo

Department of Geology, University of Jordan, Amman, Jordan

References

[1] Ishibe T, Tsuruoka H, Shimazaki K. The dependency of the b-value on the focal mechanism and on the hypocentral depth. Paper Presented at the Meeting of the Japan Geoscience Union Meeting, Chiba, Japan; 2008

[2] Gutenberg B, Richter C. Frequency of earthquakes in California. Bulletin of the Seismological Society of America. 1944;**34**(4):185

[3] Gutenberg B, Richter C. Seismicity of the Earth and Associated Phenomena. Vol. 2. Princeton University Press; 1954

[4] El-Isa ZH, Eaton DW. Spatiotemporal variations in the *b*-value of earthquake magnitude-frequency distributions: Classification and causes. Tectonophysics. 2014;**615**:1-11

[5] Imoto M. Changes in the magnitude–frequency b-value prior to large (M 6.0) earthquakes in Japan. Tectonophysics. 1991;**193**(4):311-325

[6] Frohlich C, Davis S. Teleseismic b values; or, much ado about 1.0. Journal of Geophysical Research. 1993;**98**(B1):631-644

[7] Wiemer S, McNutt S, Wyss M. Temporal and three-dimensional spatial analyses of the frequency–magnitude distribution near Long Valley caldera, California. Geophysical Journal International. 1998;**134**(2):409-421

[8] Kagan Y. Universality of the seismic moment-frequency relation. Pure and Applied Geophysics. 1999;**155**(2-4):537-573

[9] Jaume S, Sykes L. Evolving towards a critical point: A review of accelerating seismic moment/energy release prior to large and great earthquakes. Pure and Applied Geophysics. 1999;**155**(2):279-306

[10] Stein S, Wysession M. An Introduction to Seismology, Earthquakes, and Earth Structure. Oxford: Blackwell Publishing; 2003

[11] Mishra O, Kayal J, Chakrabortty G, Singh O, Ghosh D. Aftershock investigation in the Andaman-Nicobar Islands of India and its seismotectonic implications. Bulletin of the Seismological Society of America. 2007;**97**(1A):S71

[12] Ghosh A, Newman A, Thomas A, Farmer G. Interface looking along the subduction megathrust from b-value mapping near Nicoya peninsula, Costa Rica. Geophysical Research Letters. 2008;**35**:L01301. DOI: 10.1029/2007

[13] Wech A, Kenneth W, Creager C, Houston H, Vidale J. An earthquake-like magnitude-frequency distribution of slow slip in northern Cascadia. Geophysical Research Letters. 2010;**37**:L22310

[14] El-Isa ZH. Continuous cyclic variations in the b-value of the earthquake frequency-magnitude-distribution. Earthquake Science. 2013:**26**(5):301-320

[15] Mogi K. Study of elastic shocks caused by the fracture of heterogeneous materials and its relations to earthquake phenomena. Bulletin of the Earthquake Research Institute. 1962;**40**(1):125-173

[16] Mogi K. Earthquakes and fractures. Tectonophysics. 1967;**5**(1):35-55

[17] Scholz C. Microfracturing and the inelastic deformation of rock in compression. Journal of Geophysical Research. 1968;**73**(4):1417-1432

[18] Gibowicz S, Lasocki S. Seismicity induced by mining: Ten years later. Advances in Geophysics. 2001;**44**:39-181

[19] Wyss M. Towards a physical understanding of the earthquake frequency distribution. Geophysical Journal of the Royal Astronomical Society. 1973;**31**(4):341-359

[20] Fiedler G. Local b-values related to seismicity. Tectonophysics. 1974;**23**(3):277-282

[21] Udias A. Time and magnitude relations for three microaftershock series near Hollister, California. Bulletin of the Seismological Society of America. 1977;**67**(1):173

[22] Robinson R. Variation of energy release, rate of occurrence and b-value of earthquakes in the main seismic region, New Zealand. Physics of the Earth and Planetary Interiors. 1979;**18**(3):209-220

[23] Båth M. Earthquake recurrence of a particular type. Pure and Applied Geophysics. 1981;**119**(5):1063-1076

[24] Smith W. Evidence for precursory changes in the frequency-magnitude B-value. Geophysical Journal of the Royal Astronomical Society. 1986;**86**(3):815-838

[25] Wang J. b Values of shallow earthquakes in Taiwan. Bulletin of the Seismological Society of America. 1988;**78**(3):1243

[26] Ogata Y, Imoto M, Katsura K. 3-D spatial variation of b-values of magnitude-frequency distribution beneath the Kanto District, Japan. Geophysical Journal International. 1991;**104**(1):135-146

[27] Sahu O, Saikia M. The b value before the 6th August, 1988 India-Myanmar border region earthquake–A case study. Tectonophysics. 1994;**234**(4):349-354

[28] Gupta S, Singh V, Kumar A. Attenuation of coda waves in the Garhwal Himalaya, India. Physics of the Earth and Planetary Interiors. 1995;**87**(3-4):247-253

[29] Wiemer S, Benoit J. Mapping the B-value anomaly at 100 km depth in the Alaska and New Zealand Subduction zones. Geophysical Research Letters. 1996;**23**(13):1557-1560

[30] Wiemer S, Wyss M. Mapping the frequency-magnitude distribution in asperities: An improved technique to calculate recurrence times? Journal of Geophysical Research. 1997;**102**(B7):15115-15115

[31] Zuniga F, Wyss M. Most-and least-likely locations of large to great earthquakes along the Pacific coast of Mexico estimated from local recurrence times based on b-values. Bulletin of the Seismological Society of America. 2001;**91**(6):1717

[32] Gerstenberger M, Wiemer S, Giardini D. A systematic test of the hypothesis that the b value varies with depth in California. Geophysical Research Letters. 2001;**28**(1):57-60

[33] Wiemer S, Wyss M. Mapping spatial variability of the frequency-magnitude distribution of earthquakes. Advances in Geophysics. 2002;**45**:259-302

[34] Cao A, Gao S. Temporal variation of seismic b-values beneath northeastern Japan island arc. Geophysical Research Letters. 2002;**29**(9):48-41

[35] Wyss M, Stefansson R. Nucleation points of recent main shocks in southern Iceland, mapped by b-values. Bulletin of the Seismological Society of America. 2006;**96**(2):599

[36] Console R, Murru M, Catalli F, Falcone G. Real time forecasts through an earthquake clustering model constrained by the rate-and-state constitutive law: Comparison with a purely stochastic ETAS model. Seismological Research Letters. 2007;**78**(1):49

[37] Zhou P, Xia S, Sun J, Cao J, Xu H, Zhao F, Chen C. Spatial variations of b-values in the coastal area of Guangdong. Journal of Ocean University of China. 2018;**17**(1):177-185

[38] Leptokaropoulos K, Staszek M, Lasocki S, Mart́ınez-Garz'on P, Kwiatek G. Evolution of seismicity in relation to fluid injection in the North-Western part of The Geysers geothermal field. Geophysical Journal International. 2018;**212**:1157-1166

[39] Urbancic T, Trifu C, Long J, Young R. Space-time correlations of b values with stress release. Pure and Applied Geophysics. 1992;**139**(3):449-462

[40] Singh S, Carton H, Tapponnier P, Hananto N, Chauhan A, Hartoyo D, Bayly M, Moeljopranoto S, Bunting T, Christie P, Others. Seismic evidence for broken oceanic crust in the 2004 Sumatra earthquake epicentral region. Nature Geoscience. 2008;**1**(11):777-781

[41] Schorlemmer D, Wiemer S, Wyss M. Earthquake statistics at Parkfield: 1. Stationarity of b values. Journal of Geophysical Research. 109, B12308, DOI: 10.1029/2004JB003235

[42] Schorlemmer D, Wiemer S, Wyss M. Variations in earthquake-size distribution across different stress regimes. Nature. 2005;**437**(7058):539-542

[43] Wyss M, Sammis C, Nadeau R, Wiemer S. Fractal dimension and b-value on creeping and locked patches of the San Andreas fault near Parkfield, California. Bulletin of the Seismological Society of America. 2004;**94**(2):410

[44] Wyss M, Matsumura S. Verification of our previous definition of preferred earthquake nucleation areas in Kanto-Tokai, Japan. Tectonophysics. 2006;**417**(1-2):81-84

[45] Nuannin P, Kulhanek O, Persson L. Spatial and temporal b value anomalies preceding the devastating off coast of NW Sumatra earthquake of December 26, 2004. Geophysical Research Letters. 2005;**32**(11):L11307

[46] Nuannin P. The Potential of b-Value Variations as Earthquake Precursors for Small and Large Events [Ph.D. thesis]. Uppsala University; 2006

[47] Del Pezzo E, Esposito A, Giudicepietro F, Marinaro M, Martini M, Scarpetta S. Discrimination of earthquakes and underwater explosions using neural networks. Bulletin of the Seismological Society of America. 2003;**93**(1):215

[48] Amorese D, Grasso J, Rydelek P. On varying b-values with depth: Results from computer-intensive tests for Southern California. Geophysical Journal International. 2010;**180**(1):347-360

[49] Wyss M, Lee WH. Time variations of the average earthquake magnitude in Central California. In: Proceedings of the conference on tectonic problems of the San Andreas fault system; 1973, edt. R. Kovach & A. Nur. Stanford University Geol. Sci. pp. 24-42

[50] Main I, Meredith P, Jones C. A reinterpretation of the precursory seismic b-value anomaly from fracture mechanics. Geophysical Journal International. 1989;**96**(1):131-138

[51] Gupta H, Singh H. Earthquake swarms precursory to moderate to great earthquakes in the Northeast India region. Tectonophysics. 1989;**167**(2-4):285-298

[52] Wyss M, Booth D. The IASPEI procedure for the evaluation of earthquake precursors. Geophysical Journal International. 1997;**131**(3):423-424

[53] Mori J, Abercrombie R. Depth dependence of earthquake frequency-magnitude distributions in California: Implications for rupture initiation. Journal of Geophysical Research. 1997;**102**(15):081-015

[54] Power J, Wyss M, Latchman J. Spatial variations in the frequency-magnitude distribution of earthquakes at Soufriere Hills volcano, Montserrat, West Indies. Geophysical Research Letters. 1998;**25**(19):3653-3656

[55] Warren N, Latham G. An experimental study of thermally induced microfracturing and its relation to volcanic seismicity. Journal of Geophysical Research. 1970;**75**(23):4455-4464

[56] Amelung F, King G. Earthquake scaling laws for creeping and non-creeping faults. Geophysical Research Letters. 1997;**24**(5):507-510

[57] Beauval C, Scotti O. Quantifying sensitivities of PSHA for France to earthquake catalog uncertainties, truncation of ground-motion variability, and magnitude limits. Bulletin of the Seismological Society of America. 2004;**94**(5):1579

[58] Kulhanek O. Seminar on b-value. Prague Centre of Mathematical Geophysics, Meteorology, and their Applications (MAGMA); 2005

[59] Hauksson E, Jones L, Hutton K. The 1999 Mw 7.1 Hector mine, California, earthquake sequence: Complex conjugate strike-slip faulting. Bulletin of the Seismological Society of America. 2002;**92**(4):1154

[60] Guttorp P. On least-squares estimation of b values. Bulletin of the Seismological Society of America. 1987;**77**(6):2115

[61] Aki K. Maximum likelihood estimate of b in the formula log N= a-bM and its confidence limits. Bulletin of the Earthquake Research Institute Tokyo University. 1965;**43**:237-239

[62] Bender B. Maximum likelihood estimation of b values for magnitude grouped data. Bulletin of the Seismological Society of America. 1983;**73**(3):831

[63] Utsu T. A method for determining the value of b in a formula log n= a-bM showing the magnitude-frequency relation for earthquakes. Geophys. Bull. Hokkaido Univ. 1965;**13**:99-103

[64] Shi Y, Bolt B. The standard error of the magnitude-frequency b value. Bulletin of the Seismological Society of America. 1982;**72**(5):1677

[65] Felzer K. Calculating the Gutenberg-Richter *b*-Value. USGS publication; 2007

[66] Melojevic S. Power-law distributions in information science-making the case for logarithmic binning. Journal of the American Society for Information Science and Technology. 2010;**61**(12):2417-2425

10

Principles of Probabilistic Seismic Hazard Assessment (PSHA) and Site Effect Evaluation and its Application for the Volcanic Environment in El Salvador

Walter Salazar

Abstract

This book chapter explains the fundamental concepts of the probabilistic seismic hazard and site effect evaluation. It is divided into four parts: firstly, the theoretical background of the probabilistic seismic hazard methods is explained to compute the earthquake loads used in structural analysis of buildings, namely, the rigid-zone method, the free-zone methods, and the characteristic models. We emphasize the physical meaning of the seismic coefficient prescribed in the seismic code regulations and its association with the return period of ground motion and spectral ordinates. The interconnection of the return period, the recurrence interval, and the lifetime concepts are explained to clarify misconceptions among these terms in connection with the probability of exceedance of motion. Secondly, the seismic hazard methods are applied employing volcanic chain seismicity data, and preliminary seismic hazard maps for rock site are presented for flat topography conditions along El Salvador. Thirdly, the site effects in terms of the amplification of ground motion are studied using soil profiles characterized by the interbedding of lava flows and volcanic ashes. Finally, we present a summary that highlights the most important concepts explained in this book chapter.

Keywords: seismic hazard, peak ground acceleration, spectral ordinates, amplification of motion, elastic design spectrum

1. Introduction

The probabilistic seismic hazard assessment (PSHA) involves the production of contour maps that represent the levels of ground and structural shaking expected to be experienced over a particular return period (in years). Generally, the shaking is represented in terms of peak ground acceleration (PGA) that is generated by the effects of the rupture of a geological fault and the propagation and attenuation of seismic waves reaching the bedrock. The shaking can be also represented by the structural acceleration (SA), and it depends on the period of vibration and damping of the building, and it is related to the characteristics of the ground motion itself; the SA is also referred in the literature as a spectral ordinate, and the damping is set as 5% of the critical (e.g., reinforced concrete structures) in PSHA computations. The PGA is associated with a zero structural period or with an infinite rigid structure. Then the severity of the shaking computed in a PSHA is primarily calculated on rock site conditions (NEHRP B soil class with a Vs = 760 m/s) and assuming a flat topography. The amplification of ground motion due to the presence of sediments above the bedrock is taken into consideration as a site effect in a second phase of evaluation.

The conjunction of spectral accelerations is represented in a response spectrum that is indeed the seismic coefficient C_S that is always prescribed in the codes' regulations as a fraction of the gravity g (9.8 m/s^2), i.e., 1.19 g = 11.66 m/s^2 at 0.2 s period (**Figure 1a,b**). A common misunderstanding is to assume that the seismic coefficient is related only to the ground motion or PGA; such confusion comes from the fact that the shear base V of the building is related with the seismic coefficient employing the following equation:

$$V = C_S W \tag{1}$$

where W is the total weight of the building. What Eq. (1) really reflects is the reaction at the ground level in the foundation of the building due to the all inertial forces F that result at each mass story subjected to its correspondent spectral acceleration SA during an earthquake. When considering a one degree of freedom system or a one-story building, the inertia force F can be calculated by the second Newton law as

$$F = ma \tag{2}$$

where m is the mass and a is the spectral acceleration. For the one-story building case $V = F$, $a = C_S$, and $m = W/g$. Note that engineers use the weight instead of mass for structural calculations, so the seismic loads are directly calculated vanishing the gravity g constant when using Eq. (2): $F = (W/g)$*1.19 g. In the case of having a several degrees of freedom system or a multi-story building, Eq. (1) also represents the reaction at the base of the building due to all inertia forces above; its application is still valid since spectral ordinates are always computed considering a single degree of freedom system that represents the seismic demand of one mode of vibration in the structure; in other words, the shear base in Eq. (1) retains the total inertia forces that are distributed later among all the stories along the building height when considering multi-degree of freedom systems. When considering the first overtones or harmonics in the calculations, the seismic demand of each mode of vibration can be combined employing classical methods of structural dynamics based on the same response spectrum (**Figure 1b**) derived by seismic hazard computations.

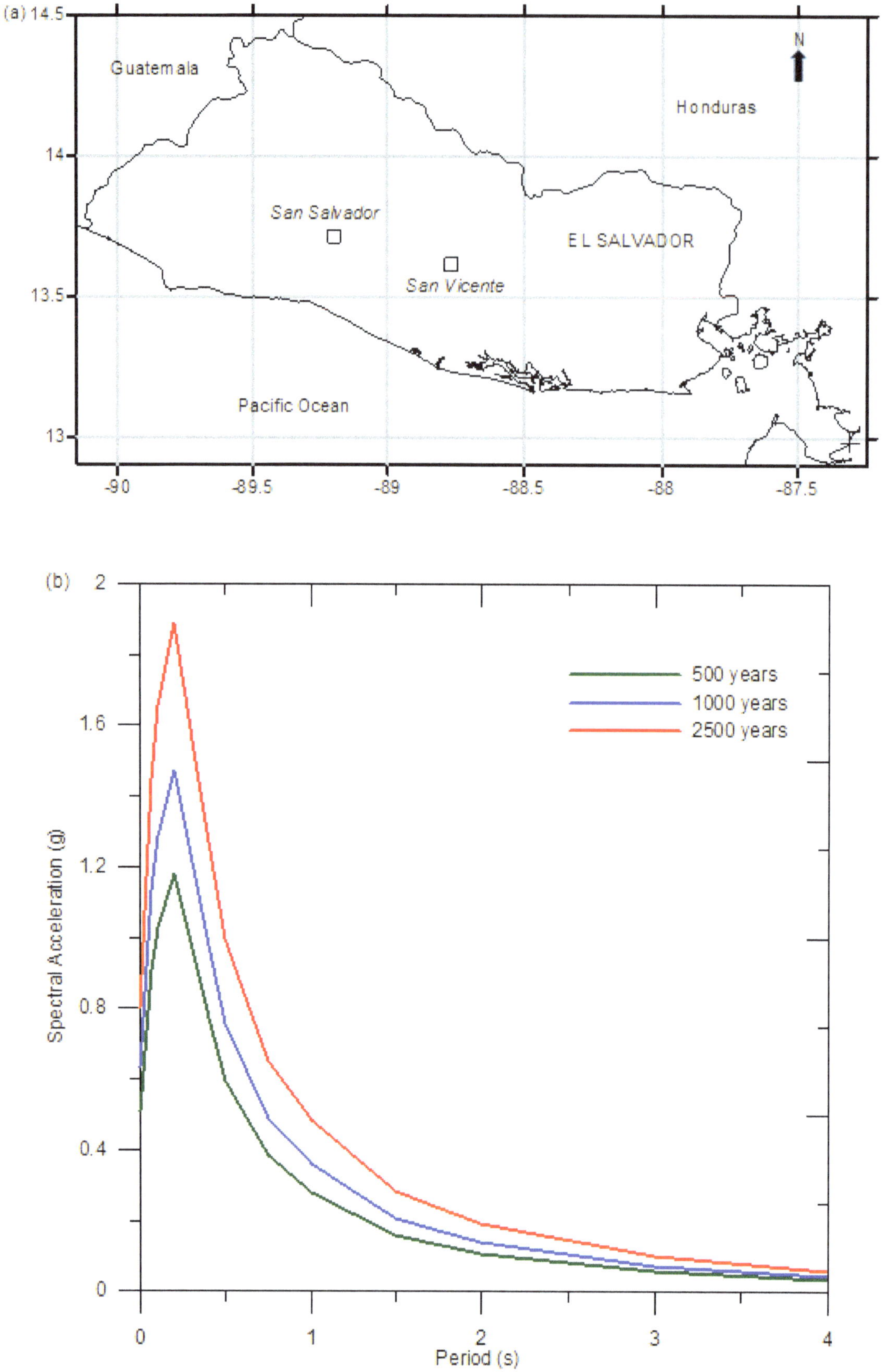

Figure 1. (a) Geographical location of El Salvador. (b) Elastic design response spectrum for 5% of critical damping at San Vicente City setting several years of return period.

2. Principles of probabilistic seismic hazard assessment (PSHA)

In PSHA calculations, the return period *RP* is associated with the expected level of shaking C_S at one place and not with the recurrence interval *RI* of a single earthquake related with a specific magnitude in a seismogenic source. Despite the concepts of the return period *RP* and the recurrence interval *RI* are different—although sometimes confused in the literature—they complement each other to produce the seismic loads that depend of the selected lifetime *L* (in years) of the building. The seismic coefficient C_S in the modern code provisions is usually calculated setting 500- or 2500-year return period which is associated with the probability of exceedance: this is the probability that the building experiment at least one time—might be two or more—a specific level of shaking in its lifetime *L*. Employing a Poisson distribution, the probability of exceedance q for a specific motion level can be expressed as

$$q = 1 - e^{-LE(z)} \tag{3}$$

where *E(z)* is the mean annual rate (number of earthquakes per year) for a specific level of motion *z*. Setting q to 2 and 10% and $L = 50$ years, *E(z)* yields 0.0004 and 0.002 earthquakes per year after evaluating Eq. (3). The return period is the reciprocal of *E(z)*; $RP = 1/E(z)$ yields 2500 and 500 years in these cases. Note that the lifetime of 50 years is presumably set as the average time span for which a person can own a building property. Note that the multiplication of *L* by *E(z)* vanishes the unit of years allowing probability results q. The usefulness of the earthquake interval concept on hazard computations is presented in Section 2.1.

Then, the main objective of seismic hazard computations is to obtain the mean annual rate $E(z)$ for a specific level of motion z, and consequently the probability of exceedance is calculated in Eq. (3) setting a lifetime usually set as $L = 50$ years. For a specific period of vibration, a seismic hazard curve is calculated at several motion levels, and the design value of acceleration for a specific period of vibration is obtained at 2 or 10% of probability of exceedance via interpolation of the hazard curve (**Figure 2**). When considering another period of vibration in the hazard computations, it is possible to construct a response spectrum related to the engineering project at the return period of interest. Finally, the seismic hazard maps are constructed using a grid of points smoothing the acceleration values obtained at each point that constitutes the grid over the region under consideration.

In the next sections, the main concepts for three PSHA classical methods, namely, the rigid-zone method, the free-zone methods, and the characteristic model are discussed.

2.1. The rigid-zone method

The rigid-zone method is the standard Cornell-McGuire approach [1, 2] based on a Poissonian probability distribution for which the occurrence of an event is independent of others. The first step in PSHA is to quantify the level of seismicity in the region of study answering three questions: (i) where do earthquakes happen? (ii) what magnitude do they have? and (iii) how do they frequently occur? Several *ith* seismogenic sources must be declared as polygons that

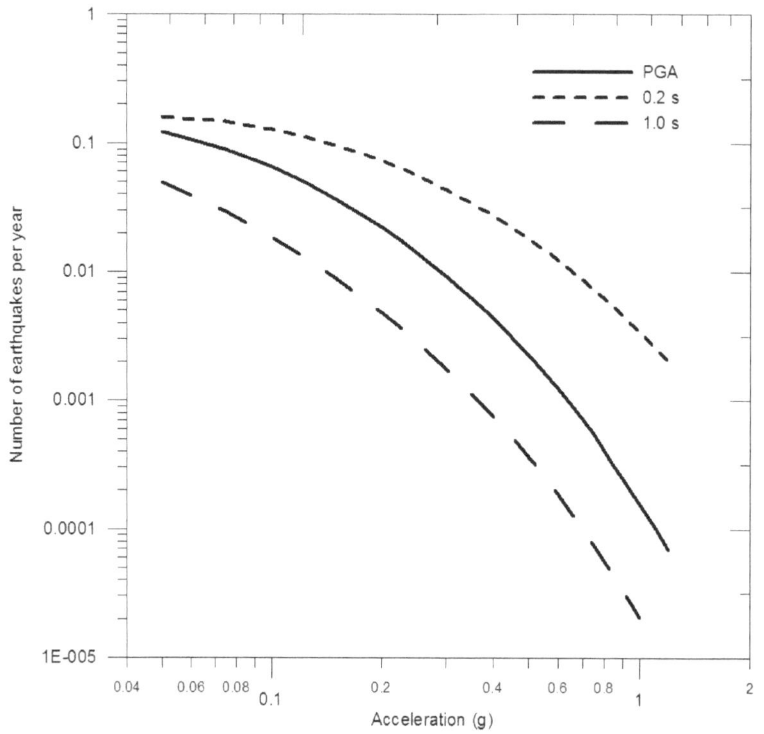

Figure 2. Seismic hazard curves for San Vicente City (**Figure 1**a) depicting the number of earthquake per year vs. the peak ground acceleration and spectral accelerations for 0.2 and 1.0 s structural period of vibration.

enclose a group of earthquakes in particular geographical location that have similar geological characteristics, namely, the tectonic environment, the focal mechanism, the focal depth, and a maximum magnitude. As a result, that area constitutes a rigid zone (**Figure 3**). It is called the rigid zone because it is assumed that the level of seismicity is constant all over the area inside the polygon and cannot change with time and dimensions. Since the hazard analysis is based on a Poisson probability distribution, the compiled earthquake catalog must be declustered, that is, the foreshocks and the aftershocks must be removed from the list of events, and only the main shocks must remain in the catalog; this is usually accomplished employing decluster algorithms [3]. To quantify the seismicity in a specific seismogenic source, a Gutenberg-Richter (G-R) recurrence relationship is defined as

$$log N_i = A - BM \tag{4}$$

where N_i represents the cumulative number of earthquakes per year above magnitude M, and A and B set two constants derived from regression analysis; such constants are independent of the period of vibration of buildings. The number of earthquakes N_i within different magnitude ranges are usually computed employing the bounded G-R relationship based on [4]. The constant -B is the measure of the relative abundance of large to small shocks, and it includes the negative sign indicating the negative slope that must result from the regression analysis. Note that Eq. (4) must be derived after a completeness analysis of the earthquake catalog [5, 6]; what it means is that the cumulative number of earthquakes must be divided by the number of years for which a particular magnitude bin is completed. It is worth mentioning that the

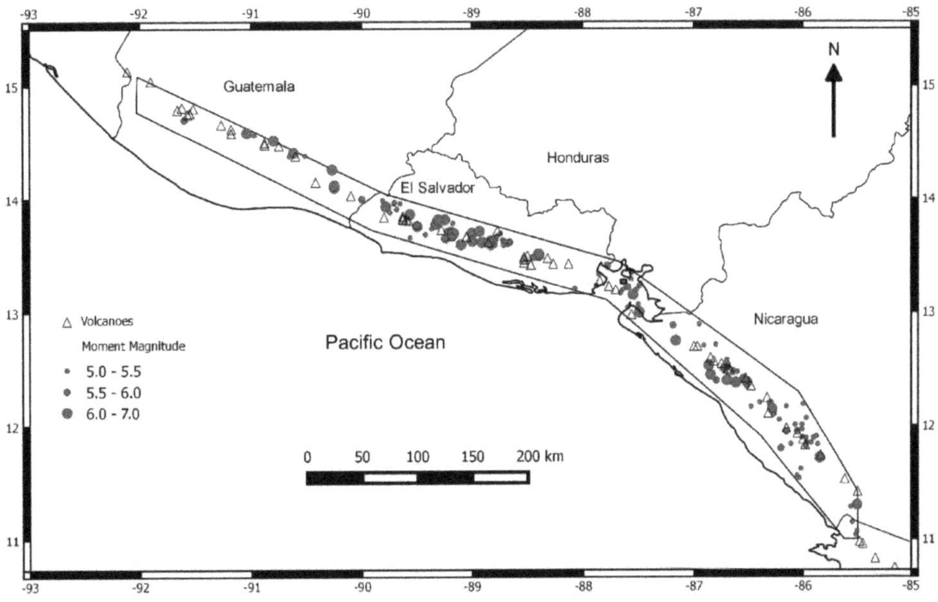

Figure 3. Epicenters (solid circles) and geometrical delimitation (polygon) of volcanic chain earthquakes in El Salvador and neighboring countries 1528–2009. The active quaternary volcanoes are depicted with white triangles. In this Figure the epicenters of main events, foreshocks, and aftershocks are included.

reciprocal value of N_i yields a recurrence interval *RI* and does not yield the return period for a certain magnitude *M*. For example, if $A = 5.2$ and $B = 1$ for a *M* 4.0 and *M* 7.0, N_i yield 16.0 and 0.016 earthquakes/year, and the recurrence interval are ¾ and 63 years, respectively. Then the return period is associated with the level of motion at a particular site, and the recurrence interval is related with the size of an earthquake [7].

Although this information is valuable in terms of occurrence characteristics of the earthquakes sources, such recurrence interval is still not associated with a specific ground or structural motion needed to design a building. In the first instance, a ground motion prediction equation (GMPE) must be developed in order to relate the expected shaking at one place of interest depending on the magnitude and distance from the source at rock site conditions. Such GMPE must be compatible with the same tectonic environment of the seismogenic area *ith* to develop Eq. (4). A general form of a GMPE is

$$\log(y) = a + bM + cR + d\log R \pm \sigma_{\log(y)} \tag{5}$$

where y is the PGA or spectral acceleration SA in g, a and b are regression constants related to the source, and c and d are constants related with the anelastic and geometrical attenuation respectively (path term). The logarithm nature of this equation is presented in Section 3. *M* is the moment magnitude, *R* is the distance from the project site to any point of the seismogenic source *i*, and $\sigma_{log(y)}$ is the standard deviation of the regression in terms of the logarithm value of y. Employing Eq. (5) and setting $a = 1.68$, $b = 0.30$, $c = -0.01$, and $d = -1.0$, the estimated PGA for an earthquake of $M = 7.0$ at a distance $R = 10$ km gives 479 cm/s² = 0.49 g. Despite that such value could be used for the design of a short period structure, it is not associated with any recurrence interval RI of this earthquake size and with any lifetime L and return period RP. Moreover, there is no information of other magnitude and distance pairs that can affect the project site; besides, the standard deviation σ is not being taken into account in the ground motion estimation at this stage.

The mean annual rate $E(z)$ for a specific motion level "z" can be obtained combining the seismicity evaluation and the GMPEs employing the following integrals considering all possible magnitudes and distances probabilities:

$$E(z) = \sum_{i=1}^{Ns} v_i \int_{R=0}^{\alpha} \int_{M_{min}}^{M_{max}} f_i(M) f_i(R) P(Z > z \mid M, R) dR dM \tag{6}$$

where Ns = number of seismogenic sources; v_i = mean rate of occurrence of earthquakes above the minimum magnitude for the "*ith*" source; $f_i(M)$ = probability density distribution of magnitude within the "*ith*" source, which is obtained using the G-R relationship; $f_i(R)$ is the probability density distribution of epicentral distance "R" between various locations within source "*ith*" and the site where hazard is estimated; and $P(Z > z \mid M,R)$ = probability that a given earthquake of magnitude "M" and source-site distance "R" will exceed motion level "z", which is obtained employing the selected GMPE and the standard deviation σ.

Events with a lower magnitude M_{min} (i.e., 5.0) might not cause damage to structures, so seismic hazard computation is performed establishing a lower bound limit in all cases. The mean rate of occurrence v_i is evaluated using Eq. (4) at M_{min} and represents all earthquakes per year that could happen in the seismogenic source above the lower bound limit nearby the project location, including the maximum credible magnitude M_{max} that can produce each seismogenic source *i*th; M_{max} is considered as the upper-bound physical limit in hazard computations. The units of $E(z)$ are in accordance with the units of v_i and must be multiplied by the results of the integrals that contain the unitless quantities represented by the probabilities of occurrence of magnitudes, distances, and the level of shaking z produced by all possible magnitude-distance pairs that affect the site of interest. The multiplication of $f_i(M)$, $f_i(R)$, and $P(Z > z \mid M,R)$ obeys to the fact that the occurrences of these probabilities are physically and statistically independent.

The probability distribution of magnitude M or the cumulative probability function $CDF(M)$ can be expressed as

$$CDF_i(M) = \frac{N_{Mmin} - N_M}{N_{Mmin}} \tag{7}$$

The $CDF_i(M)$ is the quotient between the numbers of earthquakes of a magnitude M that exceeds the number of earthquakes of minimum magnitude M_{min} after evaluation in Eq. (4). However, the derivation of Eq. (7) with respect to the magnitude M is used in seismic hazard calculations, yielding the probability density distribution after Neperian logarithm transformation of the B value in the G-R relationship [8]:

$$f_i(M) = \beta e^{-\beta(M-Mmin)} \tag{8}$$

where $\beta = (\ln 10)B$ and e is the Neperian number. Note that the probability density function is suitable for PSHA since it represents the chances of success of the occurrence of particular size M of an earthquake. In computational practice, Eq. (8) is discretized over the same bin magnitude classes (e.g., $M1$ 5.5 to $M2$ 6.0) used to develop the G-R recurrence relationship (Eq. 4); then the probability $P(M)$ of a magnitude to fall in that bin can be written as

$$P(M) = \int_{M1}^{M2} f_M \, dM \tag{9}$$

where the value of M is taken as the average of the bin magnitudes $M1$ and $M2$.

When performing PSHA calculations, the polygon that represents the rigid zone is usually discretized in squares of 1 km side which represent all possible earthquake sources. The probability $f_i(R)$ that an earthquake source could happen at a specific distance R from the project site is the quotient between R and all the possible distances R_j:

$$P(R) = f_i(R) = \frac{R_j}{\sum_{j=1}^{n} R_j} \tag{10}$$

where n is the number of elements used to discretized the seismogenic source. Note that all the earthquake sources can have magnitudes between the lower- and upper-bound limits with related probabilities of occurrence explained above. So, $f_i(M)$ and $f_i(R)$ are independent of the level of motion being evaluated; $f_i(R)$ is dependent on the location of the project relative to the seismogenic source *ith* into consideration. The probability $P(Z > z \mid M,R)$ that a given earthquake of magnitude M and source-site distance R will exceed motion level z is evaluated using a standard normal variable G:

$$G = \frac{\log(z) - \log(y)}{\sigma_{\log(y)}} \tag{11}$$

To estimate $log\ (y)$ in Eqs. (5) and (11), the value of M to be used in Eq. (5) is taken as the average of $M1$ and $M2$ in Eq. (9). $P(Z > z \mid M,R)$ is then calculated as

$$P(Z > z \mid M, R) = 1 - \phi_{(G)} \tag{12}$$

where $\varphi_{(G)}$ is the probability density function derived from a log-normal Gaussian distribution evaluated for the G value obtained in Eq. (11). For computational purposes, the exact integrals in Eq. (6) are substituted by discrete sums yielding:

$$E(z) = \sum_{i=1}^{Ns} v_i \sum_{R_{min}}^{R_{max}} \sum_{M_{min}}^{M_{max}} P(M)P(R)P(Z > z \mid M, R) \tag{13}$$

The probabilities in Eq. (13) are calculated after selecting a segmentation of magnitude and source-site distance pairs. Usually such segmentation is done for every 0.25–0.5 magnitude units of magnitude and 1–10 km distance. R_{min} and R_{max} constitute the minimum and maximum source-site distances.

2.2. The free-zone methods

Woo [9] proposed an alternative approach to compute the PSHA in order to remove the uncertainties inherent in the definition of the rigid seismogenic sources above. The arbitrariness in delineating zone boundaries may be artificially expanded to dilute activity around a site or on the other hand artificially contracted to concentrate activity around a site. Besides, the assumption that the activity rate is constant in a seismogenic source is rarely tested statistically [9]. In this method, the seismic hazard is calculated directly based on the characteristics of the earthquake catalog rather than defining rigid zones considering a spatial uniformity seismicity; then earthquake sources are not defined according to a geological and tectonic criteria. The Woo method is called the kernel estimation method since it implements spatial smoothing directly to earthquake epicenters contained in the catalog depicting the spatial nonuniformity of seismicity in the region of study. Then the earthquake catalog must contain the main events, the foreshocks, and aftershocks when performing hazard calculations.

The first step in the method consists in defining a grid of nodes that constitute the point earthquake sources around the project site. Each point of the grid would have an activity rate that is magnitude-distant dependent based on fractal characterization. A multivariate probability density function K is expressed in terms of magnitude and distance yielding:

$$K(M,x) = \frac{n-1}{\pi H^2}\left[1+\left(\frac{R}{H}\right)^2\right]^n \tag{14}$$

where n is the exponent of the power law (usually between 1.5 and 2.0), R is the source-site distance, and H is the bandwidth for normalizing distances that is a function of magnitude M representing the average distance between epicenters of the same magnitude bin:

$$H = c\, e^{dM} \tag{15}$$

where c and d are constants to be derived by a regression analysis. By summing over all cataloged events with historical epicenters x_i, the cumulative activity rate density is computed at a general point x in the grid for each magnitude bin class from the minimum magnitude to the highest value, as follows:

$$\lambda(M,x) = \sum_{i=1}^{n_o} K(M, x-x_i)/T(x_i) \tag{16}$$

$T(x_i)$ is the effective observation time period which is equivalent to the completeness period, and n_o is the number of earthquakes in the catalog; the contribution of each event is inversely weighted by its completeness period. Once $\lambda(M,x)$ is calculated for each node in the mesh, a

GMPE is employed, and the seismic hazard is calculated by summing over each node source for an specific motion z. Note that the probability $P(Z > z \mid M,R)$ will be evaluated using Eqs. (11) and (12) and multiplied by $\lambda(M,x)$ to calculate $E(z)$ and the correspondent probability of exceedance q in Eq. (3) setting a specific lifetime L. The observational uncertainties in event magnitude and epicentral location can be accounted for in this method employing Gaussian error distributions [10]. The basic principle underlying the Woo method is that the epicenter of each past shallow event is smoothed geographically to generate a spatial probability distribution for event recurrence. As a conclusion, the method intends to represent both the fractal clustering of moderate magnitude epicenters and the haphazard migration of major events [9].

Frankel [11] also proposed a free-zone method used for the US National Hazard Maps developed by the US Geological Survey (USGS). In this method a region is partitioned into cells by a grid of 0.1° (11 km on a side). Within the *i*th grid cell, a count is made of the number of events with a magnitude above a reference value M_{ref}. This cumulative count of events is converted to incremental values from $M_{ref} + \Delta M$ or for a specific magnitude bin. The grid of n_i values is smoothed spatially in a simple Gaussian way with correlation distance c_d. The smoothed value of n_i is obtained as follows:

$$n_i = \frac{\sum_j n_j e^{-\Delta^2 ij/c_d^2}}{\sum_j e^{-\Delta^2 ij/c_d^2}} \tag{17}$$

where Δ_{ij} is the distance between the *i*th and *j*th cells. For a specific site, the values of n_i are binned by their distance from the site, and a value of N_r is calculated as the total value of n_i for cells within a certain distance increment of the site.

The mean annual rate $E(z)$ for a specific motion level "z" at the project site is determined as follows:

$$E(z) = \sum_r \sum_m 10^{\left[\log\left(\frac{N_r}{T}\right)+B(M_m-M_{ref})\right]} P(Z > z \mid M, R) \tag{18}$$

where r and m are the counters for the distance and the magnitude bins, respectively. It is worth mentioning that the term $10^{\left[\log\left(\frac{N_k}{T}\right)+B(M_i-M_{ref})\right]}$ is the value of 10^A in Eq. (4) for the completeness period T; $P(Z > z \mid M, R)$ is calculated as in the previous methods. Since a Poisson probability distribution is used in this method, the earthquake catalog must be declustered removing the foreshocks and aftershocks retaining only the main events for PSHA calculations.

2.3. The characteristic earthquake model

The characteristic earthquake model is employed when detailed fault information is collected by a geologist in the field and where there is a clear evidence of the slip via geodetic surveys (GPS) and/or paleoseismological investigations [12–14]. Instead of declaring an area as a rigid seismogenic source or employing fractal geometry concepts in free-zone methods, a fault is defined employing the following input data:

i. The slip rate (mm/year)

ii. A *B* value (usually set as 1.0)

iii. Precise geographical coordinates of a series of points that define the fault trace

iv. The dip, the sense of slip, and depth of the fault

v. The minimum magnitude M_{min} (e.g., 6.0) to be considered in the analysis

Two alternate characteristic models are applied actually in a PSHA:

i. Considering only the maximum magnitude M_{max} and its associated activity rate or number of earthquakes per year

ii. Classical Gutenberg-Richter (G-R) relationships

For both models the first step in a PSHA is to calculate the number of earthquake per year *E(z)* for the M_{max} in model (i) and for the M_{min} in the model (ii).

The idea of using two alternate models is that it is unknown whether future earthquakes will be large fault-filling ruptures of smaller ruptures which only occupy a relatively small portion of the fault [15]. Generally a weight of 0.5 is assigned to each model if the mean recurrence time for M_{max} in the first model is less than the return period of motion.

2.4. Probabilistic seismic hazard application for the volcanic chain earthquakes in El Salvador

In this section, the free-zone method, the rigid-zone methods, and the characteristic models are applied for the volcanic chain earthquakes in El Salvador. For the Cornell-McGuire method, an own program was developed and used the KERFRACT program for the Woo method. The programs developed by the USGS [15] were employed for the Frankel and characteristic models. The seismic activity of this source is concentrated in upper part of the crust with focal depth less than 25 km and within a continuous belt of 20–30 km width along the active volcanoes in Central America (**Figure 3**). The events can appear in clusters. Faults with right and left lateral focal mechanisms are located parallel and perpendicular to the volcanic axis, respectively. The last destructive earthquakes inside the volcanic chain in El Salvador dated on October 10, 1986, M5.7, and February 13, 2011, M 6.5 [16].

Seismic hazard maps for PGA were developed employing a $0.1 \times 0.1^{\circ}$ grid assuming a flat topography setting 500-year return period of ground motion in terms of PGA. The homogenized earthquake catalog compiled by Salazar et al. [17] was used for the moment magnitudes above 5.0 (M_{min}) and the GMPE developed by Sadigh et al. [18] for shallow crustal earthquakes and a truncation value of 3σ in the hazard computations. We updated the work of Bommer et al. [19] which also presents PSHA maps for El Salvador in terms of normalizing PGA to the maximum computed value; here the absolute values of PGA are presented based on new seismological information [17] applying the methods explained above.

A M_{min} of 5.0 and M_{max} of 6.9 were set when using all methods with a constant seismogenic depth of 5 km. The G-R relationship for the rigid-zone method is presented in **Figure 4** after the completeness analysis. A tri-linear trend that reflects interesting features of the seismicity along the volcanoes is observed. The first part (B1) suggests a large number of small earthquakes related to big shocks, the middle part yields a very low B2 value reflecting the effects of clusters or the abundance or moderate size events, and the third part (B3) reflects the small number of destructive events related to small shocks. The tri-linear trend of the G-R relation yields the probability density function when applying Eqs. (7) and (8). The sum of all magnitude bin probabilities of the tri-linear trend yields 76, 9, and 15% for the first (B1), second (B2), and third (B3) segments, respectively, which sum must be 100%. Although, it is interesting to note that the probability of occurrence of an earthquake of a specific magnitude M 5.6 and M 6.3 is the same of about 3%, presumably, the contribution to the seismic hazard from both events in terms of magnitude probability is the same despite the size differences of these events (**Figure 5**). Generally, the greater the magnitude, the lower the probability. The probability of occurrence for magnitudes between 5.7 and 6.1 yield similar (2%) due to the low B2 value in the G-R relationship.

Carr et al. [20] suggest that the volcanoes in Central America cluster into centers, whose spacing is random but averages about 27 km. **Table 1** and **Figure 6** present the average distance of earthquakes within magnitude bins of 0.25; the all average distance for all size bins yields the same 27 km as the volcano centers suggested by Carr et al. [20]. What it means is that earthquake epicenters and volcano centers have the same spatial distribution among them. Regarding with the

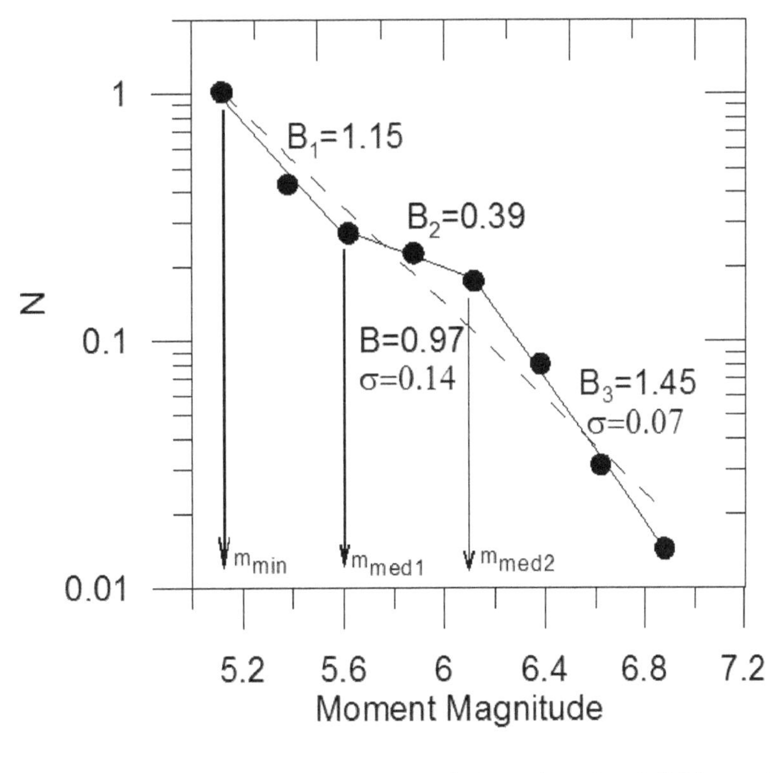

Figure 4. Gutenberg-Richter (G-R) relationships for the volcanic chain earthquakes. Solid line and solid circles: tri-linear trend employed in the rigid-zone Cornell-McGuire method. Broken line: linear regression employed in the free-zone Frankel method with B value of 0.97; see Eq. (4).

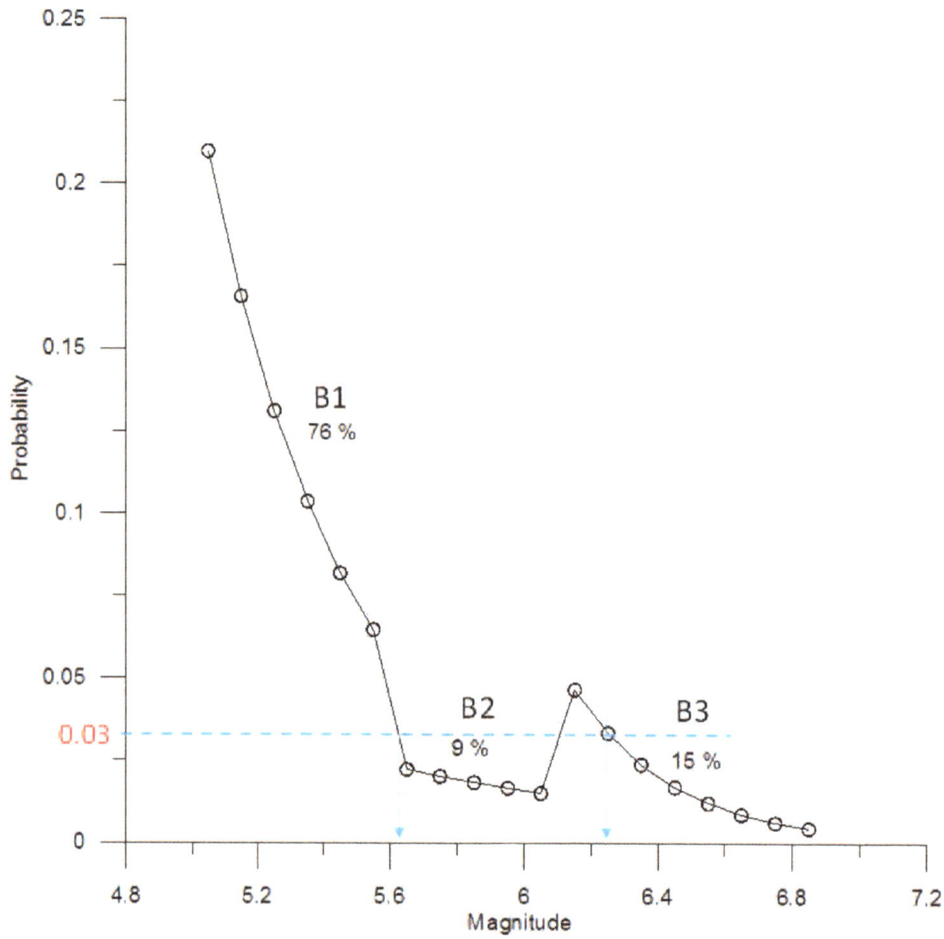

Figure 5. Magnitude probability density function for the volcanic chain earthquakes based on the tri-linear G-R relationships depicted in **Figure 4**; see Eq. (8). The percentages are the sum of all magnitude bin probabilities for each segment of the tri-linear trend with a total of 100%.

Woo free-zone method, the kernel only depends on the epicentral distance rather than magnitude, so the kernel function must be unique for the volcanic chain events (**Figure 7**) and independent of the size of the earthquakes; then $c = 27$ and d must be zero yielding $H = 27$ km in Eq. (15).

We model the characteristic earthquake for the San Vicente Fault employing the following parameters:

i. The slip rate: 4.1 mm/year [21] based on paleoseismological investigations

ii. A B value equal to 1.0

iii. The geographical coordinates that define the fault trace are taken from [21]

iv. The dip = 90°, the sense of slip as a pure right lateral and depth of the fault = 15 km

v. The minimum magnitude M_{min} as 6.0 to be considered in the analysis.

We present in **Figures 8–11** the seismic hazard maps employing the four methods explained above. As a first observation, it is noted that the rigid-zone method yields a uniform level of ground motion along the volcanoes and that the free-zone methods can capture the

Magnitude range	Number of earthquakes	Distance (km)
5.00 ≤ M < 5.25	68	15.84
5.25 ≤ M < 5.50	13	35.19
5.50 ≤ M < 5.75	20	22.33
5.75 ≤ M < 6.00	14	24.79
6.00 ≤ M ≤ 6.25	22	19.44
6.25 ≤ M ≤ 6.50	11	22.64
6.50 ≤ M ≤ 6.75	5	47.98
6.75 ≤ M ≤ 7.00	2	27.75
Total	155	Average H = 27

Table 1. Average distances for magnitude bins for volcanic chain earthquakes.

nonuniformity of hazard spatial distribution especially at the western part of the country. However, at San Salvador City (see location in **Figure 1a**), the PGA yields about 0.45–0.5 g for all methods. At San Vicente City (see location in **Figure 1a**) when applying the rigid

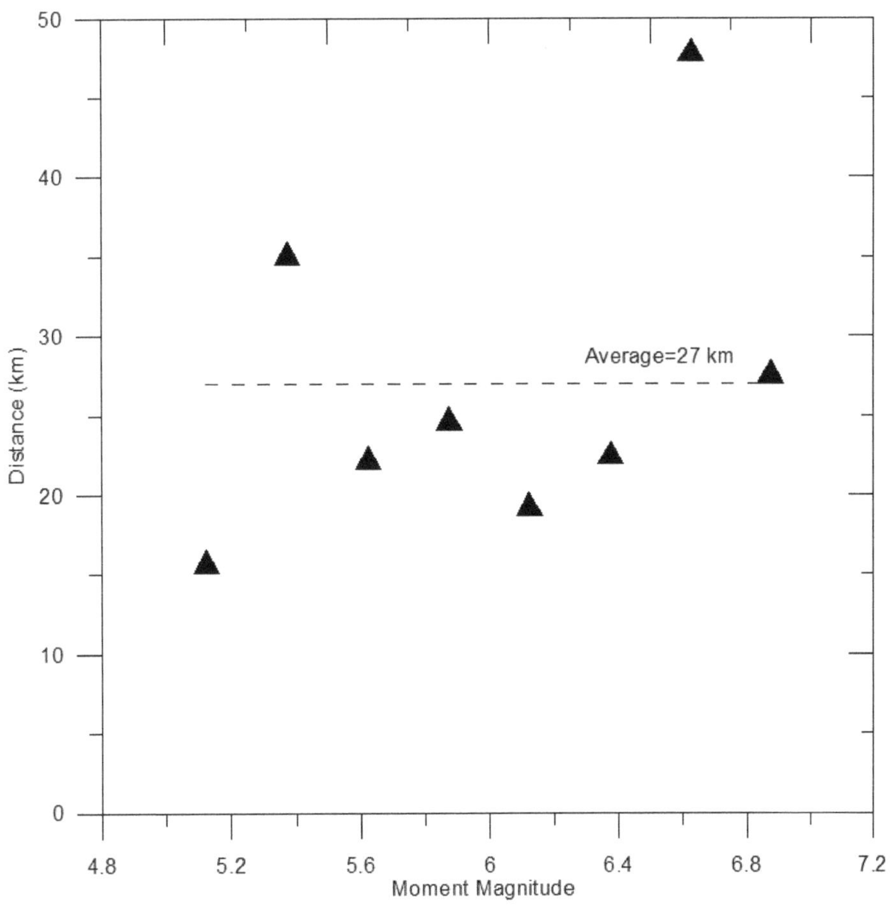

Figure 6. Bandwidth for normalizing epicentral distances in El Salvador volcanic chain and neighboring countries where H = 27 km in Eqs. (14) and (15). Carr et al. [20] suggest the volcanoes in Central America clusters into centers, whose spacing is random but averages about 27 km. It is noted that the all average distance for all earthquake epicenters also yields the same 27 km as the volcano centers (see **Table 1**).

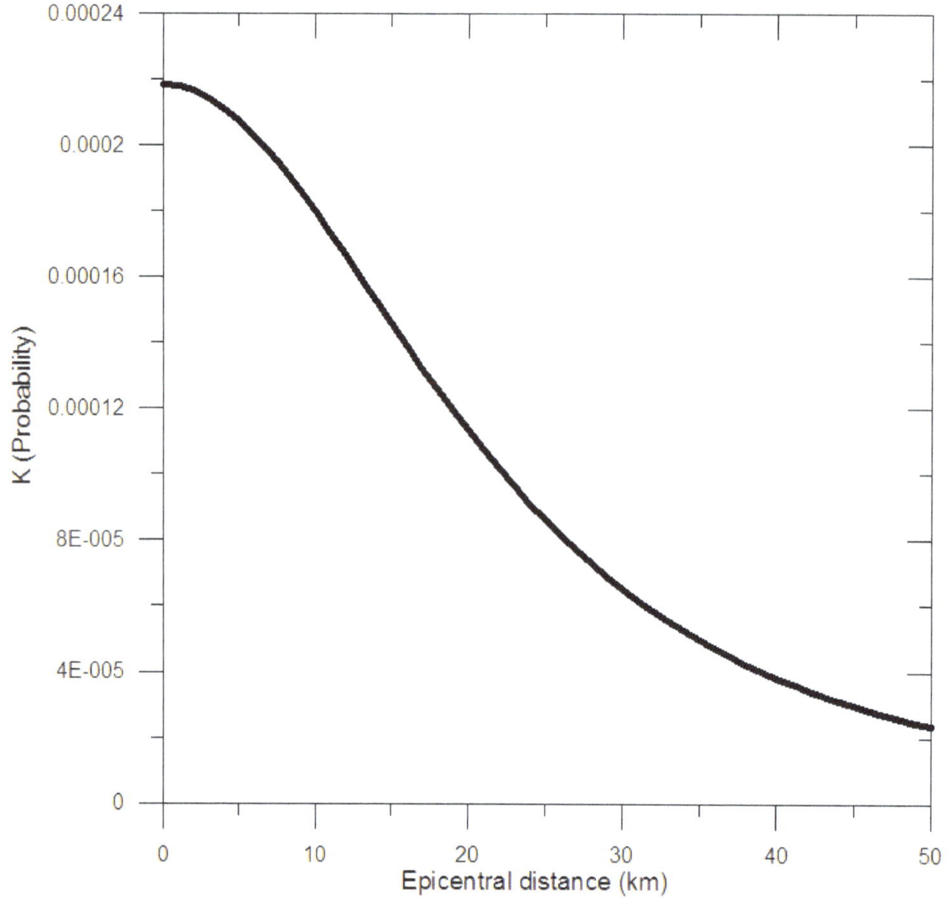

Figure 7. Kernel function for upper-crustal earthquakes in the volcanic chain. Note that the kernel probability function K is unique for H = 27 m in Eqs. (14) and (15).

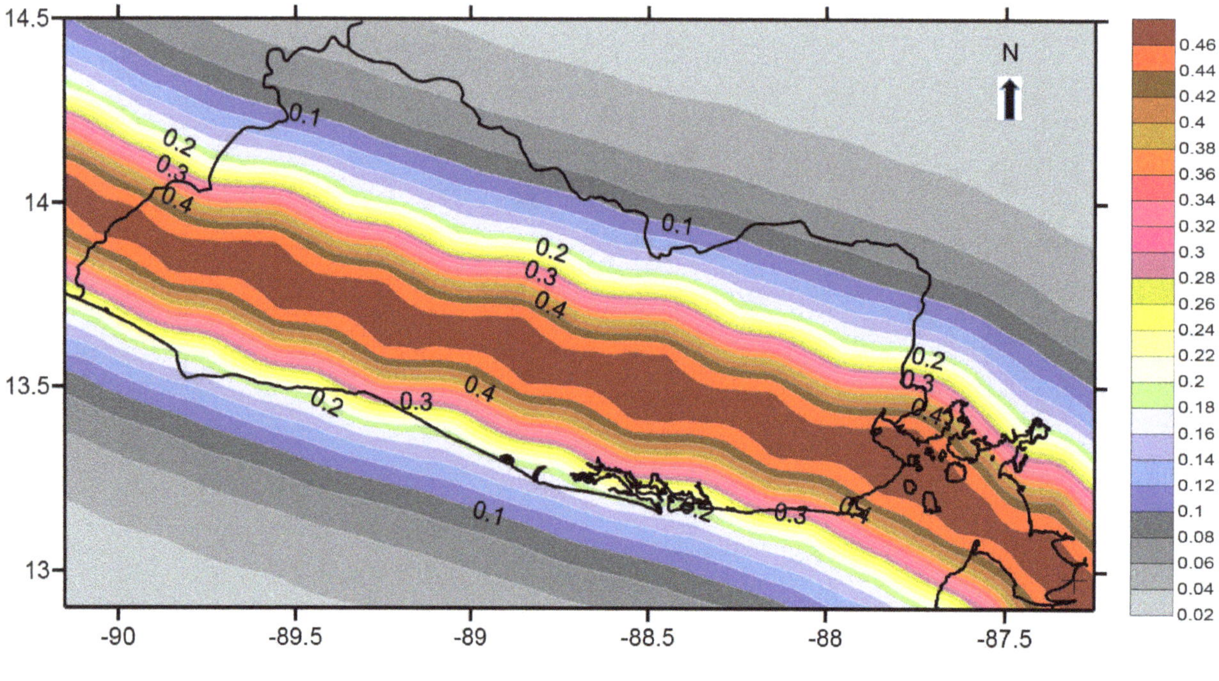

Figure 8. Seismic hazard map for the peak ground acceleration (PGA) as a fraction of gravity "g" employing the rigid-zone Cornell-McGuire method setting 500 years of return period.

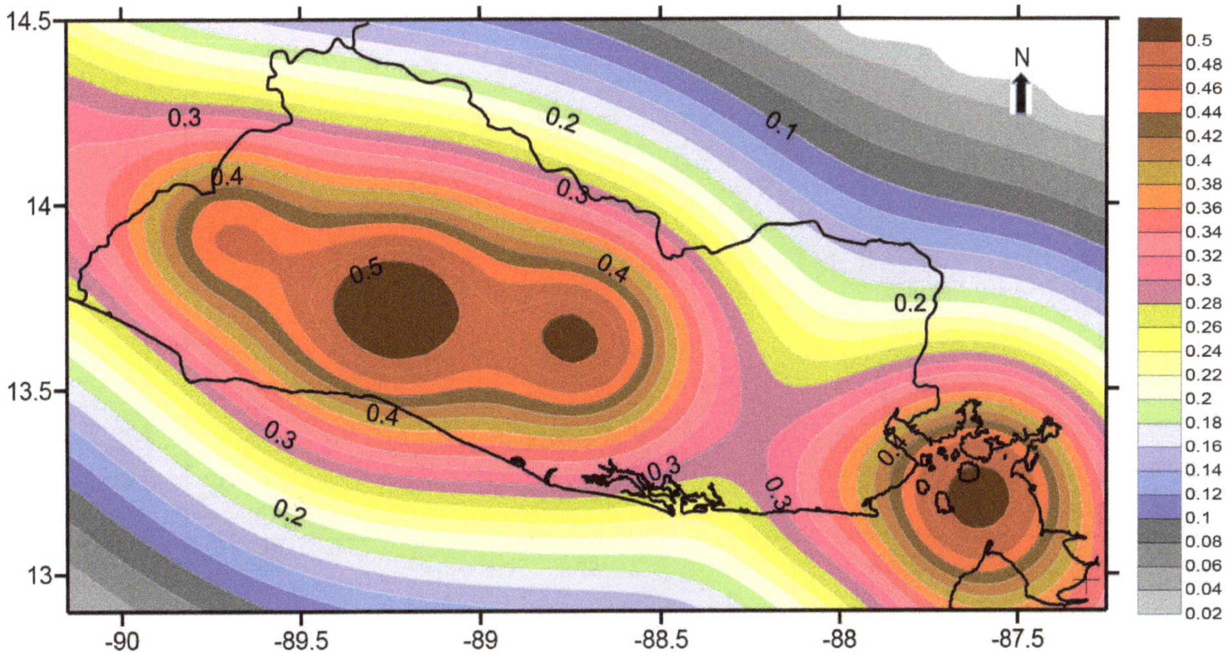

Figure 9. Seismic hazard map for the peak ground acceleration (PGA) as a fraction of gravity "g" employing the free-zone Woo method setting 500 years of return period.

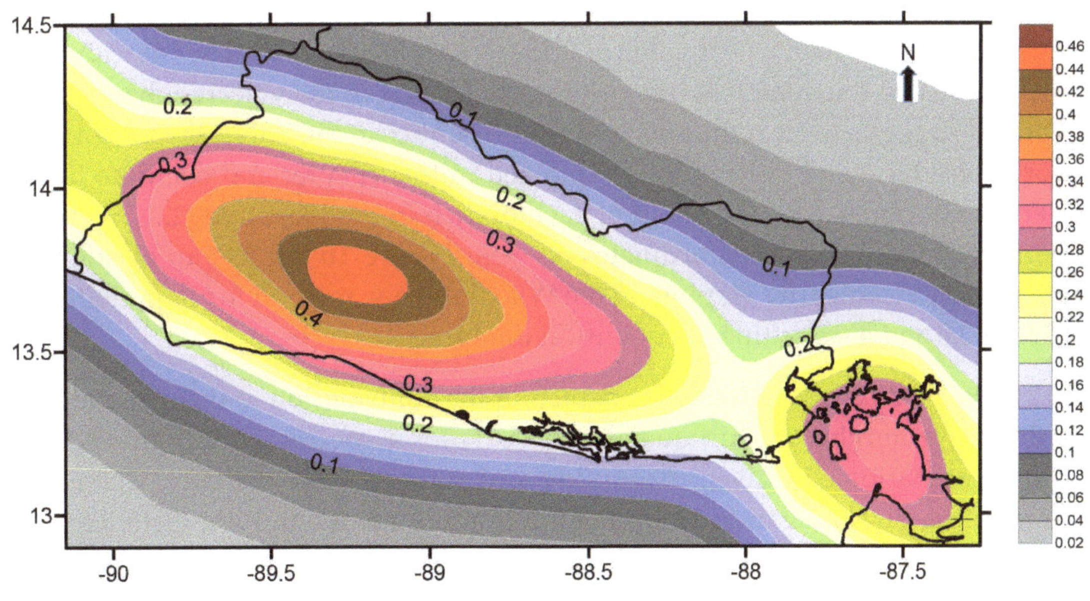

Figure 10. Seismic hazard map for the peak ground acceleration (PGA) as a fraction of gravity "g" employing the free-zone Frankel USGS method setting 500 years of return period.

zone, the Woo free zone, and the characteristic model, the PGA yields 0.45 g; however, a lower value of about 0.35–0.40 g yields for the Frankel free-zone method. This difference is attributed to the fact that the Woo method employs the unclustered catalog and the Frankel method employs the declustered one that it only contains the main events; it is noticed also

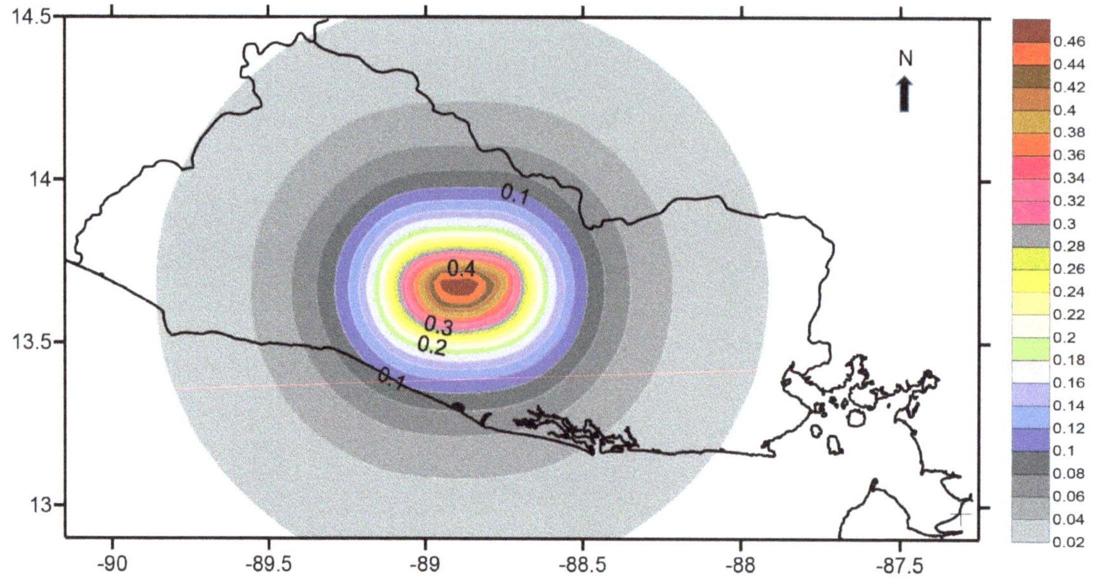

Figure 11. Seismic hazard map for the peak ground acceleration (PGA) as a fraction of gravity "g" employing the characteristic model at San Vicente Fault setting 500 years of return period and 4.1 mm/year of slip rate.

that for the Frankel method, we employed a unique B value of 0.97 for a G-R relationship (**Figure 4**); however, such hazard estimates are not significantly different from each other. When applying the characteristic model employing the paleoseismological slip rate at San Vicente City of 4.1 mm/year, a near value of 0.45 g is obtained as well. The coincidence of the PGA values revealed the quality of our earthquake catalog in the volcanic chain. The actual state of the art suggests to treat the epistemic uncertainties in the PSHA through a logic-tree framework considering all the methods explained in this chapter and compute the covariance or error associated with the result of each branch in the logic tree [22–26]. The PSHA could be performed for spectral ordinates as well; **Figure 1b** depicts an elastic response spectrum for San Vicente City employing the Woo free-zone method for 13 periods of vibration between 0.0 and 4.0 s setting 5% of critical damping.

Figure 12 shows a disaggregation chart depicting the contribution to the mean annual rate of exceedance (or number of earthquakes per year) for a specific ground motion value at a site due to scenarios of a given magnitude and source-to-site distance. Then it is possible through the disaggregation analysis to determine the magnitude-distance pair with the largest contribution to the hazard [27]. This analysis is performed setting a PGA of 0.46 g observed at San Vicente City (**Figure 15**) when applying the rigid-zone method. It is interesting to note that small earthquakes of M 5.0 have the largest contribution and that an earthquake of M 6.2 has the same contribution of an earthquake of M 5.4 to the seismic hazard in the volcanic chain of El Salvador. We attribute such especial features due to the tri-linear trend observed in the seismicity evaluation represented by the G-R recurrence relationships and its correspondent magnitude probability density function presented in **Figure 5**.

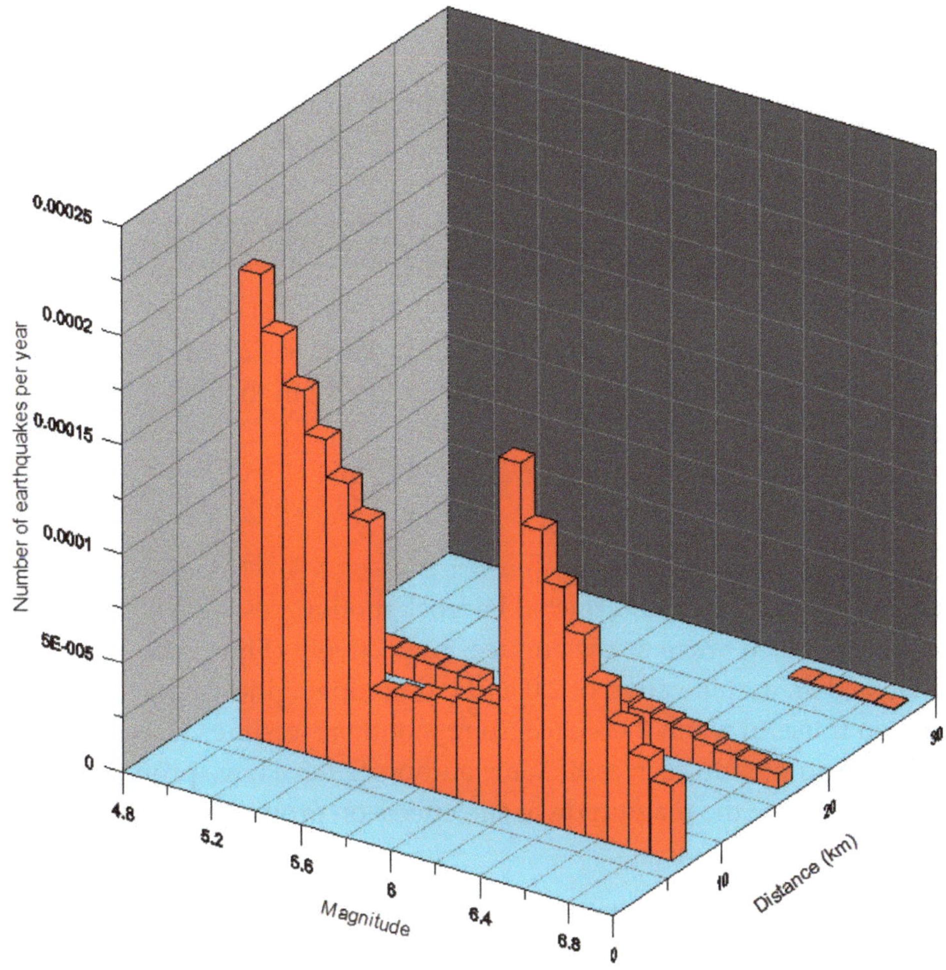

Figure 12. Example of magnitude-distance pair disaggregation charts for the peak ground acceleration (PGA) of 0.46 g at 500-year return period employing the rigid-zone Cornell-McGuire method at San Vicente City (see location in **Figure 1a**). Note that an earthquake of M 6.2 has the same contribution to seismic hazard of an earthquake of M 5.4 in the volcanic chain of El Salvador.

3. Site effects related with the interbedding of sediments and lava flows

Different studies carried out after the occurrence of several destructive seismic events have confirmed that the presence of sedimentary deposits exerts a particularly strong influence on the distribution of damages. These studies have also corroborated that the spatial distribution of damages correlates on the magnitude of the amplifications of the arriving seismic waves caused by the upper sedimentary layers. Consequently, we can regard the distribution of the ground-shaking intensity as a phenomenon closely related to the filtering effects of the soil profile above the bedrock level [28].

It is convenient to introduce the most general and widely accepted formulation of the ground motion phenomenon, departing from assuming that in the frequency domain, the S-wave can be expressed in terms of Fourier amplitude spectra $A_{ie(f)}$ recorded at a site i due to an earthquake e as

$$A_{ie}(f) = S_e(f)\, P_{ie}(f)\, G_i(f) \tag{19}$$

where *f* denotes the frequency domain and *Se(f)* represents the source, $P_{ie}(f)$ the path, and $G_i(f)$ the site effect term (**Figure 13**). When applying logarithm to Eq. (19) yields

$$\log A_{ie}(f) = \log S_e(f) + \log P_{ie}(f) + \log G_i(f) \tag{20}$$

where the amplitude term $A_{ie(f)}$ is related to PGA or and spectral ordinate SA, the source term can be converted to $a + bM$ and the path term to $cR + d\log R$ in Eq. (5). The convolution of the source and path effects yields the input bedrock motion B_n; in the first instance the site effect term can be defined as

$$G_i(f) = \frac{B_m + A_m}{B_n} = \frac{2\,B_m}{B_n} \tag{21}$$

where B_n and A_n are the input and refracted bedrock motion, respectively, and B_m and A_m are the incident and refracted wave at the surface, respectively (**Figure 13**). Note that the free-surface effects yield an amplification of 2.0 at the outcrop engineering bedrock since $B_n = A_n$. Such formulation is useful when there are strong motion recordings (accelerograms) for both sediment and outcrop sites. In such cases some authors [29–33] have proposed several empirical techniques to evaluate the magnitude of site effects on ground motion without relying on specific soil profile information.

The estimation of the magnitude of site effects applying theoretical methods requires knowledge of the physical properties of the layers composing the soil profile, namely, shear wave velocity, density, and thickness of the layers of the upper 30 meters. In most practical cases involving thick sedimentary deposits, knowledge of the thickness and physical properties of the upper 30 meters of the soil profile appears to provide a rather poor representation of the problem.

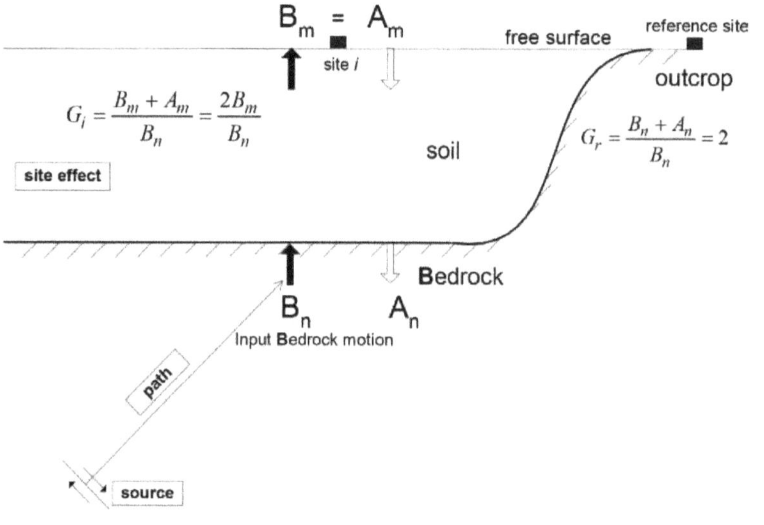

Figure 13. Illustration of theoretical transfer functions at soil sites and outcrop site (reference site) taking into account the free-surface amplification. B_n and A_n: input and refracted bedrock motion, respectively. B_m and A_m: incident and refracted wave at the surface, respectively.

The amplification function employing 1-D SH wave propagation can be obtained when solving the following partial equation of wave motion:

$$\frac{\partial^2 u}{\partial t^2} = v_s^2 \frac{\partial^2 u}{\partial v_v^2} \tag{22}$$

where u is the horizontal displacement, v_s is the shear wave velocity, t is time, and v is the vertical direction [8].

A theoretical 1D SH amplification function is derived solving Eqs. (21) and (22) for a soil profile volcanic environment at the capital of El Salvador, San Salvador. The work of Salazar and Seo [34] is revisited analyzing a volcanic ash and rigid lava flows' interbedding patterns

Figure 14. Interbedding of lava flow and volcanic ashes in San Salvador City (see location in **Figure 1a**).

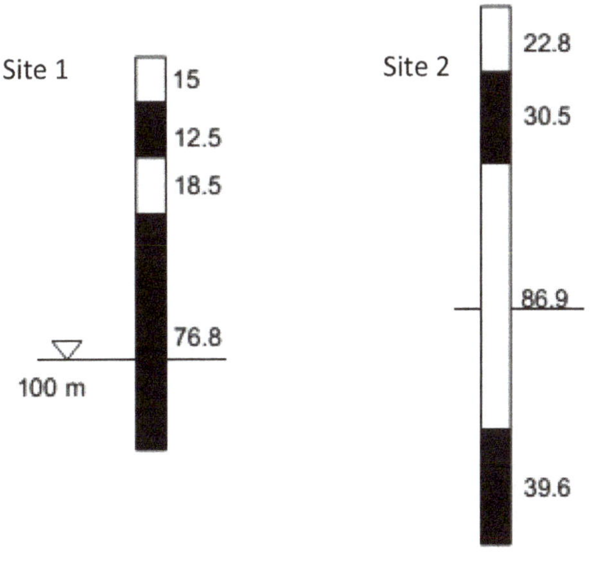

Figure 15. Profiles of lava flow and volcanic ashes interbedding in San Salvador City; see location in **Figure 1a** (after Salazar & Seo (2002)). The value of thickness of the volcanic ash (white part) and the lava flow (black part) is written to the right of the profiles in meters.

(**Figures 14** and **15**). A value of $v_s = 225$ m/s is used for the volcanic ashes in the upper part (*Tierra Blanca*) and a value of $v_s = 500$ m/s for the soils under laying the stiff lava flow (consolidated tuffs). In the case of the stiff lava flow, we used a $v_s = 2100$ m/s. The interbedding of soft and rigid materials is the consequence of different volcanic eruptions through historic times. The results of the 1D amplification functions are presented in **Figure 16**.

For the site 1, the highest amplification is observed at the second mode of vibration, and it does not take place at the first mode as can be usually expected, so in this case, the predominant period is the second mode of soil vibration at 0.4 s rather that the first one. For the site 2, similar amplification factors of about 10–12 are observed at the first mode and the subsequent

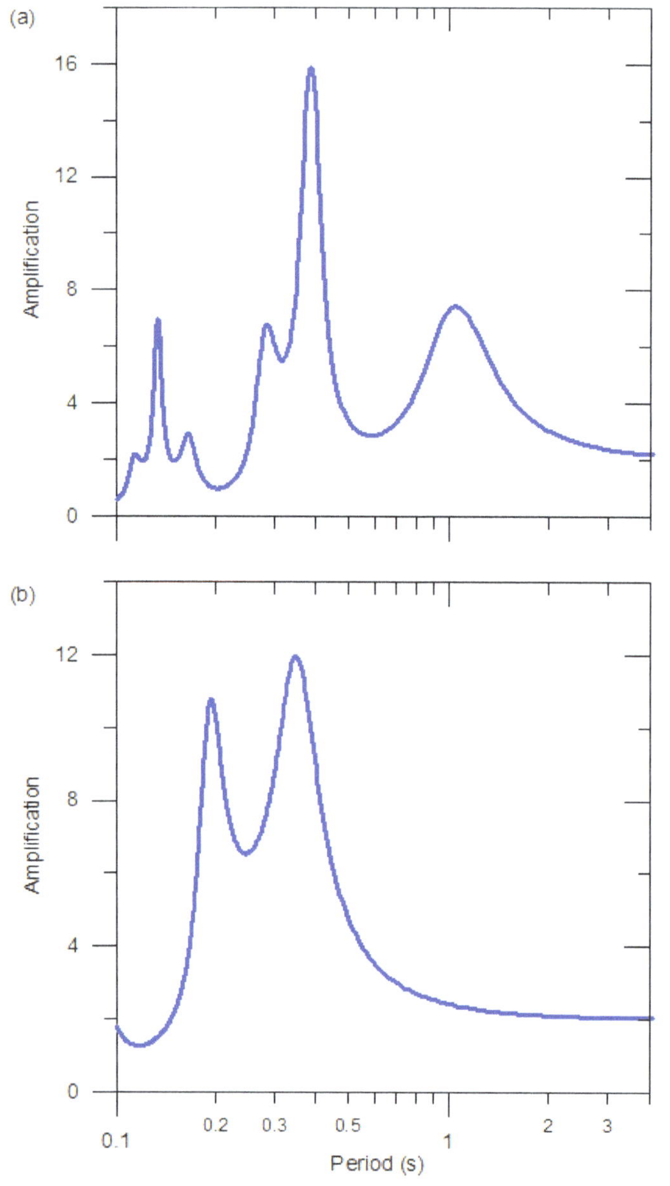

Figure 16. 1D SH wave transfer functions for interbedding of lava flow and volcanic ashes in San Salvador City; see location in **Figure 1a**. (a) Site 1 and (b) Site 2 in **Figure 15**. Note that the second mode is predominant at 0.4 s rather than the first one for the Site 1 and that similar amplification factors of about 10–12 are obtained for the first and second modes for Site 2.

overtone. Note that for long period components above 2.0 s the amplification factors yield a value of 2.0 confirming that the earthquake motion at these periods yields no amplification. These amplification features might have serious implications in the distribution of damages since several modes can be highly excited during an earthquake affecting medium- and low-rise buildings in San Salvador City. Moreover, if the building foundations are being constructed on the first lava flow encountered as a recommendation from a standard penetration test (SPT), the site would be characterized as a rock yielding a lower seismic coefficient despite that volcanic ashes might underlay the lava several tens of meters.

4. Summary

To finalize this book chapter, the most important concepts are summarized below when performing a PSHA:

i. The necessary input data to start a PSHA is the conformation of a homogeneous earthquake catalog in terms of moment magnitude and a strong motion data base. The earthquake catalog serves to compute the classical G-R relationships that depend on the completeness analysis of the catalog itself for a specific seismogenic source when applying the rigid-zone method. New methods are oriented to capture the uncertainty of the seismicity rate in the G-R relationships [35]. The free-zone methods test directly the earthquake catalog itself containing in the case of the Woo method, the foreshocks, the main events, and the aftershocks, and its application also depends of the analysis of completeness of the catalog. On the other hand, the strong motion data base serves to conform ground motion prediction equations (GMPEs) for both, the peak ground acceleration and spectral ordinates, or to test the applicability of another GMPE derived for similar tectonic environments.

ii. In all seismic hazard methods, the first objective is to compute the number of earthquakes per year that a single site can expect during the lifetime of the structure. The combination of a lifetime and the probability of exceedance yield the return period (in years) that is employed to establish a design seismic load of the building. Then, the return period, the recurrence interval, and the lifetime of the structure are different concepts that connect each other through probability concepts [7]; however, it is a must to employ GMPEs to estimate the seismic coefficient prescribed in the building code regulations to take into account all the magnitude-distance pairs that can affect the project site. Actual research frontiers for state of the art in PSHA suggest to test the hazard curves against record strong motions and Mercalli intensity observations [36, 37].

iii. The rigid-zone methods have the advantage to assessing a seismicity level when there is a geological evidence that shocks can occur within specific area; however, one must have caution to spread such seismicity in arbitrary manner. Instead, the free-zone methods are based solely on the quality of the earthquake catalogs, and they do not depend on a priori geological information neither of a particular seismogenic source delimitation.

iv. The characteristic models are usually applied when there is clear information about geological faults, namely, the slip rate (mm/years), the geographical trace coordinates, the dip, and the lock depth of the fault. However, in the Woo free-zone method, it is possible to establish a background seismicity setting in a certain area, a recurrence interval for the maximum credible magnitude that could be directionally dependent applying anisotropy features, which is indeed an equivalent approach of the characteristic model.

v. A robust seismic hazard map must be the result of a rigorous computation employing a logic-tree formulation based on all methods explained in this chapter and the combination of several GMPEs [23, 38–40], possible maximum magnitudes and slip rates. Besides, it is advisable to check the covariance or the associated error or dispersion of the final results [17, 24, 25]. When employing a logic tree, the source-to site distance definitions and the ways to use the two horizontal components of motion in the GMPEs must be compatible among them [41–45]. The effect of truncation of ground motion distributions at a specific number above the median (i.e., 3σ) might have a substantial influence in the results when related to a very low probability of exceedance [46, 47].

vi. The PSHA is generally performed at rock site conditions (NEHRP B soil class with a Vs = 760 m/s) and a flat topography. However, the final response spectrum must take into account the site effects through the amplification of ground motion due to the presence of sediments above the bedrock; however, the topographic effects might be also influence the final motion for design at highland sites. The 1D SH wave amplification function (or transfer function) can be obtained via theoretical methods based on shear wave velocity test at the sites or employing nondestructive geophysical methods as the microtremor arrays and refraction methods.

vii. When studying the amplification of ground motion, the compilation of a strong motion data base [48] and the mobile microtremor measurement surveys [29] can be combined to validate the transfer function obtained via inversion analysis of earthquake data. As a result, the elastic response spectra calculated in the PSHA at rock site condition can be amplified using suitable amplification factors for the final design of a building [49].

viii. Before applying any method for PSHA or site effect computation, the inherent geological characteristics of the earthquakes and the soil profiles in the region of study must be checked carefully. We have demonstrated in this chapter that the volcanic chain earthquakes in El Salvador and the intercalation of rigid lava flows and volcanic ashes yield special patterns of seismicity [50, 51] and vibration of soil deposits, respectively, that must be included in the future seismic code regulations and design foundation guidelines for the country.

Acknowledgements

I firstly thank editorial board for inviting me to write this book chapter and to be a part of this important global educational project. I openly thank the book editor Valentina Svalova for the detailed review that helps to improve the manuscript. This article was sponsored by the Catholic University of El Salvador UNICAES in terms of research grants for international publishing. I thank Gordon Woo from RMS London who allows me to use his Fortran code KERFRACT. I also thank the assistance of Stephen Harmsen at the USGS who guides me to use the USGS Fortran codes available at GitHub: https://github.com/usgs/nshmp-haz-fortran. Vanessa Rivera at UNICAES helped to correct the manuscript.

Author details

Walter Salazar

Address all correspondence to: walter.salazar@catolica.edu.sv

Catholic University of El Salvador UNICAES, Santa Ana, El Salvador

References

[1] Cornell CA. Engineering seismic risk analysis. Bulletin of the Seismological Society of America. 1968;**58**:1583-1606

[2] McGuire RK. FORTRAN computer program for seismic risk analysis, U.S. Geological Survey; Open-File Report; 1975. pp. 76-67

[3] Gardner JK, Knopoff L. Is the sequence of earthquakes in Southern California, with aftershocks removed, Poissonian? Bulletin of the Seismological Society of America. 1974;**64**:1363-1367

[4] McGuire R, Arabasz J. An introduction to probabilistic seismic hazard analysis. In: Ward SH, editor. Geotechnical and Environmental Geophysics. Vol. 1. Oklahoma: Society of Exploration Geophysicist Tulsa; 1990. pp. 333-353

[5] Stepp J. Analysis of completeness of the earthquake catalogue sample in the Puget sound area. In: Harding ST, editor. Seismic Zoning, National Oceanic and Atmospheric Administration. US Department of Commerce. Technical Report ERL 267-ESL30. Boulder; 1973

[6] Tinti S, Mulargia F. Completeness analysis of a seismic catalogue. Annali di Geofisica. 1985;**3**:407-414

[7] Bommer J, Abrahamson N. Why do modern probabilistic seismic hazard analysis often lead to increased seismic hazard estimates? Bulletin of the Seismological Society of America. 2006;**96**:1967-1977

[8] Kramer S. Geotechnical Earthquake Engineering. Upper Saddle River, New Jersey: Prentice-Hall; 1996. pp. 653 ISBN: 0-13-374943-6

[9] Woo G. Kernel estimation methods for seismic hazard area source modelling. Bulletin of the Seismological Society of America. 1996;**86**(2):353-362

[10] Getsenberg M, Fry B. Preface to the focus section on the joint Japan-Taiwan-New Zealand National seismic hazard collaboration. Seismological Research Letters. 2016; **87**(6):1236-1239

[11] Frankel A. Mapping seismic hazard maps in the Central and Eastern United States. Seismological Research Letters. 1995;**66**(4):8-21

[12] Petersen M, Frankel A, Harmsen S, Mueller C, Haller K, Wheeler R, Weson R, Zeng Y, Boyd O, Perkins D, et al. Documentation for the 2008 update of the United States National Seismic Hazard Maps. U.S. Geological Survey. Open File Report 2014-1091; 2008. p. 243

[13] Petersen M, Moschetti M, Powers P, Mueller C, Haller K, Frankel A, Zeng Y, Razaeian S, Harmsen S, Boyd O, et al. Documentation for the 2014 update of the United States National Seismic Hazard Maps. U.S. Geological Survey. Open File Report 2008-1128; 2014. p. 61

[14] Petersen M, Harmsen S, Jaiswal K, Rukstales K, Luco N, Haller K, Mueller C, Shumway A. Seismic hazard, risk, and design for South America. Bulletin of the Seismological Society of America. 2018;**108**(2):781-800. DOI: 10.1785/0120170002

[15] Harmsen S. USGS Software for Probabilistic Seismic Hazard Analysis (PSHA) Draft Document, United States Geological Survey; 2010

[16] Salazar W, Seo K. Earthquake disasters of January 13th and February 13th 2001, El Salvador. Seismological Research Letters. 2003;**74**(4):420-439

[17] Salazar W, Brown L, Hernández W, Guerra J. An earthquake catalogue for El Salvador and neighboring Central American countries (1528-2009) and its implication in the seismic Hazard. Journal of Civil Engineering and Architecture. 2013;**7**(8):1018-1045 (Serial No. 69)

[18] Sadigh K, Chang C-Y, Egan JA, Makdisi F, Youngs RR. Attenuation relationships for shallow crustal earthquakes based on California strong motion data. Seismological Research Letters. 1997;**68**(1):180-189. ISSN: 1934-7359

[19] Bommer J, McQueen C, Salazar W, Scott S, Woo G. A case of study of the spatial distribution of seismic hazard (El Salvador). Natural Hazards. 1998;**18**:145-166

[20] Carr M, Feigenson M, Patino L, Walker J. Volcanism and geochemistry in Central America: Progress and problems. In: Eiler J, editor. Inside the Subduction Factory. Washington, DC: American Geophysical Union; 2013. DOI: 10.1029/138GM09

[21] Canora C, Martínez-Díaz J, Insua-Arévalo J, Alvarez-Gómez J, Villamor P, Alonso-Henar J, Capote-Villar R. The 1719 El Salvador earthquake: An M > 7.0 event in the Central American volcanic arc? Seismological Research Letters. 2014;**85**(4):784-793

[22] Coppersmith K, Youngs R. Capturing uncertainty in probabilistic seismic hazard assessments within intraplate tectonic environments. In: Proceeding of the Third U.S. National Conference on Earthquake Engineering. Vol. 1; 1986. pp. 301-312

[23] Bommer J, Scherbaum F, Bungum H, Cotton F, Sabetta F, Abrahamson N. On the use of logic trees for ground-motion prediction equations in seismic-hazard analysis. Bulletin of the Seismological Society of America. 2005;**95**(2):377-389

[24] Cramer C, Wheeler R, Mueller C. Uncertainty analysis for seismic hazard in the Southern Illinois Basin. Seismological Research Letters. 2002;**73**:792-805

[25] Salazar W, Brown L, Mannette G. Probabilistic seismic hazard assessment for Jamaica. Journal of Civil Engineering and Architecture. 2013;**7**(9):1118-1140. ISSN 1934-7359. (Serial No. 70)

[26] Marzocchi W, Jordan T. A unified probabilistic framework for seismic hazard analysis. Bulletin of the Seismological Society of America. 2017;**107**(6):2738-2744

[27] Chapman M. A probabilistic approach to ground-motion selection for engineering design. Bulletin of the Seismological Society of America. 1995;**85**(3):937-942

[28] Salazar W, Sardina V, Cortina J. A hybrid inversion technique for the evaluation of source, path and site effects employing S-wave spectra for subduction and upper-crustal earthquakes in El Salvador. Bulletin of the Seismological Society of America. 2007;**97**(1B):208-221

[29] Salazar W, Mannete G, Reddock K, Ash C. High resolution grid of H/V spectral ratios and spatial variability on microtremors at port of Spain, Trinidad. Journal of Seismology. 2017;**21**:1541-1557. DOI: 10.1007/s 10950-017-9681-1

[30] Nogoshi M, Igarashi T. On the amplitude characteristics of microtremors (part I). Journal of the Seismic Society of Japan. 1970;**23**:281-303

[31] Rodríguez V, Midorikawa S. Applicability of the H/V spectral ratio of microtremors in assessing site effects on seismic motion. Earthquake Engineering and Structural Dynamics. 2002;**31**:261-279

[32] Borchedt R. Effects of local geology on ground motion in San Francisco Bay. Bulletin of the Seismological Society of America. 1970;**60**:29-61

[33] Nakamura Y. A method for dynamic characteristics estimation of subsurface using microtremor on the ground surface. Quarterly Report of RTRI. 1989;**30**(1):25-33

[34] Salazar W, Seo K. Spectral and amplification characteristics in San Salvador City (El Salvador) for upper-crustal and subduction earthquakes. 11th Japan Earthquake Engineering Symposium, P065; 2002. pp. 329-334

[35] Stromeyer D, Grüntal G. Capturing the uncertainty of seismic activity rates in probabilistic seismic-hazard analysis. Bulletin of the Seismological Society of America. 2015;**105**:580-589

[36] Anderson J, Biasi G. What is the basic assumption for probabilistic seismic hazard assessment? Seismological Research Letters. 2016;**81**(2A):323-326

[37] Wyss M. Testing the basic assumption for probabilistic seismic-hazard assessment: 11 failures. Seismological Research Letters. 2015;**86**(5):1405-1411

[38] Isuck A, Bjerrum L, Kamchybekov M, Abdrakhmatov K, Lindholm C. Probabilistic seismic hazard assessment for the area of Kyrgyzstan, Tajikistan, and eastern Uzbekistan, Central Asia. Bulletin of the Seismological Society of America. 2018;**108**:130-144

[39] Gülerce Z, Vakilinezhad M. Effect of seismic source model parameters on the probabilistic seismic-hazard assessment results: a case study for the North Anatolian fault zone. Bulletin of the Seismological Society of America. 2015;**105**:2808-2822

[40] Haas M, Agnon A, Bindi D, Parolai S, Pittore M. Data-driven seismic-hazard models prepared for a seismic risk assessment in the Dead Sea region. Bulletin of the Seismological Society of America. 2016;**106**:2584-2598

[41] Boore D, Watson-Lamprey J, Abrahamson N. Orientation-independent measures of ground motion. Bulletin of the Seismological Society of America. 2006;**96**:1502-1511

[42] Haji-Soltani A, Pezeshk S. Relationships among various definitions of horizontal spectra accelerations in Central and Eastern North America. Bulletin of the Seismological Society of America. 2018;**108**:409-417

[43] Tavakoli B, Sedaghati F, Pezeshk S. An analytical effective point-source-based distance-conversion approach to mimic the effects of extended faults on seismic hazard assessment. Bulletin of the Seismological Society of America. 2018. DOI: 10.1785/0120170171

[44] Monelli D, Pagaini M, Weatherwill G, Danciu L, García J. Modeling distributed seismicity for probabilistic seismic-hazard analysis: Implementation and insights with the OpenQuake Engine. Bulletin of the Seismological Society of America. 2014;**104**:1636-1649

[45] Atik L, Abrahamson N, Bommer J, Scherbaum F, Cotton F, Kuehn N. The variability of ground-motion prediction models and its components. Seismological Research Letters, Seismological Society of America. 2010;**81**(5):794-801

[46] Strasser F, Abrahamson N, Bommer J. Sigma: Issues, insights, and challenges. Seismological Research Letters, Seismological Society of America. 2009;**80**(1):40-56

[47] Bommer J, Abrahamson N, Strasser F, Pecker A, Bard P-Y, Bungum H, Cotton F, Fäh D, Sabetta F, Scherbaum F, Studer J. The challenge of defining upper bounds on earthquake ground motions. Seismological Research Letters, Seismological Society of America. 2004;**75**(1):82-95

[48] Salazar W, Torres R, Mixco L, Rivera V, Burgos E, Rivas J, Brown L. Strong motion networks, digital signal processing and data base for El Salvador earthquakes: 1966-2017. Seismological Research Letters. 2018;**89**(2A):483-494. DOI: 10.1785/0220170198

[49] Haji-Soltani A, Pezeshk S. A comparison of different approaches to incorporate site effects into PSHA: A case study for a liquefied natural gas tank. Bulletin of the Seismological Society of America. 2017;**107**:2927-2947

[50] Ansari A, Firuzi E, Etemadsaeed L. Delineation of seismic sources in probabilistic seismic-hazard analysis using fuzzy cluster analysis and Monte Carlo simulation. Bulletin of the Seismological Society of America. 2015;**105**:2174-2191

[51] Gerstenberg M, Rhoades D, McVerry G. A hybrid time-dependent probabilistic seismic-hazard model for Canterbury, New Zealand. Seismological Research Letters. 2016;**87**(6):1311-1318

11

Structure-Soil-Structure Interaction between Underground Structure and Surface Structure

Huai-feng Wang

Abstract

A numerical study is made on the dynamic through-soil interaction between underground station and nearby pile supported surface structure on viscous-elastic soil layer, under vertically incident S wave. This paper focuses on the dynamic interaction and the interactive influence on seismic response of adjacent surface structure and underground structure. To this end, ANSYS has been further developed for calculation in frequency domain, in which hysteretic damping can be considered, so that structure-soil-structure interaction (SSSI) can be investigated via direct methodology. Discussion is made on the influence of arrangement of structures, distances between structures, shaking direction of seismic wave, shear wave velocity and damping of soil, scale and burial depth of underground structure, storey number, stiffness, style and pile length of surface structure on SSSI, in terms of horizontal acceleration of surface structure and horizontal relative displacement of underground structure. Maximum acceleration and displacement is also presented for 12 seismic inputs. Arrangement and shaking direction are two of the most important factors. The response can be either amplified or attenuated according to the distance, related to dynamic properties of the overall system. The underground structure, surrounded by building with the fundamental frequency approximate to that of free field, is heavily affected.

Keywords: structure-soil-structure interaction (SSSI), soil-structure interaction (SSI), underground structure, surface structure, seismic response

1. Introduction

With the increasing narrowing of urban ground development space, plentiful underground space structures have been constructed under the building groups and arterial roads in urban bustling business districts. Underground structures, such as subway station and underground plaza, have become increasingly common in urban construction, and they are often close to the surrounding high-rise buildings. In the metropolitans, like Kobe in Japan, structures, such as buildings, stations and tunnels, are built closely to each other on the soft soil deposit. Under such circumstances, the dynamic interaction among structures is bound to occur through the radiation energy emitted from a vibrating structure to other structures. Hence, the dynamical characteristics and earthquake response characteristics of a structure are closely related to those of the adjacent structures.

Structure-soil interaction (SSI) is an important research topic in earthquake engineering, which has attracted extensive attention for several decades. As a branch of SSI, structure-soil-structure interaction (SSSI) is a key research topic during recent years. It is an interdisciplinary field of endeavor, which lies at the intersection of soil and structural mechanics, soil and structural dynamics, earthquake engineering, geophysics and geomechanics, material science, computational and numerical methods, and diverse other technical disciplines.

Researches on SSI originate from the vibration problem between machines and foundations, which belongs to the category of linear harmonic vibration. The influence of impedance matrix of foundations and SSI on the machine operation is studied. In the 1960s, studies on anti-seismic safety of nuclear power station hastened researches on the issue of SSI under earthquake action. Owing to the development of numerical calculation theory and computer technology, especially the application of finite element method and boundary element method, effective measures have been provided for seismic analysis on complicated engineering structures considering the influence of SSI. Finite element method is used to handle the problems with irregular field, while boundary element method can process the problems with infinite field conveniently. Due to the finite element-boundary element coupling method, the solution scope of SSI is further expanded. Problems, with complicated shape, flexibility, burial depth of foundation, sheet separation between foundation and base, soil layers, local topography, nonlinear characteristic can be handled via numerical discretization method. In recent years, the issue of SSI under nonlinear analysis conditions has already become the mainstream direction of research work. When various theoretical analysis methods are further studied, model test and prototype test have also attracted increasing attention from scholars, to become the new hotspot of SSI study. With the successful outcome about SSI, various kinds of theoretical methods and experimental installations are used to promote the study of SSSI.

The history and current status of SSSI were elucidated by the author in another paper [1]. A simple introduction will be given here. To the writer's knowledge, it is Luco and Contesse [2] in 1973 to come up with the structure-soil-structure interaction designation for this area of study first. The steady-state response of two parallel infinite shear walls placed on rigid foundations for a vertically incident SH wave is obtained and compared with corresponding values resulting from consideration of only one structure. During this period, analytical method and semi-analytical numerical method were mainly used to study the issue of SSSI, and most of them were based on elastic half-space theory. Shallow foundation or surface foundation was adopted. Besides, system with single-particle or columnar block was used to simulate the surface structures, and some work gone so far as to simulate the foundation only, without considering the upper structures. Undoubtedly, these researches have laid a solid theoretical foundation for follow-up studies, but still cannot solve practical SSSI problems.

With the rapid development of computer technology, calculation theory, and calculation software, numerical simulation and analysis became the most extensive method to study SSSI. In the 1980s, many researchers began to conduct a study through numerical methods. In 1982, Fu and Yu [3] analyzed the dynamic response of two-dimensional SSSI system placed on elastic half-space by applying substructure method. By simplifying the surface structure into a multi-particle string system, Tian and Yu [4] studied the interaction of two multi-particle systems on the elastic half-space by solving the dynamic impedance function of soil via boundary element method for the first time. In addition, Jiang et al. [5] analyzed the interaction of two multi-particle string systems on homogeneous soil layer via finite element method. According to the results, when the distance between structures is greater than the scope of 2.5 times the structure width, the influence of SSSI can be ignored.

Later, the simulation for surface structures became elaborate. Dou and Yang [6] studied the adjacent 6-storey and 21-storey three-span two-dimensional frame structures by applying finite element method. Padron et al. [7] and Alamo et al. [8] utilized finite element-boundary element coupling method to study the influence of SSSI on transversal, vertical and torsional deformation and shearing force of several 1-storey frames under S wave and Rayleigh wave in frequency domain and time domain. In addition, Fariborz and Ali [9] analyzed the seismic response of contiguous 15-storey and 30-storey steel-frame structures. In order to obtain the dynamic response of two adjacent 12-storey frame-shear wall structures under earthquake action, Li et al. [10] conducted three-dimensional finite element model to analyze adjacent high-rise structures-pile-soil interaction system. Alam and Kim [11] explored the seismic response of two adjacent 3-storey frames considering the spatial effect of seismic oscillation. Moreover, Ghandil et al. [12] studied the applicability of equivalent linearization in SSSI system by simulating soil nonlinearity via equivalent linearization.

The lumped parameter method is a common method used to analyze SSI and SSSI, in which soil is simulated by spring, mass, and damper, or an equivalent impedance function [13]. In 1998, Mulliken et al. [14] firstly made use of this method to present efficient discrete models with frequency-independent masses, springs, and dampers. Each model has modes of vibration considered independent degree of freedom for predicting the dynamic interaction between adjacent rigid surface foundations, which are supported by a homogeneous, isotropic, and linear elastic half-space. This finding is achieved through a proposed modification of the Wilson-θ method; thus, the time-lagging effects due to wave propagation are also considered. Besides, the basic foundation interaction model is extended to the evaluation of coupled building-foundation systems. Moreover, Alexander et al. [15] applied this method to study SSSI effect. After that, Naserkhaki and Pourmohammad [16] probed into the work of Mulliken et al. [14].

Tests are important means for scientists and engineers to improve humans' knowledge about the nature law. In 1980, Mizuno [17] firstly clarified actual phenomena of SSSI by a series of tests such as forced vibration tests, micro tremor or measurements and earthquake observations for a full-scale building and a model structure. In recent years, more tests have been conducted on civil structures. Yin [18] and Kang [19] analyzed the dynamic interaction of underground structure and surface structure through shaking table test and numerical analysis. Besides, Pan et al. [20] conducted a forced vibration tests on two steel frames on the shear soil box. According to the results, in SSSI system, modes of vibration with approximate two frequencies and opposite phase exist, and the frequency of mode corresponding to SSI system is between the two frequencies. In addition, Trombetta et al. [21] conducted a centrifuge test on the 1-storey steel frame and 2-storey shear wall model, which shows that the SSSI effect is relatively obvious when the seismic excitation is low. By simplifying the surface structure into a single-particle system, Xiong et al. [22] conducted a shaking table test on five adjacent single-particle systems on the shear soil box, but the test results are different from the results of Pan et al. [20] and Trombetta et al. [21] in the aspects of dynamic characteristics and seismic response respectively, showing the complexity of SSSI. By simulating the soil with foam and rubber respectively, Aldaikh et al. [23] and Cacciola et al. [24] conducted shaking table tests on several adjacent single-particle systems. Owing to the limitation of test equipment, especially shaking table test, multi-storey structures are often investigated, and soil layer with the depth of 100 m and even several hundred meters cannot be simulated.

Obviously, less attention is paid to the problem of the interaction of adjacent surface structure and underground structure through the underlying and surrounding soil. Many researchers have done a lot of work about dynamic response of underground structure, diffraction and scatter of wave, and effect of seismic field by underground structure, but they have ignored adjacent structure and interaction between them, such as Lee [25, 26], Lin [27], Hatzigeorgiou [28], Abate and Massimino [29]. Guo and Chen [30] studied the influence of subway station on the dynamic response of a nearby 6-storey frame structure by adopting the two-dimensional model. Meanwhile, Yang et al. [31], Li and Zhou [32], and Farghaly [33] explored the influence of subway tunnel on the dynamic response of a nearby frame structure with more than 10 storeys. Abate and Massimino [34–36] investigated the effects of the tunnel on the response of the soil and/or of the building and vice versa. Via full-coupled FEM modeling, a cross-section of the underground network in Catania including an aboveground building was modeled to study its behavior during the expected scenario earthquake (MS = 7.0–7.4).

Indeed, more work needs to be done in order to assess the effects of interaction between surface structure and underground structure on general three-dimensional problems with different structural and subsoil configurations, and its associated risks.

In the following sections, the problem is first stated and the parameters and properties are defined. Then, the calculation method and model are briefly outlined. Afterwards, a set of numerical results, in the frequency domain, is presented to assess the effects of building on the seismic response of underground station in terms of its horizontal relative displacement and the effects of underground station on the seismic response of building in terms of its magnification factor of horizontal acceleration. Maximum relative displacement and acceleration responses are also presented for 12 seismic inputs.

2. Problem definition

The system under investigation is composed of one or more common multi-storey frame structures (GS), which is founded on fixed-head pile groups embedded into homogeneous soil layer, and a neighboring typical underground subway station (US), three-dimensionally distributed.

Note that **Figures 1** and **2** are two-dimensional representations of the three-dimensional model used herein. The frame construction plane is shown in **Figure 1** and the cross-section of subway station is presented in **Figure 2**. The mechanical and geometrical properties of surface structure are defined by the following parameters: storey height 3.6 m; number of storey $N_F = 10$; cross section of frame column 600 × 600 mm; cross section of frame beam 250 × 600 mm; cross section of foundation beam 300 × 800 mm; diameter of pile 600 mm; length of pile $H_{pile} = 18m$; thickness of floor slab 150 mm; thickness of base plate 300 mm; concrete modulus 3 × 10^4 MPa; Poisson ratio of concrete 0.2; damping ratio $\zeta_{GS} = 0.05$. The fundamental period of surface structure is 0.71 s along lateral axis, and 0.67 s along longitudinal axis. The mechanical and geometrical properties of underground structure are defined by the following parameters: burial depth of underground station $H_{bd} = 2m$; length of underground station 72 m; concrete modulus of underground structure 3.45 × 10^4 MPa; Poisson ratio of concrete 0.2; damping ratio $\zeta_{US} = 0.05$.

On the other hand, the most important parameters to define the soil dynamic behavior are as follows: mass density $1900kg/m^3$; shear wave velocity $c_s = 300m/s$; soil height $H = 60m$; damping ratio $\zeta = 0.05$; Poisson ratio 0.3.

An analysis is made on the dynamic behavior of several problems (of which the different relative arrangements of structures are shown in **Figure 3**, termed as "AR0"–"AR10") under vertically incident S waves (producing motions on the X axis, termed as "SX", or Y axis, termed as "SY") from rigid bedrock. The response of underground structure and surface structure in the group is compared with that of single-underground structure-soil system or single-surface structure-soil system (termed as "AR0") respectively, in order to find out whether or not SSSI effects between two or more structures can be of importance. Note that in all configurations the distance D between adjacent structures is measured in parallel to X axis and Y axis, and is the same between all structures in the same problem.

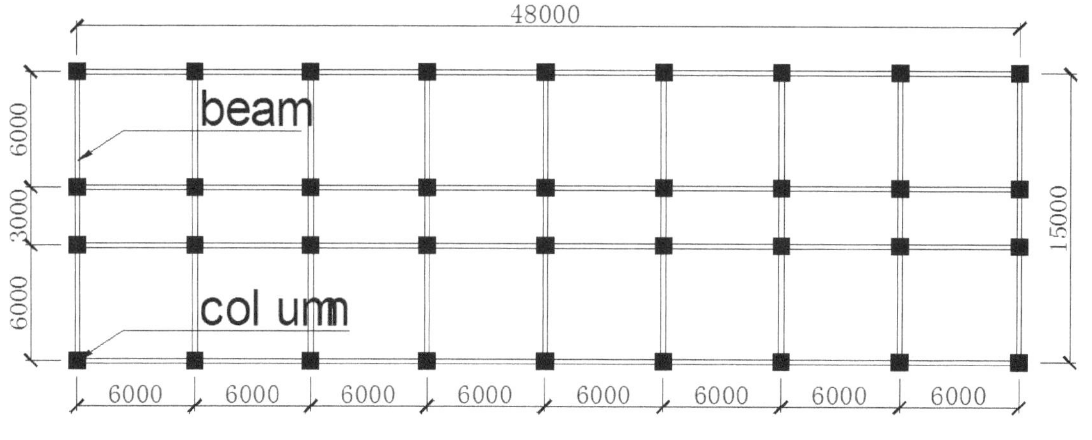

Figure 1. Typical plan of surface structure (unit: Mm).

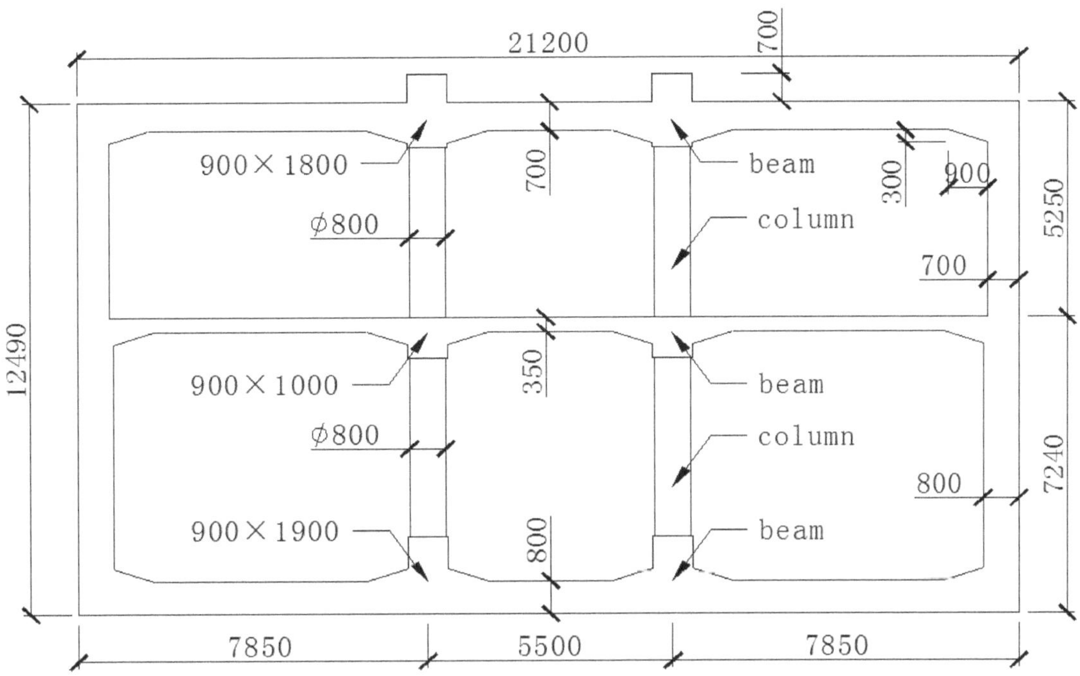

Figure 2. Cross-section of underground station (unit: Mm).

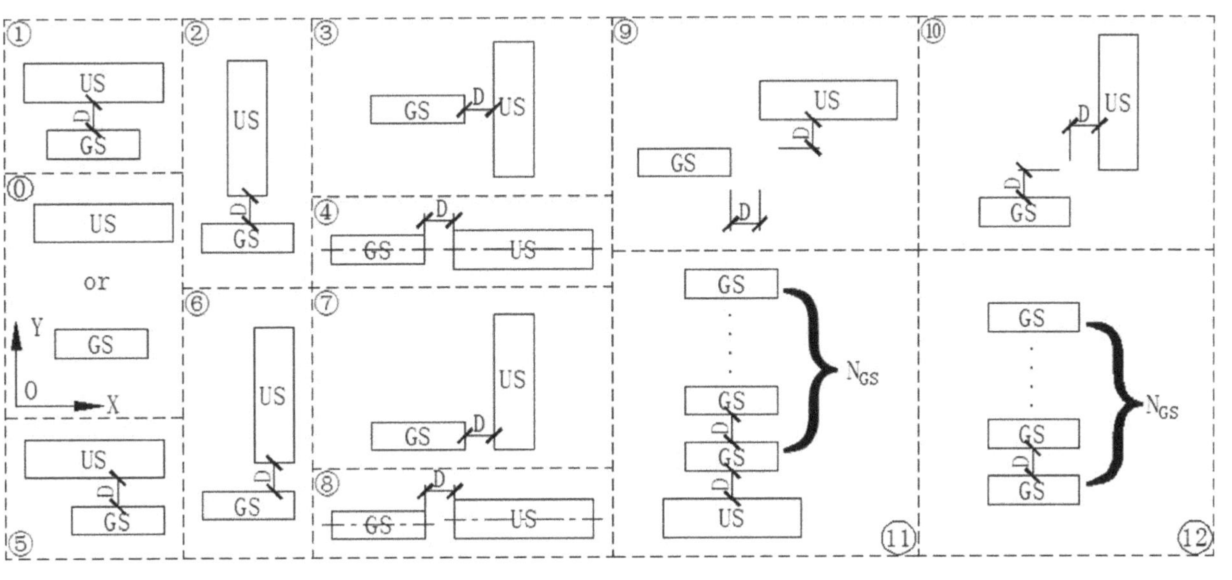

Figure 3. Relative arrangements of structures.

3. Finite element model

A commercial software product for finite element analysis, ANSYS, has been further developed and enhanced for calculation in frequency domain, in which hysteretic damping can been considered for both the soil and the structures by defining a complex soil shear modulus $\mu = \mathrm{Re}[\mu](1+2i\zeta)$ and a complex structural stiffness $k = \mathrm{Re}[k](1+2i\zeta_{GS})$ or $k = \mathrm{Re}[k](1+2i\zeta_{US})$.

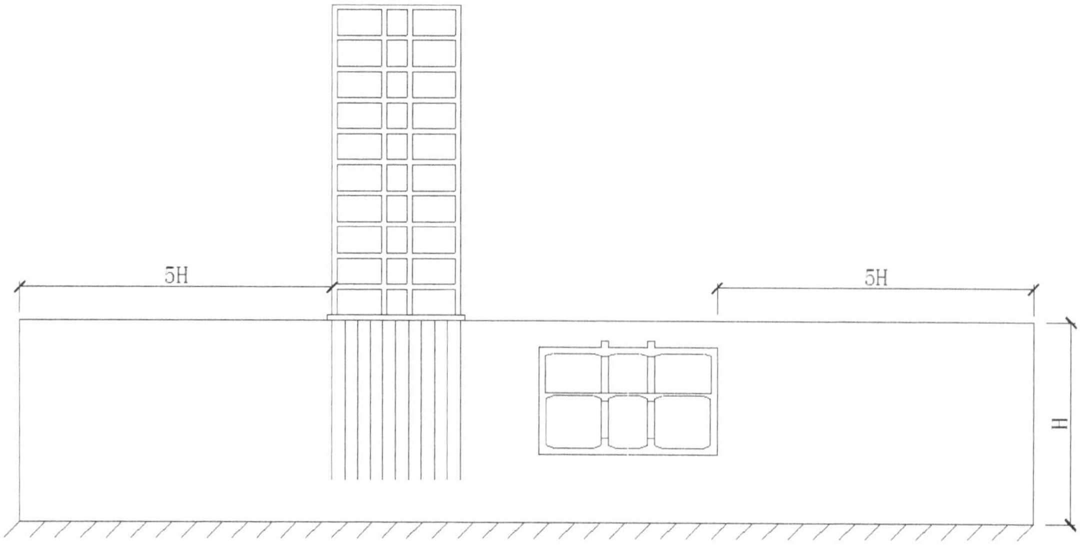

Figure 4. Arrangement of structures and soil (AR1).

The soil, considered as a consecutive, isotropic, homogeneous, linear, viscous-elastic medium, is modeled by solid elements (SOLID45). Fully bonded contact conditions are assumed between the soil and the underground station, and between soil and piles. Beams, and columns of the structures are modeled as Bernoulli beams via beam elements (beam188) and floor slabs and walls are modeled by shell element (shell63). Piles, whose heads can be fixedly connected to the base plate of surface structure, are modeled as vertical Bernoulli beams via pipe elements (pipe16).

Fixed constrains are applied at the bottom of the soil to simulate rigid bedrock where the seismic waves come. In view of the radiation damping of semi-infinite space, the scope of soil is set large enough, extending $5H$ long from structure scope (**Figure 4**). And in view of the transmission of energy in soil, the finite element mesh is set small enough, limited to $1/8\lambda_{min}$, being $\lambda_{min} = c_s/f_{max}$ the minimum premeditated wavelength. Here $f_{max} = 20Hz$ is the maximum premeditated frequency of seismic wave.

The seismic inputs include harmonic wave, whose amplitude is 1 and frequency ranges from 0 to 12.5 Hz, and nine ground motion records from bedrock and three artificial seismic waves with different exceedance probabilities.

4. Numerical results

The influence of SSSI on the dynamic response of subway station is expounded in this section. Here, an analysis is made on the horizontal seismic behavior of underground structure in terms of its dimensionless displacement u, which is defined as $\left|\frac{\Delta u_{US}}{T^2 a_{bedrock}}\right|$, where $\Delta u_{US} = u_{top} - u_{bottom}$ is the relative displacement, paralleling to seismic excitation orientation, of underground structure at the top and bottom, and $a_{bedrock}$ is the peak horizontal acceleration at the bedrock and $T = 4H/c_s$ is the fundamental period of free field. The horizontal seismic behavior of ground

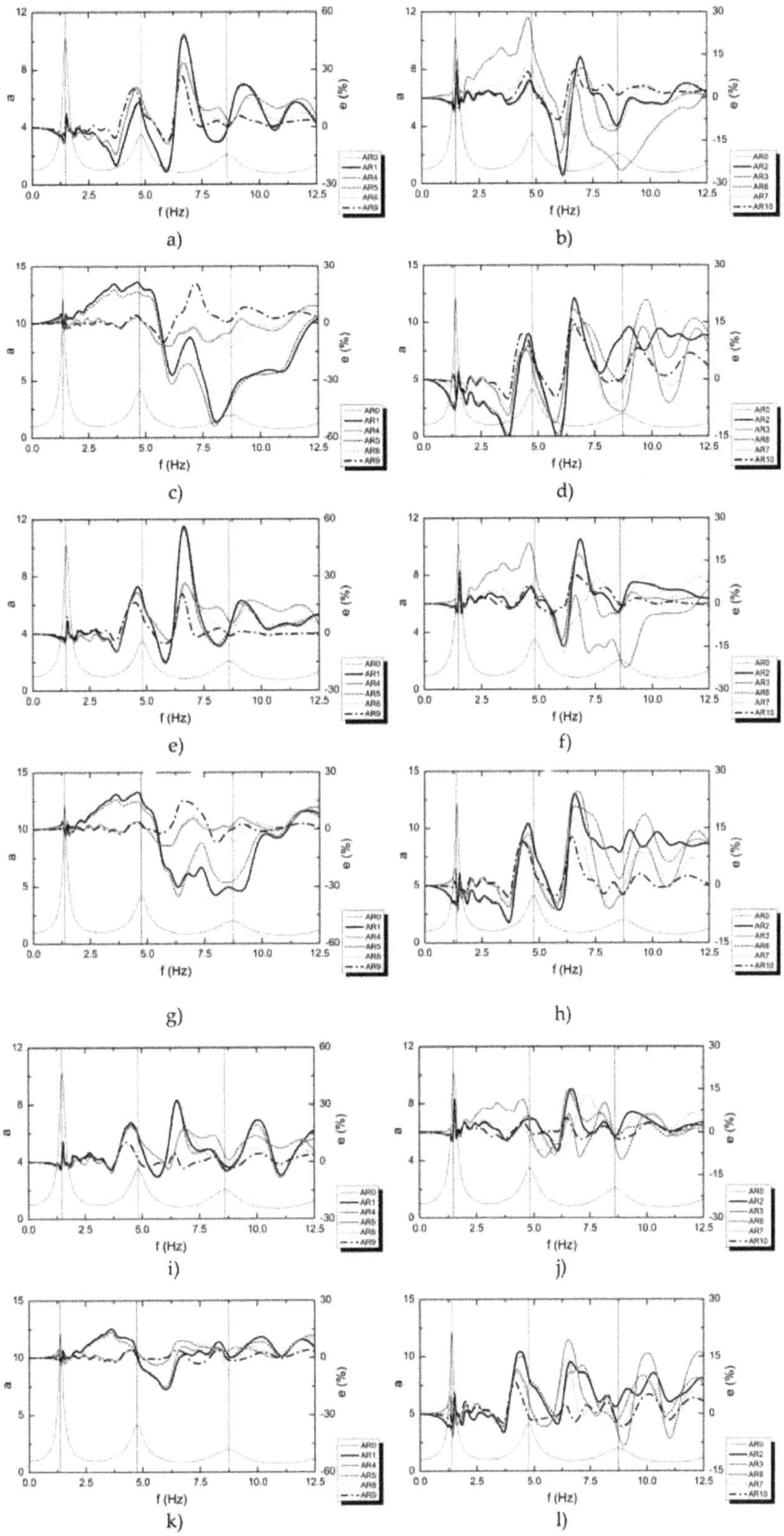

Figure 5. Interaction between surface structure and underground structure in terms of harmonic response spectra and influence coefficient *e* for different arrangements under S waves with different excitation orientation, a) AR0/1/4/5/8/9, SX and d = 0.1, b) AR0/2/3/6/7/10, SX and d = 0.1, c) AR0/1/4/5/8/9, SY and d = 0.1, d) AR0/2/3/6/7/10, SY and d = 0.1, e) AR0/1/4/5/8/9, SX and d = 0.5, f) AR0/2/3/6/7/10, SX and d = 0.5, g) AR0/1/4/5/8/9, SY and d = 0.5, h) AR0/2/3/6/7/10, SY and d = 0.5, i) AR0/1/4/5/8/9, SX and d = 1.0, j) AR0/2/3/6/7/10, SX and d = 1.0, k) AR0/1/4/5/8/9, SY and d = 1.0, l) AR0/2/3/6/7/10, SY and d = 1.0.

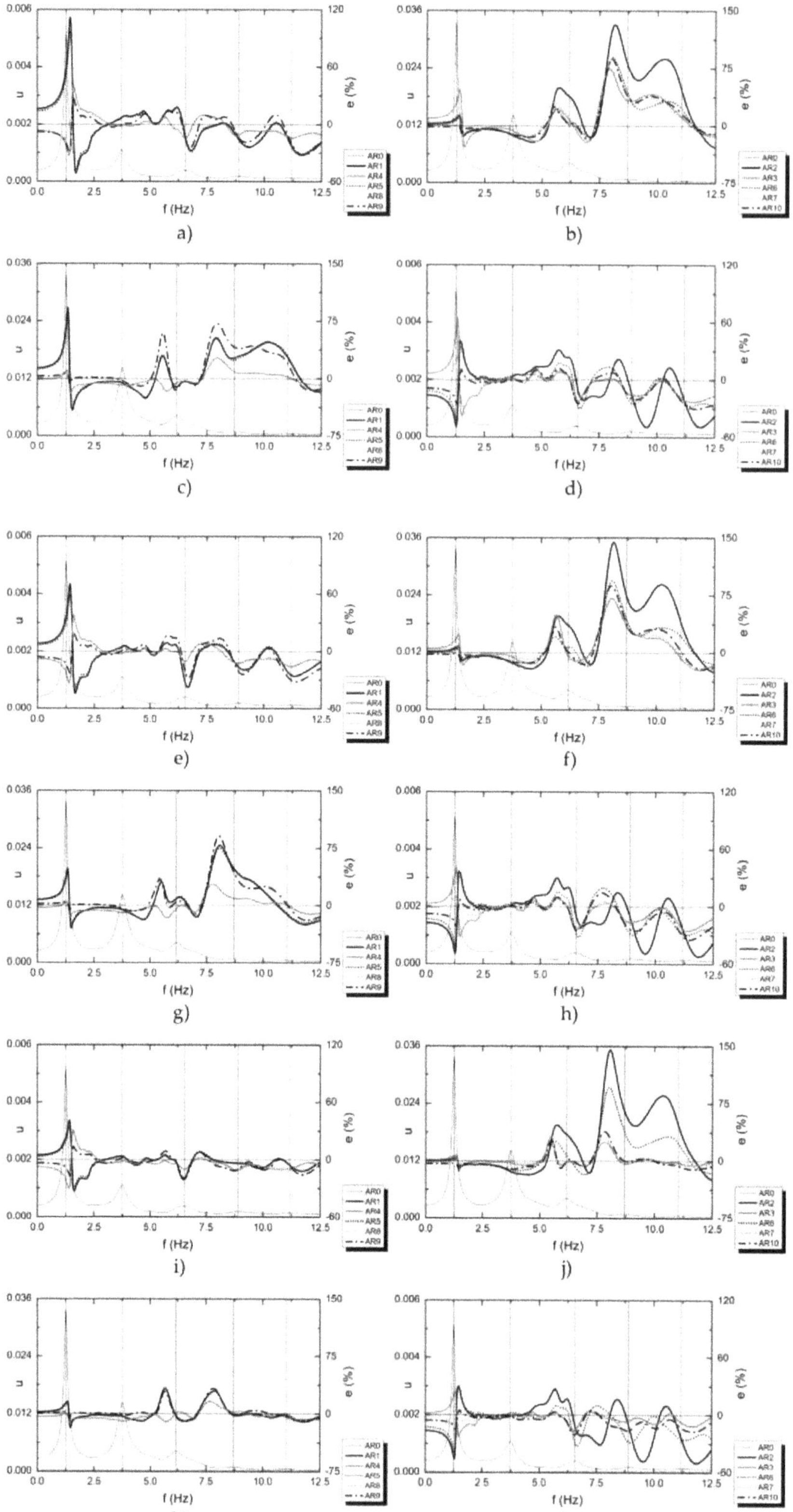

Figure 6. Interaction between surface structure and underground structure in terms of harmonic response spectra and influence coefficient e for different arrangements under S waves with different excitation orientation, a) AR0/1/4/5/8/9, SX and d = 0.1, b) AR0/2/3/6/7/10, SX and d = 0.1, c) AR0/1/4/5/8/9, SY and d = 0.1, d) AR0/2/3/6/7/10, SY and d = 0.1, e) AR0/1/4/5/8/9, SX and d = 0.5, f) AR0/2/3/6/7/10, SX and d = 0.5, g) AR0/1/4/5/8/9, SY and d = 0.5, h) AR0/2/3/6/7/10, SY and d = 0.5, i) AR0/1/4/5/8/9, SX and d = 1.0, j) AR0/2/3/6/7/10, SX and d = 1.0, k) AR0/1/4/5/8/9, SY and d = 1.0, l) AR0/2/3/6/7/10, SY and d = 1.0.

structure is analyzed in terms of its dynamic magnification factor a, defined as $|a_{GS}/a_{ff}|$, where a_{GS} and a_{ff} are respectively the horizontal acceleration, parallel to seismic excitation orientation, of surface structure and of free-field at the ground surface. The product of this value with the structural mass and the corresponding free-field horizontal acceleration at ground surface level yields the amplitude of shear force at each storey of the structure.

In order to show the SSSI effect explicitly, the influence of surface structure on underground structure is illustrated with influence coefficient e, defined as $(u_{AR1-11} - u_{AR0})/u_{AR0}$, where u_{AR0-11} is the aforementioned dimensionless displacement (u) of underground structure, arranging according to AR0–AR11 (**Figure 3**). Meanwhile, the influence of underground structure on surface structure is illustrated by influence coefficient e, defined as $(a_{AR1-10} - a_{AR0})/a_{AR0}$ or $(a_{AR11} - a_{AR12})/a_{AR12}$, where a_{AR0-12} is the dynamic magnification factor (a) of surface structure, arranging according to AR0–AR12 (**Figure 3**).The dimensionless distance d is D/B, where B is the width of surface structure.

4.1. Steady-state response

In this section, the following figures show the harmonic dynamic response (thin line), the aforementioned dimensionless displacement u of single-underground structure-soil systems (AR0) or the dynamic magnification factor (a) of single-surface structure-soil systems (AR0) together with the influence coefficient e (thick lines) of surface structure-soil-underground structure systems (AR1–11) under vertically incident S waves (SX or SY).

Figure 5 shows the dynamic response (thin line) of single-surface structure-soil system (AR0) together with the influence of adjacent surface structure on the response of underground structure (thick lines) under vertically incident S waves while **Figure 6** shows the dynamic response (thin line) of single-underground structure-soil system (AR0) together with the influence of adjacent surface structure on the response of underground structure (thick lines) under vertically incident S waves. Consideration is given to 10 different arrangements of structures (AR1–10), including one underground structure and one surface structure, and three distances between adjacent structures (d=0.1, 0.5 and 1.0). Shaking direction is assumed to be either parallel or perpendicular to the axes of structures (SX or SY).

A very slight shift at the peak of harmonic response spectra of underground structure takes place, slighter than that of surface structure-soil-surface structure system in many previous works [1]. The peak of harmonic response spectra of underground structure drifts slightly too, even slighter than that of surface structure, which can be ignored. It can be seen that the horizontal response of a surface structure and an underground structure may vary significantly due to the presence of underground structure or neighboring surface structure. The influence coefficient e, much depending on excitation frequency, fluctuates around zero with excitation frequency, and the dynamic response can be positive or negative. The influence of SSSI varies for different arrangements, distances between structures and shaking directions, and the dynamic response of structure may even increase or decrease, depending on the configuration.

The underground structure has conspicuous influence on the surface structure on many configurations, and the maximum impact can be up to 50%. But it has little influence ($e \leq 12\%$) if structures arrange in accordance with AR4 or AR8 under S waves producing motions on the Y.

That the central axis of two structures coincide with each other (AR1–4) or not (AR5–8) has very little influence on the interaction of structures. The most impact appears when AR3 contrast with AR7 under S waves producing motions on the X. The influence coefficient e is very depending on excitation frequency. When the excitation frequency is lower than 1 Hz, the influence coefficient e is almost close to zero for each arrangement. Namely that, the adjacent underground structure has little influence on the surface structure. When the excitation frequency is higher than 1 Hz, the influence coefficient e fluctuates with excitation frequency, and the dynamic response can be plus or minus. For all arrangement, the values of the lateral response at the top of surface structure around fundamental frequency are impact but not much. And that e are almost within 10% except the conditions with seismic motion on the X axis, $d = 0.1$ and arrangement according to AR3 or AR10.

The surface structure has a conspicuous influence on the underground structure on many configurations, and the maximum impact can reach up to 150%. It has smallest influence ($e \leq 27\%$) when structures arrange in accordance with AR4 or AR8 under S waves producing motions on the Y axis. This condition is consistent with that of influence of underground structure on surface structure, with influence coefficient not exceeding 12%. Whether the central axis of two structures coincide with each other (AR1–4) or not (AR5–8) has a little influence on the interaction of structures. There is little difference between AR1 and AR5 and between AR4 and AR8 under S waves producing motions on the X axis or Y axis. The greatest impact appears when AR2 contrast with AR6 under high frequency S waves producing motions on the X axis ($\geq 6.0Hz$) and Y axis ($\geq 3.5Hz$) and when AR3 contrasts with AR7 under low frequency S waves producing motions on the X axis ($\leq 6.0Hz$) and Y axis ($\leq 3.5Hz$).

For all arrangements, the horizontal response (dimensionless displacement) of underground structure around fundamental frequency of surface structure on shaking direction is impacted sharply. And the curves of influence coefficient e present peaks here. As the fundamental frequency of surface structure is approximate to that of underground structure-soil system, it is hard to determine the frequency near which the peaks appear. But this is clear in **Figures 9, 11–13** for the peaks move with the fundamental frequency of surface structure but not that of underground structure-soil system. As the distance increases, the peaks decline rapidly for AR1 and AR3, and slowly for AR2 and AR4. Experiments and observations from real cases have shown that underground structure exhibits a significantly different seismic response from surface structures as they do not respond in resonance with the ground motion, but rather on the basis of the response of the surrounding soil. This special behavior occurs, first because the mass effect is much smaller in underground than in surface structures and second, because the damping in underground structures is very high due to the energy radiating from them into the surrounding ground. Thus, the fundamental frequency of underground structure-soil system is approximately equal to that of free field. When the fundamental frequency of surface structure is approximate to that of free field, the surface structure might produce a more obvious influence on adjacent underground structure.

The influence of distance between structures on the interaction is investigated in **Figure 7**. **Figure 7(a)** shows the dynamic response (thin line) of surface structure-soil system (AR0) together with the influence of SSSI (thick lines) on the surface structure while **Figure 7(b)** shows the dynamic response (thin line) of underground structure-soil system (AR0) together with the influence of SSSI (thick lines) on the underground structure, for AR1 under S waves producing motions on the Y axis, for different distance (d). Combining with **Figures 5** and **6**, for all frequencies, the interaction between structures reduces with the increase of distance, fluctuating but not monotonous. Obviously, with the increase of distance, the energy of scattered wave from structure reduces and the interaction between structures fades away. This is also the explanation of the phenomenon illustrated in **Figure 8**. **Figure 8** presents the dynamic response (thin line) of single-structure-soil system (AR0) and the influence coefficient e (thick lines) for AR1 under S waves producing motions on the Y axis, for different soil damping (ζ = 0.05, 0.10, 0.15, 0.20). With the increase of soil damping, the absolute value of influence coefficient e decreases for all frequencies. And the curves fluctuate more gently.

Another important attribute of subsoil is shear wave velocity. So **Figure 9** presents the dynamic response (thin line) of single-structure-soil system (AR0) and the influence coefficient e (thick lines) for AR1 under S waves producing motions on the Y axis, now for different shear wave velocities of soil (c_s = 200, 300, 400m/s). At low frequency points, those shapes of influence

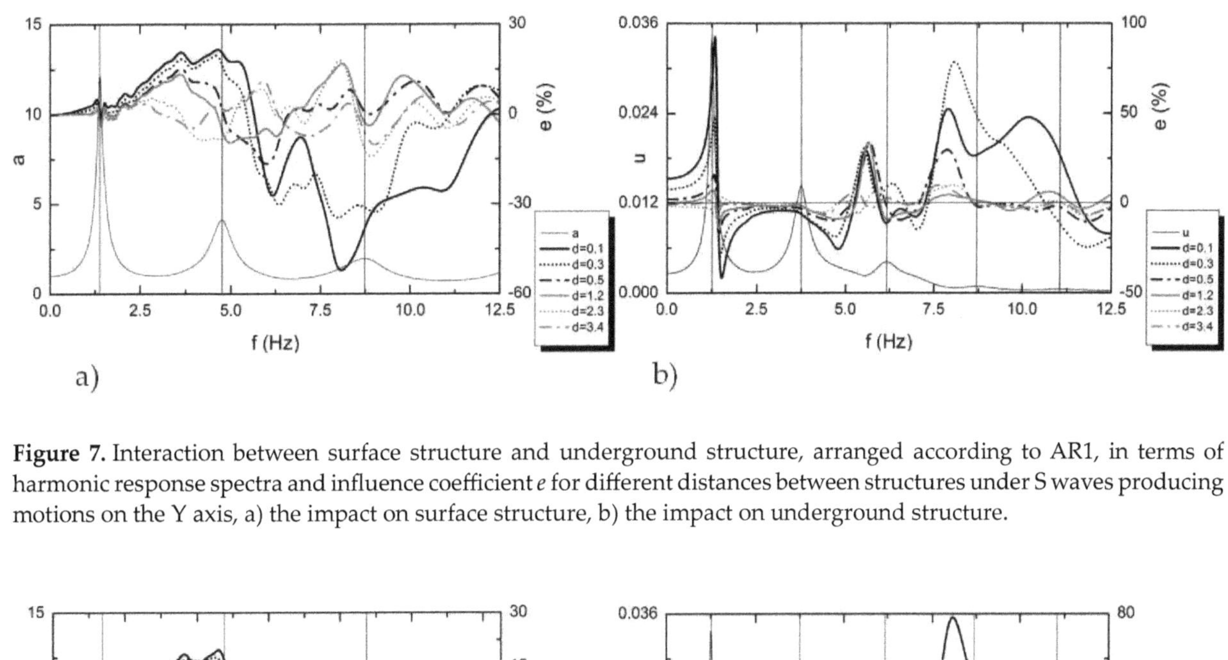

Figure 7. Interaction between surface structure and underground structure, arranged according to AR1, in terms of harmonic response spectra and influence coefficient e for different distances between structures under S waves producing motions on the Y axis, a) the impact on surface structure, b) the impact on underground structure.

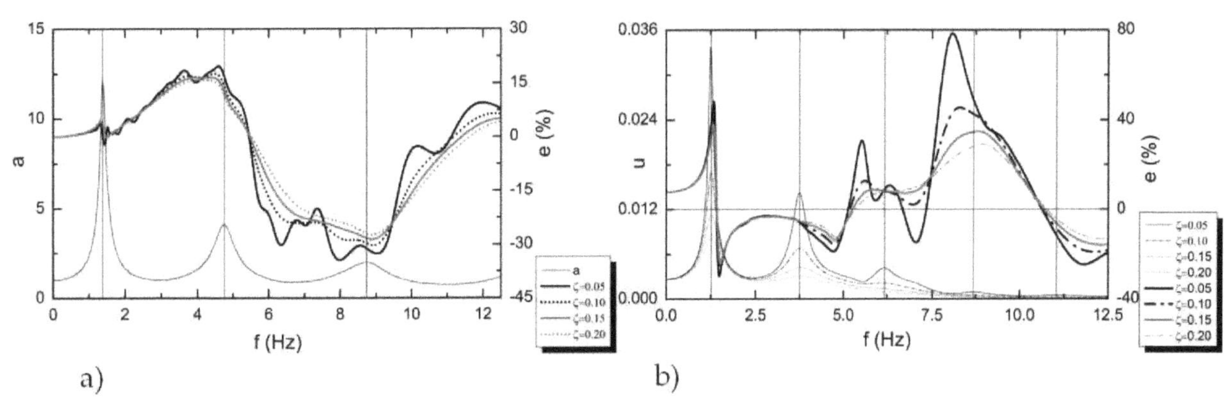

Figure 8. Interaction between surface structure and underground structure, arranged according to AR1, in terms of harmonic response spectra and influence coefficient e for different soil damping under S waves producing motions on the Y axis (d = 0.5), a) the impact on surface structure, b) the impact on underground structure.

coefficient curves are similar, but at high frequency points, the increase of shear wave velocity leads to the increase of absolute value of influence coefficient e.

Figure 10 presents the dynamic response (thin line) of single-structure-soil system (AR0) and the influence coefficient e (thick lines) for AR1 under S waves producing motions on the Y axis, now for different lengths of piles of the surface structure (H_{pile} = $6m, 12m, 18m, 24m$) and surface foundation (H_{pile} = $0m$). Those curves of influence coefficient e almost coincide at all frequencies, which mean that the influence of pile can be ignored here. However, with the decrease of length of pile, the impact of underground structure on surface structure around fundamental frequency increases. And that e can be up to 15% for surface foundation.

Figures 11 and **12** show the dynamic response (thin line) of single-structure-soil system (AR0) and the influence coefficient e (thick lines) for AR1 under S waves producing motions on the Y axis, for frame with different stiffness and frame-shear wall structures with different stiffness respectively. The frame-shear wall structure is based on the aforementioned frame structure with the addition of four L-shape shear walls on each storey. Meanwhile, consideration is given to three different kinds of stiffness (original stiffness multiplied by 0.8, 1.0 and 1.2,

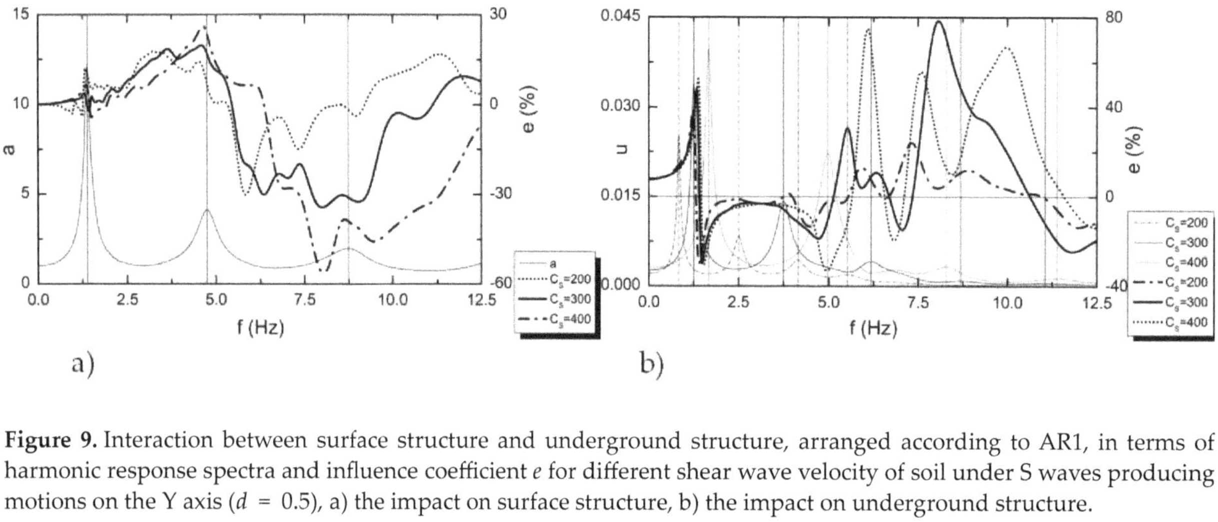

Figure 9. Interaction between surface structure and underground structure, arranged according to AR1, in terms of harmonic response spectra and influence coefficient e for different shear wave velocity of soil under S waves producing motions on the Y axis (d = 0.5), a) the impact on surface structure, b) the impact on underground structure.

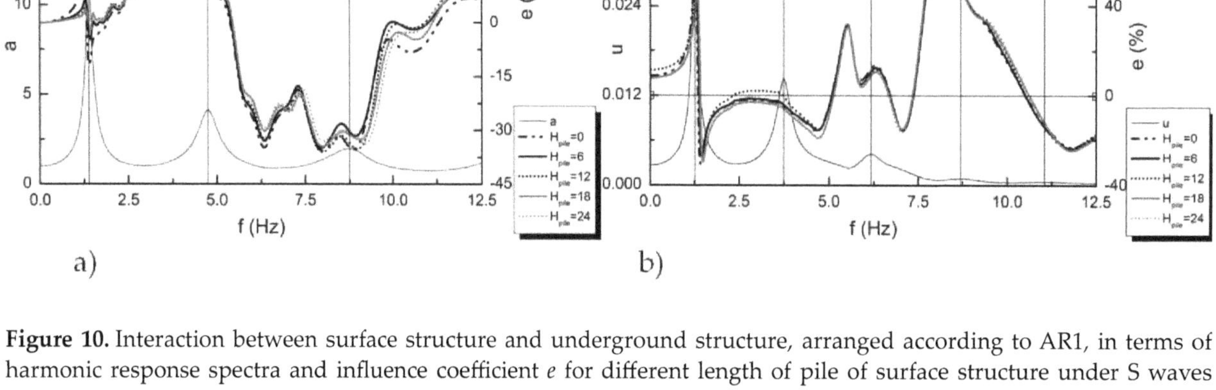

Figure 10. Interaction between surface structure and underground structure, arranged according to AR1, in terms of harmonic response spectra and influence coefficient e for different length of pile of surface structure under S waves producing motions on the Y axis (d = 0.5), a) the impact on surface structure, b) the impact on underground structure.

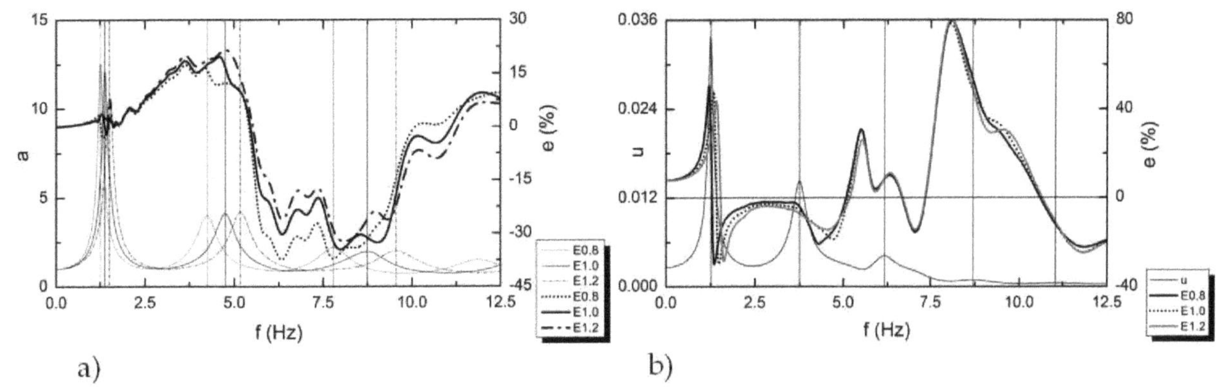

Figure 11. Interaction between surface structure and underground structure, arranged according to AR1, in terms of harmonic response spectra and influence coefficient *e* for different stiffness of surface structure under S waves producing motions on the Y (frame structure, $d = 0.5$), a) the impact on surface structure, b) the impact on underground structure.

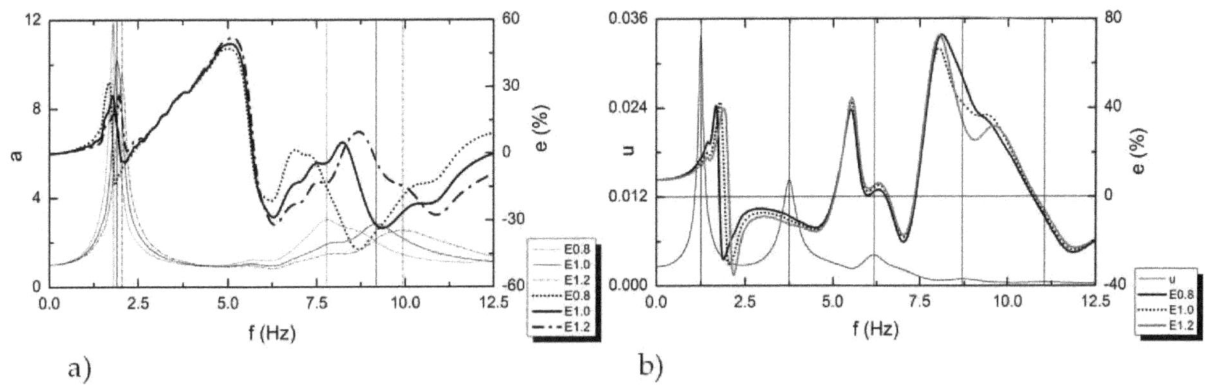

Figure 12. Interaction between surface structure and underground structure, arranged according to AR1, in terms of harmonic response spectra and influence coefficient *e* for different stiffness of surface structure under S waves producing motions on the Y axis (frame-shear wall structure, $d = 0.5$), a) the impact on surface structure, b) the impact on underground structure.

termed as "E0.8", "E1.0" and "E1.2") of the surface structure. The fundamental periods, along lateral axis, of frame structures are 0.77 s for "E0.8"and 0.67 s for "E1.2"; and that of frame-shear wall structures are 0.56 s for "E0.8", 0.53 s for "E1.0" and 0.50s for "E1.2".

The moderate adjustment of stiffness does not have much influence on the interaction at all frequency points. Meanwhile, through contrast on **Figures 11** and **12**, style of surface structure, frame structure or frame-shear wall structure, has no great influence on SSSI when focus on the response of underground structure. The shapes of those influence coefficient curves are similar except that the first peaks drift with fundamental frequencies of surface structures. However, style of structure has unexpectedly impact on SSSI when focus on the response of surface structure. But this impact does not just owing to the difference of stiffness.

Figure 13 shows the dynamic response (thin line) of single-structure-soil system (AR0) and the influence coefficient *e* (thick lines) for AR1 under S waves producing motions on the Y axis, for different storey numbers ($N_F = 6, 10, 14$) of the surface structure. The fundamental periods, along lateral axis, of the frame structure are 0.42 s for "$N_F = 6$"and 1.11 s for "$N_F = 14$". The variation of number of storey has some influence on the interaction between structures, but the shapes of influence coefficient curves are similar by and large. There are some perturbations at those frequency points around the natural frequency of surface structure, and the smaller the number of storey, the sharper the perturbation is. Expectedly, except that the first peaks of influence coefficient curves of SSSI on underground structure shift with fundamental frequencies of surface structures, the variation of storey number has no great influence on the interaction between structures. Meanwhile, the larger the storey number is, the sharper the first peak will be.

Figure 14 presents the dynamic response (thin line) of single-structure-soil system (AR0) and the influence coefficient e (thick lines) for AR1 under S waves producing motions on the Y axis, for different burial depths of the underground station (H_{bd} = $2m, 3m, 4m$).Those shapes of influence coefficient e curves of SSSI on surface structure are similar, but at some frequency points, such as 1–4 Hz, the increase of burial depth leads to a little increase of interaction, and at the other frequency points, such as 5.5–7.5 Hz, the increase of burial depth leads to a little decrease of interaction. At the same frequency, the increase or decrease of interaction is monotonous. Except at high frequency points ($\geq 9.0Hz$), those curves of influence coefficient e on underground structure almost coincide at most of the frequency points, which means that burial depth generates little impact on the interaction.

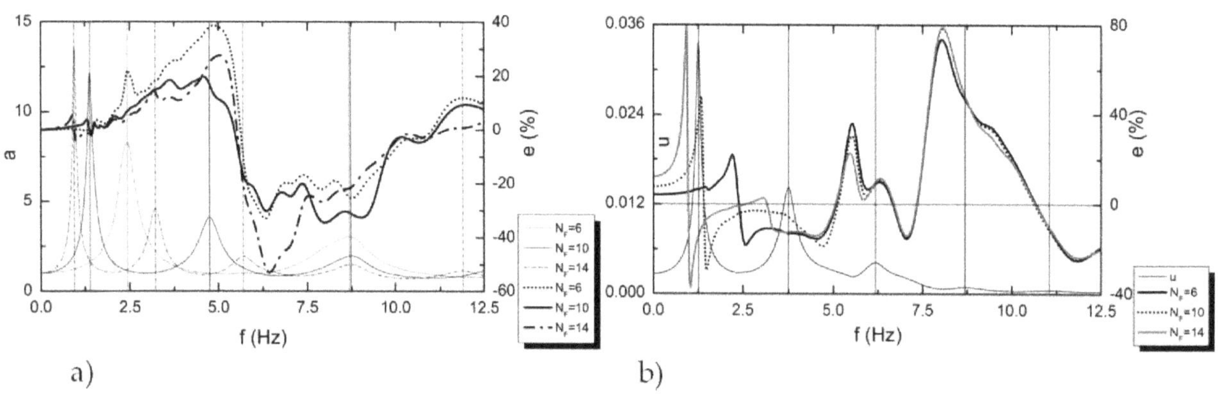

Figure 13. Interaction between surface structure and underground structure, arranged according to AR1, in terms of harmonic response spectra and influence coefficient e for different number of storey of surface structure under S waves producing motions on the Y axis (d = 0.5), a) the impact on surface structure, b) the impact on underground structure.

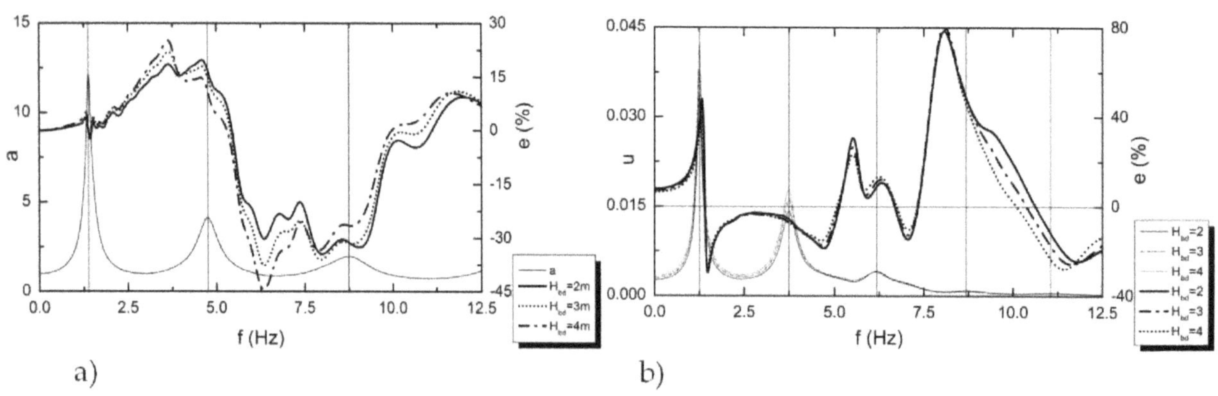

Figure 14. Interaction between surface structure and underground structure in terms of harmonic response spectra and influence coefficient e for different burial depth of underground station under S waves producing motions on the Y axis (d = 0.5), a) the impact on surface structure, b) the impact on underground structure.

Figure 2 depicts the profile of a typical three-span underground subway station, which is used for analysis in this paper. At some smaller subway stations, however, a two-span profile is widely adopted. So the middle span of the profile depicted in **Figure 2** is cut off, and the underground structure is turned into a two-span station for comparison. **Figure 15** displays the dynamic response (thin line) of single-structure-soil system (AR0) and the influence coefficient e (thick lines) for AR1 under S waves producing motions on the Y axis, for different numbers of spans ($N_S = 2, 3$) of underground station. At the lower frequency points (0–4.5 Hz), number of spans hardly impacts the interaction. But at the higher frequency points (4.5–12.5 Hz), different numbers of spans lead to the misalign of influent coefficient e. Even so, the influence of surface structure on two-span and three-span underground station is not much different.

Now, an analysis is made on a case of several similar surface structures and an underground structure arranged as shown in **Figure 3** (AR11). For comparison, the case of just only surface structures (AR12) is analyzed too. **Figure 16** presents the dynamic response (thin line) of single-structures-soil system (AR0) together with the influence coefficient e (thick lines) of surface structures-soil-underground structure systems (AR11) under S waves producing

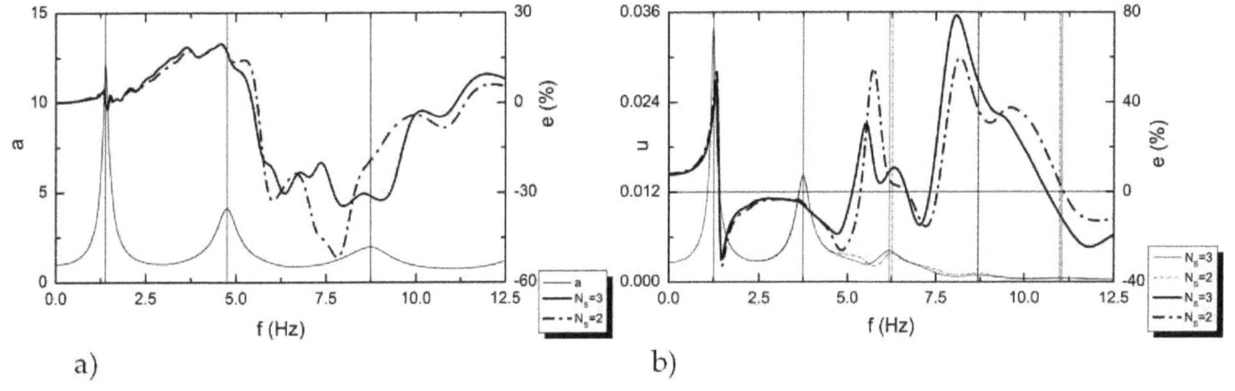

Figure 15. Interaction between surface structure and underground structure in terms of harmonic response spectra and influence coefficient e for different number of span of underground structure under S waves producing motions on the Y axis ($d = 0.5$), a) the impact on surface structure, b) the impact on underground structure.

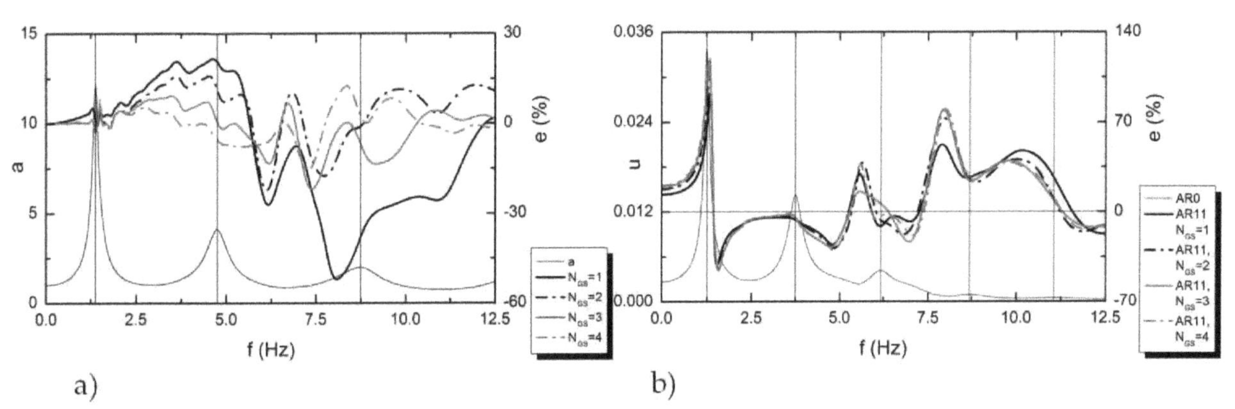

Figure 16. Interaction between surface structure and underground structure in terms of harmonic response spectra and influence coefficient e for different numbers of surface structures under S waves producing motions on the Y axis ($d = 0.1$), a) the impact on surface structure, b) the impact on underground structure.

motions on the Y axis, for different numbers of surface structures (N_{GS} = 1, 2, 3, 4).The lateral response of the surface structure, which is furthest away from the underground station, is checked here. It is exposed that the presence of underground structure can not only has influence on the contiguous surface structure but also the other neighboring ground structures. Expectedly, because the energy of scattered wave fade away, the influence falls down when the number of ground structures grows up. It is exposed that those neighboring surface structures, which are not next to the underground structure, also generate an impact on the underground structure by affecting the contiguous surface structure. More buildings will cause greater impact, particularly at the first peak of influence coefficient. As the energy of scattered wave fades away, however, the influence of building number tends to converge when it grows up.

4.2. Earthquake response

In this section, nine ground motion records from bedrock and three artificial seismic waves with different exceedance probabilities are selected in order to measure the influence of surface structure on underground structure. Results in the time domain are obtained by Fourier transformation, via the fast Fourier transform (FFT) algorithm.

Figures 17 and **18** show the influence of SSSI under vertically incident S waves. Obviously, the absolute value of influence coefficient e of SSSI on surface structure is within 15% and that of SSSI on underground structure is within 40%. Although, in harmonic analysis, the influence coefficient e can reach up to 50% (SSSI on surface structure) or 150% (SSSI on underground structure) at some frequency points, the corresponding dynamic response (a or u) or the corresponding seismic energy might still be small, and consequently, the interaction between structures is not so conspicuous. The absolute value of influence coefficient e of underground structure on surface structure is almost within 5 percent for 11 seismic waves. However, the influence coefficient e, the dynamic magnification factor a and the seismic energy (showed by **Figure 19**) around 4.5 Hz is great. This is why the interaction under GGP-100 seismic wave is greater than the others. For AR2 and AR4, generally, the influence of surface structure on

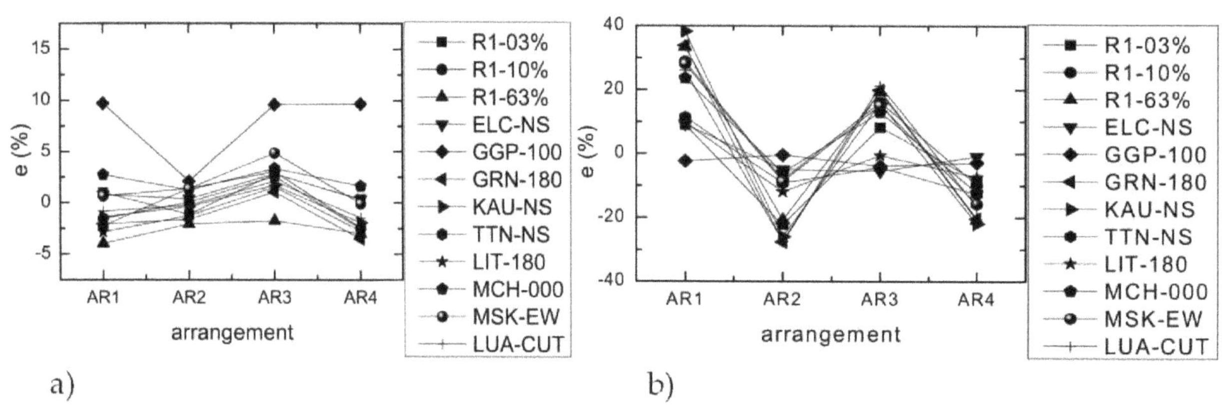

Figure 17. Interaction between surface structure and underground structure in terms of influence coefficient e for different arrangements under S waves producing motions along longitudinal axis (d = 0.5), a) the impact on surface structure, b) the impact on underground structure.

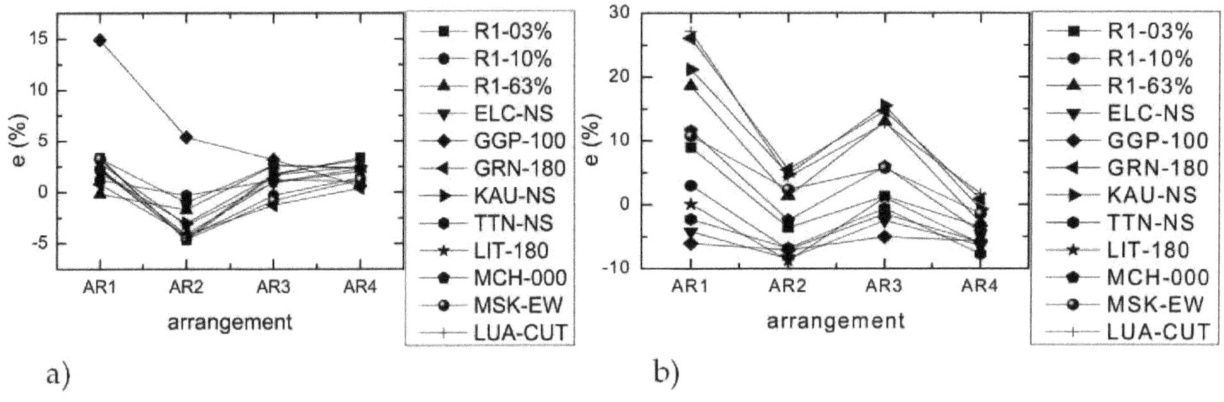

Figure 18. Interaction between surface structure and underground structure in terms of influence coefficient e for different arrangements under S waves producing motions along lateral axis (d = 0.5), a) the impact on surface structure, b) the impact on underground structure.

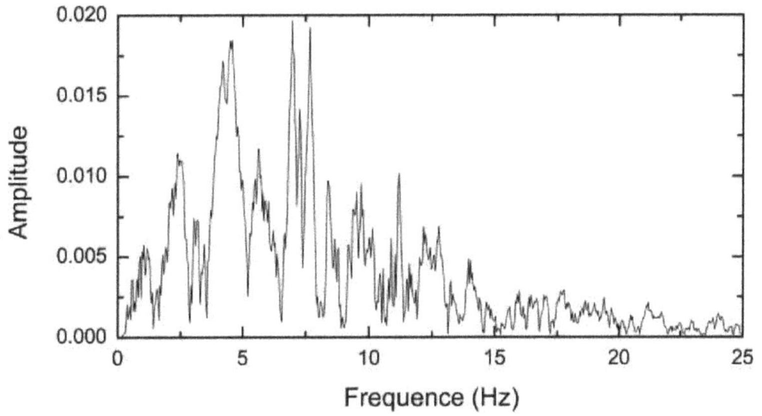

Figure 19. Fourier spectrum of seismic wave (GGP-100).

underground structure is beneficial or negligible under S waves producing motions along longitudinal axis; it is also not so conspicuous ($|e| < 10\%$) under S waves producing motions along lateral axis. For AR1 and AR3, however, the influence coefficient e can reach up to 40% which deserves more attention in seismic dynamic analysis.

5. Compare and contrast

The similarity and difference of the impact on surface structure and on underground structure are briefly summarized in this section.

The presence of adjacent structure leads to the peak of harmonic response spectra of structure shifted. But the shift of underground structure is slighter than that of surface structure, which can be ignored. The horizontal responses of surface structure and underground structure around fundamental frequency of surface structure at shaking direction are both conspicuously impacted and the influence coefficient curves show peaks here. But the peaks of underground structure are much more outstanding. The influence of adjacent structure on underground structure reaches minimum if structures arrange in accordance with AR4 or AR8 under S waves producing motions on the Y axis. Whether the central axes of the two structures coincide with each other (AR1–4) or not (AR5–8) has a small influence on the response of underground structure and has a relatively larger influence on the response of surface structure, especially in contrast between AR3 and AR7. In general, the impact of each parameter on the influence coefficient curve of surface structure on underground structure is smaller than that of underground structure on surface structure.

6. Conclusion

A three-dimensional numerical procedure for the dynamic analysis of pile supported surface building and underground station was used in this work, to solve the problem of through-soil interaction between multi-storey frame structure and neighboring underground structure. The commercial software (ANSYS), based on finite element method, has been further developed and enhanced for calculation in frequency domain, in which hysteretic damping has been considered for both soil and structures, and linear assumptions have been put forward for the rigorous three-dimensional modeling. In this way, SSSI phenomena are implicitly included in the model.

Transfer functions of horizontal acceleration of surface structure and that of relative horizontal displacement of underground structure under harmonic wave, and maximum horizontal acceleration of surface structure and maximum relative horizontal displacement response of underground structure under real or artificial seismic wave, are presented, in order to assess the influence of SSSI phenomena on the structural seismic response of neighboring surface structure and underground structure subject to vertically incident S wave.

Arrangement and shaking direction of seismic wave are two of the most important factors, which have conspicuous influences on the interaction. In general, when structures arrange according to AR1 under S waves producing motions on the X axis and according to AR2 under S waves producing motions on the Y axis, the impact of surface structure on underground structure is more noticeable. And when structures arrange AR3 under S waves producing motions on the X and according to AR1 under S waves producing motions on the Y, the impact of underground structure on surface structure is more noticeable. Depending on the distance between adjacent structures, the seismic response of surface structure can either increase or decrease. But the interaction surely fades away if the distance is large enough. When the fundamental frequency of surface structure is approximate to that of free field, surface structure may have more considerable influences on adjacent underground structure.

In view of the results, it is apparent that further studies about SSSI phenomena and their influence on structural seismic risk are needed, as it has been shown that nearby buildings can significantly increase the seismic response of a structure.

Acknowledgements

The authors would like to gratefully acknowledge the support from National Natural Science Foundation of China under Grant No. 51608462.

Author details

Huai-feng Wang

Address all correspondence to: whf_tj@163.com

School of Civil Engineering and Architecture, Xiamen University of Technology, Xiamen, China

References

[1] Lou M et al. Structure-soil-structure interaction: Literature review. Soil Dynamics and Earthquake Engineering. 2011;**31**(12):1724-1731

[2] Luco JE, Contesse L. Dynamic structure-soil-structure interaction. Bulletin of the Seismological Society of America. 1973;**63**(4):1289-1303

[3] Fu CS, Yu ZD. Analysis of dynamic behaviour of structure including the effects of an adjacent building. Shanghai Mechanics. 1982;**3**(3):28-39

[4] Tian BD, Yu ZD. Dynamic analysis of structure-soil-structure interaction. Journal of Tongji University. 1987;**15**(2):157-168

[5] Jiang XL, Yan ZD, Li XZ. Analysis of earthquake response of neighboring buildings-foundation-soil system under multi-support excitations. Earthquake Engineering and Engineering Vibration. 1997;**17**(4):65-71

[6] Dou LJ, Yang BP. Dynamic interaction of tall building with neighboring multi-storied building. Earthquake Engineering and Engineering Vibration. 2000;**20**(3):15-21

[7] Padron LA, Aznarez JJ, Maeso O. Dynamic structure-soil-structure interaction between nearby piled buildings under seismic excitation by BEM-FEM model. Soil Dynamics and Earthquake Engineering. 2009;**9**(6):1084-1096

[8] Alamo GM et al. Structure-soil-structure interaction effects on the dynamic response of piled structures under obliquely incident seismic shear waves. Soil Dynamics and Earthquake Engineering. 2015;**78**(11):142-153

[9] Fariborz N, Ali R. Nonlinear dynamic response of tall buildings considering structure-soil-structure effects. The Structural Design of Tall and Special Buildings. 2013;**22**(14): 1075-1082

[10] Li P, Yan K, Xu P. Study on dynamic interaction between soil and group of high-rise buildings under seismic excitation. China Civil Engineering Journal. 2014;**S1**:1-5

[11] Alam M, Kim D. Spatially varying ground motion effects on seismic response of adjacent structures considering soil-structure interaction. Advances in Structural Engineering. 2014;**17**(1):131-142

[12] Ghandil M, Behnamfar F, Vafaeian M. Dynamic responses of structure-soil-structure systems with an extension of the equivalent linear soil modeling. Soil Dynamics and Earthquake Engineering. 2016;**80**(1):149-162

[13] Wolf JP. Foundation Vibration Analysis Using Simple Physical Models. Englewood Cliffs: PTR Prentice Hall; 1994

[14] Mulliken JS, Karabalis DL. Discrete model for dynamic through-the-soil coupling of 3-D foundations and structures. Earthquake Engineering and Structural Dynamics. 1998; **27**(7):687-710

[15] Alexander NA, Ibraim E, Aldaikh H. Exploration of structure-soil-structure interaction dynamics. In: Proceedings of the 13th International Conference on Civil, Structural and Environmental Engineering Computing; Chania, Crete, Greece; 2011

[16] Naserkhaki S, Pourmohammad H. SSI and SSSI effects in seismic analysis of twin buildings: Discrete model concept. Journal of Civil Engineering and Management. 2012;**18**(6): 890-898

[17] Mizuno H. Effects of structure-soil-structure interaction during various excitations. In: Proceedings of 7th World Conference on Earthquake Engineering; Istanbul, Turkey; 1980

[18] Lin Y. Shaking Table Model Test of Underground Structure and Numerical Analysis. Shanghai: Tongji University; 2012

[19] Kang S. Shaking Table Model Test of Substructure-Soil-Superstructure System and Numerical Analysis. Shanghai: Tongji University; 2012

[20] Pan DG et al. Model test of the dynamic characteristics of a structure-soil-structure system. Journal of University of Science and Technology Beijing. 2014;**36**(12):1720-1728

[21] Trombetta NW et al. Nonlinear soil-foundation-structure and structure-soil-structure interaction: Engineering demands. Journal of Structural Engineering. 2015;**141**(7): 04014177

[22] Xiong F et al. Shake table tests of high-rise building group-soil interaction. Journal of Sichuan University (Engineering Science Edition). 2015;**47**(03):37-43

[23] Aldaikh H et al. Two dimensional numerical and experimental models for the study of structure-soil-structure interaction involving three buildings. Computers & Structures. 2015;**150**(4):79-91

[24] Cacciola P, Espinosa MG, Tombari A. Vibration control of piled-structures through structure-soil-structure-interaction. Soil Dynamics and Earthquake Engineering. 2015; **77**(10):47-57

[25] Lee VW, Trifunac MD. Response of tunnels to incident SH-waves. ASCE Journal of the Engineering Mechanics Division. 1979;**105**(4):643-659

[26] Lee VW, Karl J. Diffraction of SV waves by underground, circular, cylindrical cavities. Soil Dynamics and Earthquake Engineering. 1992;**11**(8):445-456

[27] Lin CH et al. Zero-stress, cylindrical wave functions around a circular underground tunnel in a flat, elastic half-space: Incident P-waves. Soil Dynamics and Earthquake Engineering. 2010;**30**(1):879-894

[28] Hatzigeorgiou GD, Beskos DE. Soil-structure interaction effects on seismic inelastic analysis of 3-D tunnels. Soil Dynamics and Earthquake Engineering. 2010;**30**(9SI):851-861

[29] Abate G, Massimino MR, Maugeri M. Numerical modelling of centrifuge tests on tunnel-soil systems. Bulletin of Earthquake Engineering. 2015;**13**(7):1927-1951

[30] Guo J, Chen JY. Influence of subway station buried in homogenous foundations on the inter-story drift ratio of adjacent structures. Journal of Hunan University (Natural Science). 2013;**40**(08):40-45

[31] Yang SY, Jiang XL, Li XG. Analysis on seismic response effect of tunnel to nearby structure. Journal of Sichuan University (Engineering Science Edition). 2007;**39**(3):41-46

[32] Li YT, Zhou ZX. Effect of tunnel on seismic response of adjacent surface building structure. Journal of Hebei University of Technology. 2014;**43**(05):97-100

[33] Farghaly AA. Seismic analysis of high rise building with deep foundation constructed near deep channel. Electronic Journal of Geotechnical Engineering. 2014;**19N**:3099-3124

[34] Abate G, Corsico S, Massimino MR. FEM modelling of the seismic behavior of a tunnel-soil-aboveground building system: A case history in Catania (Italy). In: 6th Italian Conference of Researchers in Geotechnical Engineering. Bologna, Italy: Elsevier Ltd; 2016

[35] Abate G, Massimino MR. Parametric analysis of the seismic response of coupled tunnel-soil-aboveground building systems by numerical modelling. Bulletin of Earthquake Engineering. 2017;**15**(1):443-467

[36] Abate G, Massimino MR. Numerical modelling of the seismic response of a tunnel-soil-aboveground building system in Catania (Italy). Bulletin of Earthquake Engineering. 2017;**15**(1):469-491

12

Development of an UAS for Earthquake Emergency Response and its Application in Two Disastrous Earthquakes

Chaoyong Peng, Zhiqiang Xu, Jiansi Yang, Yu Zheng, Weiping Wang, Sha Liu and Baofeng Tian

Abstract

To support humanitarian action after a disaster, we require reliable data like high-resolution satellite images for analyses aimed to define the damages of facilities and/or infrastructures. However, we cannot obtain satellite images in few days after an event. Thus, in situ surveys are preferred. Advances in unmanned aircraft system (UAS) have promoted them to become precious tools for capturing and assessing the extents and volume of damages. Safety, flexibility, low cost, and ease of operation make UAS suitable for disaster assessment. In this chapter, we developed an example of UAS for swiftly acquiring disaster information. With the selected fixed-wing UAS, we successfully performed data acquisition at specified scales. For the image analysis, we applied a photogrammetric workflow to deal with the very high resolution of the images obtained without ground control points. The results obtained from two destructive earthquakes demonstrated that the presented system plays a key role on the processes of investigating and gathering information about a disaster in the earthquake epicentral areas, like road detection, structural damage survey, secondary disaster investigation, and quick disaster assessment. It can effectively provide disaster information in hardly entered areas to salvation headquarters for rapidly developing the relief measures.

Keywords: unmanned aircraft system, emergency rescue, disaster assessment, regional panorama, earthquake

1. Introduction

For efficient mitigation of losses caused by a destructive earthquake [1–3], accurate development of relief measures, and practical improvement of the emergency rescue efficiency, we need to rapidly acquire the earthquake information and promptly attempt to evaluate the probable damages. In the case of ground transport disruption as well as communication interruption, remote sensing (RS) plays a crucial role on acquiring disaster situation for post-earthquake emergency response and evaluation. Nevertheless, once satellite remote sensing (SRS) is adopted to acquire disaster information, it is currently restricted by its spatial resolution and run cycle. Besides, the traditional manned aerial photography is also constrained by some restrictions [4], like airport and weather conditions, and some potential dangers to pilots during the flight. The unmanned aircraft system (UAS) can be used as a supplementary resort to manned aerial photography and SRS because of its unique advantages such as being real time, flexible, high image resolution, and low cost. It refers to a class of aircrafts which can easily and safely fly without the onboard presence of a pilot [5–9] and can offer several opportunities in disaster-related situations when equipped with RS instrumentations [10–12]. High-resolution images are analyzed and utilized to generate hazard maps, dense surface models, detailed building renderings, comprehensive elevation models, and other characteristics in a disaster area. These data can then be analyzed with RS methods as well as visual interpretation to coordinate rescue efforts, record the responses of a structural system to the vibrations, detect structural failures, investigate access issues, and verify experimental model of modeling disaster. In addition, the data can be gathered before a disaster for documenting immediate pre-event conditions of critical facilities and infrastructures, monitoring susceptible environmental concerns, and documenting historical conditions and sites [13]. Briefly speaking, RS based on UAS yields the best possible spatial and temporal resolutions for the respective research question or application [14].

In this chapter, we firstly introduce the utilization of a fixed-wing aircraft type called HW18 developed by HoverWings, China, which is equipped with Sigma DP2 (an optimal digital camera system). Then, we represent the field test results in the Yuxi County, Yunnan province, China, and real applications in two severe earthquakes: the 2013 Ms7.0 Lushan earthquake (occurred at Sichuan province, China) and the 2014 Ms6.5 Ludian Earthquake (occurred at Yunnan province, China). The main objectives of the application of UAS are very high-resolution monitoring of the investigated post-earthquake areas with optical RS data as well as creating detailed digital maps. Due to its very high image resolution, the UAS can be used in a broad range of applications with a superior degree of detail than SRS data. In addition, the photogrammetric potential of small-format aerial photography data makes it more suitable for applications requiring higher efficiency and spatial completeness than traditional field work. However, if we want to develop an UAS survey approach, the main prerequisite is to satisfy different requirements of resolution, scale, and accuracy [15].

2. System structure

The developed system is composed of three main parts. These parts are an UAS, a ground control station, and a system used for image processing (**Figure 1**).

2.1. UAS specifications

As the unmanned platforms continuously grow, we have many choices to select different unmanned platforms for low-altitude Earth observation, and it is difficult for us to generalize advantages or disadvantages of particular platform, as possible applications and operating conditions remarkably vary throughout the world. When we need to precisely cover small sites that require only a small number of images, tethered systems, which are manually navigated (e.g., gliders, blimps, and kites), are ideal tools. However, these systems are scarcely utilized to systematically survey larger regions, where regular overlaps along evenly spaced flight lines are preferable for an efficient processing workflow. Two well-known instances of recent development in UAS technology, the Global Positioning System (GPS) and Inertial Navigation System (INS) technologies, have been leading to the availability of a broad range of autopilot systems like drones, planes, and quadcopters, which can autonomously follow predefined flight plans.

However, as a RS platform, we need to make compromises according to hardware and software limitations of an UAS, because it is limited in its flight duration and payload capacity [16]. Thus, when developing a custom-built UAS, we must balance the platform accessibility with the data quality of low-cost sensors and the technological limitations inherent of small-scale platforms. Such weight and cost limitations necessitate a tangible reduction in quality of manufacturing process of the sensor. Through the use of cheaper fabrication methods and materials, the absence of on-board processing features, or limited data storage capacity, we can readily achieve the reductions.

In this work, we employed a fixed-wing aircraft-type HW18 developed by HoverWings, China (**Figure 1**), with weight of approximately 2.3 kg (without payload), a length of 110 cm, and a wingspan of 120 cm. It is equipped with Sigma DP2, a digital camera system. The aviation fuel-powered system is based on an airplane model. At a ground speed of 70–110 km/h and with full payload (the maximum takeoff weight is 5.2 kg), the duration of flight is more than 100 minutes. In addition, the plane can be hand-launched and does not require catapult (**Figure 2**). **Figure 3** represents the operating principle of the UAS. During flight and takeoff, the presented UAS is autonomously controlled by an autopilot system and its GPS/IMU components. We use the Pix4UAV software [17] to design flight paths followed by the UAS, and

Figure 1. Different parts of the presented UAS.

it can automatically create an accuracy report in order to investigate the result quality. After finishing the flight plan, the UAS returns to the starting place and remains rotary in a preset altitude. At the starting point, we set a center for the so-called bounding box which is a circular unit with a diameter of 150 m. When a pilot adopts an assisted flying mode, the maximum distance between the starting point and the UAS is limited by the bounding box. The UAS will automatically and immediately turn to remain within its given extent whenever it hits the limit of the bounding box.

The system also features an assisted flying mode, where an autopilot software is used to support the pilot for UAS control. Destabilization due to wind or height loss while flying turns is automatically amended. Within the software, turn radii are limited for avoiding stress concentrations which might lead to the UAS being damaged. Besides, steering is simplified. The pilot can directly turn the UAS left or right through software-controlled interaction of roll-aileron and yaw rudder. Limits to maximum values of roll angles and pitch are set in the software. The pilot can safely land and steer a plane through this flying mode while constricting the navigation region to a preset range, which is especially useful for difficult terrains where a fully autonomous landing may be impossible. In addition, it increases the flexibility during the survey when the pilot wants to cover extra targets not previously set in the flight plans. However, using the assisted flying mode, a pilot cannot be relieved from his/her responsibility because he/she must be able to safely land and steer a plane. Thus, it is vital to pass the training courses offered by HoverWings prior to independent surveys.

The UAS installed a 14 MP digital fixed lens camera, which is called Sigma DP2 [18], as the optical onboard sensor. It is a high-end compact digital camera, with a 14-megapixel Foveon X3 sensor (2652 × 1768 × 3 layers), a 2.5″ LCD, a fixed 24.2 mm f/2.8 lens (equivalent to 41 mm), and a pop-up flash. In addition, it is one of the few "compact" cameras featuring a sensor with a size equivalent to APS-C, which can produce high-quality images with the help of a digital single-lens reflex (DSLR) camera. This camera with DSLR-comparable quality is much lighter (260 g excluding battery and card) and smaller (113.3 × 59.5 × 56.1 mm including lens) than that of a traditional single-lens reflex cameras. These characteristics make it to be highly appropriate for unmanned platforms. In addition, this camera has several important features

Figure 2. Hand-launching the UAV.

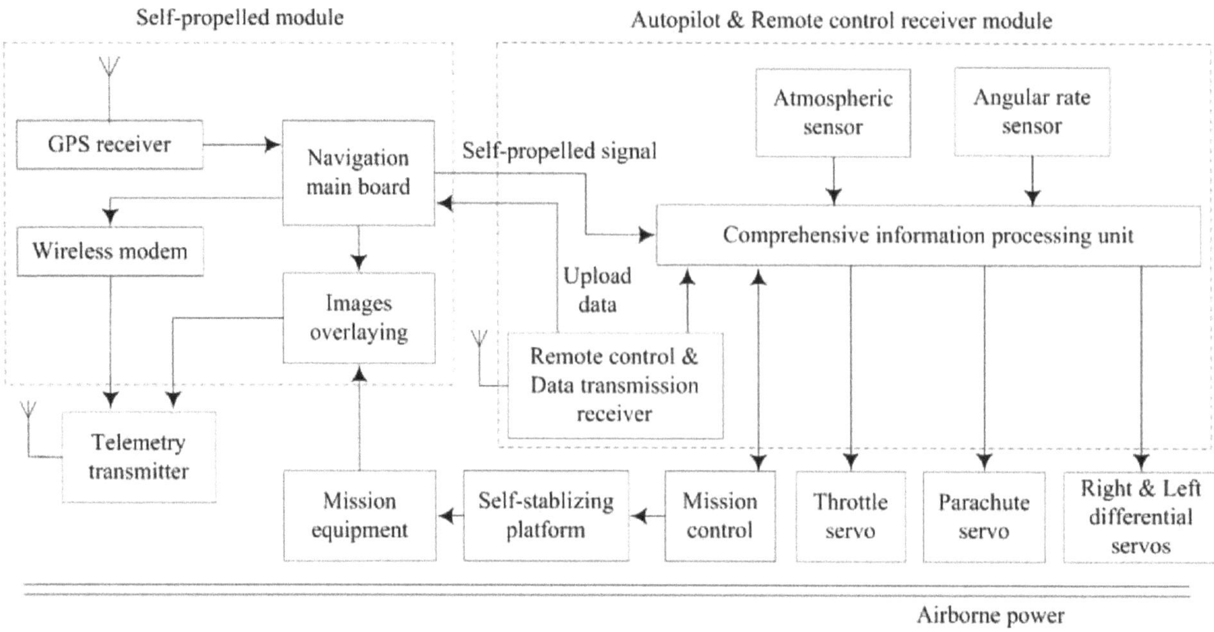

Figure 3. Working principle of the UAV.

for photogrammetric analysis of the images: large image sensor, multiple recording modes (e.g., JPEG, movie, and lossless compression RAW data), single focal length, as well of lack of an image stabilizer. For safety, we conceal the entire camera in the body of the plane and make it vertically point to the ground through an opening at the bottom of the plane. An on-board computer is used to trigger the camera exposure in a regular interval defined by a flight-planning software. Therefore, we can take images in a continuous mode during the flight.

Generally, for achieving regular stereoscopic overlap along the flight line, the flight plans are designed as parallel lines. The roll of the UAS may reach large angles up to 60° at the end of each line, where the image axis is off-nadir. Thus, we must consider a rather large overshoot resulting in a remarkably larger area covered by the predefined flight plan than the actual image acquisition area. In addition, due to restricted triggering control, we can avoid large amounts of oblique images at ends of the flight line. Maximum pitch angles (e.g., 10° off-nadir) are predefined through a flight-planning software.

2.2. Ground station

A ground station is used to remotely control the UAS, receive telemetry data, and display images transferred in real time. The processing flow of this ground station is shown in **Figure 4**. We can see from **Figure 4** that the ground station includes three components: a ground monitoring station, a telemetry receiving antenna, and a remote control transmitter. The embedded PC104 is used as the hardware of the ground motion, and it contains a digital video recorder, a 17″ LCD, a spherical mouse, and a membrane keyboard. The software system installed on it includes functions like Windows-based software modules for video processing, navigation, information processing, and map generation. We can display the video interface, the navigation interface, and the task management and status bar interface on the LCD and switch them in real time.

2.3. Image processing system

When referred to applications that include measuring and mapping, it is imperative to contain georeferencing and geometric correction of the acquired images. However, a highly precise geometric correction requires effort, time, excellent ground control, and a digital elevation model. To our UAS application, it may not really require such efforts, and based on the relief and image features, simpler solutions may be quite sufficient.

The location of an image within a given ground coordinate system can be reconstructed through two methods. The first one is to adopt Ground Control Points (GCPs) with known locations in a reference system and appearing on the images. The second one computes the exterior orientation position (X, Y, Z) and rotations (yaw, pitch, and roll of the platform or κ, ϕ, and ω of the image) of the images during the flight. Although the captured data of the latter are inaccurate and cannot be used for direct georeferencing relative to image scales and resolution, these data can be used as initial values of the system for approximate orientation. Why we adopt the second approach for image location reconstruction is that it is impossible for us to deploy GCPs for emergency rescue immediately after an earthquake and the demand for high data timeliness is of utmost priority.

Figure 5 shows the photogrammetry-based workflow for collecting and processing images acquired in a disaster. The exterior orientation values are taken from the log-files recorded during the survey flight for initial direct georeferencing of the images. The log files contain values for different angles (κ, ϕ, and ω) as well as GPS values (X, Y, and Z) for each image center. During image processing, we refine the values for image orientation in iterative triangulation

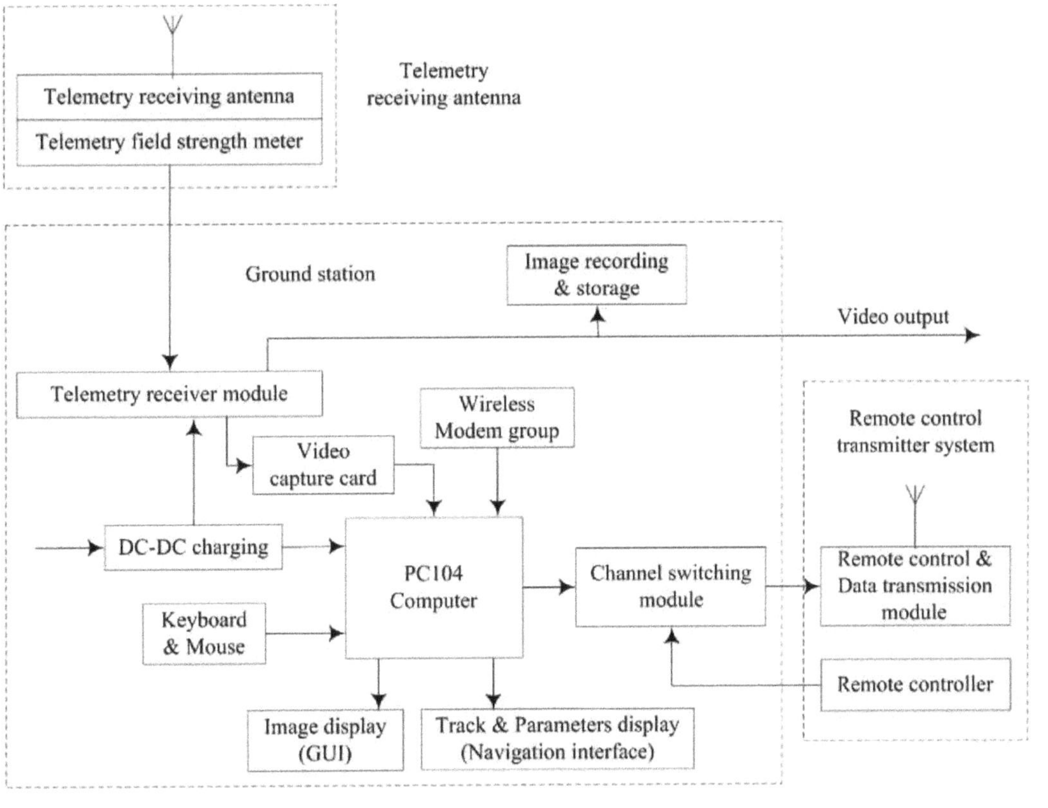

Figure 4. Processing flow of the ground station.

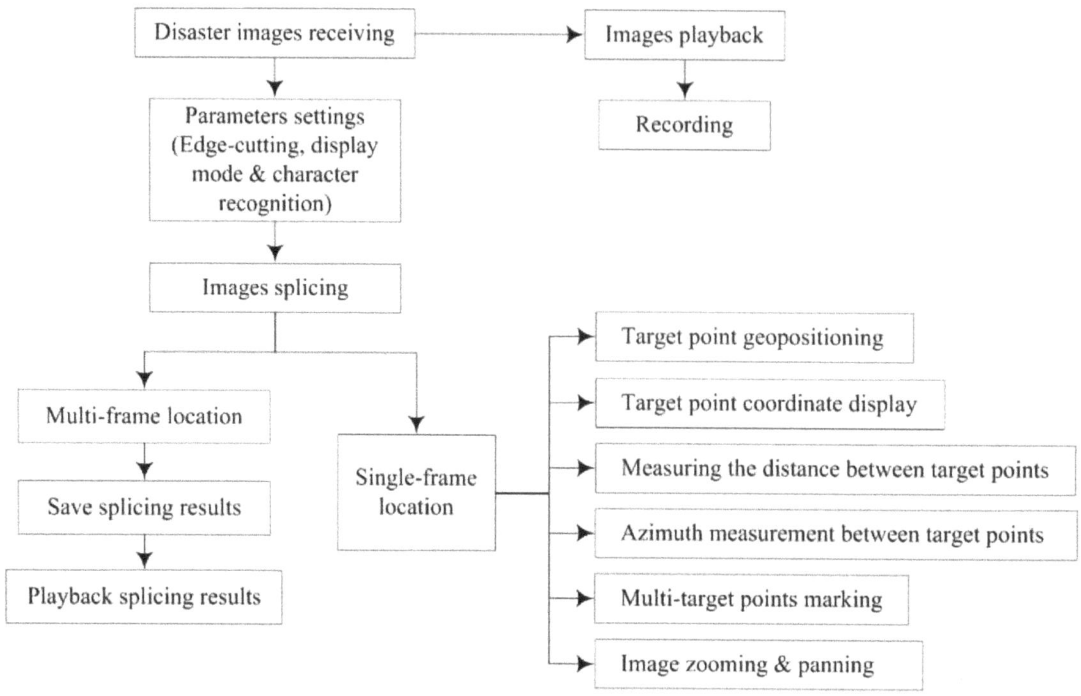

Figure 5. Working flow for collecting and processing of disaster images.

computations. Besides, calibration parameters associated with camera, which were derived from self-calibration based on a well-suited study site, are also used for image orientation.

A certain workflow is adapted to process small-format aerial photographs (SFAPs) in advance. Firstly, we screen all SFAPs obtained during survey flying and select well-suited SFAPs with a certain stereoscopic overlap rate for further processing. Those information in the log files can enable us to visualize the image location and distribution based on a GIS software. Therefore, we can select well-suited SFAPs for further processing in a much quicker manner based on an attribute-based selection of points, e.g., by excluding images deviating from flight altitude or images exceeding a predefined threshold for maximum deviance from a nadir position during image acquisition, which are taken during the starting and landing phases.

3. Field test

After completing the system development, field tests were carried out in Yuxi County, Yunnan province, China, for acquiring optimal parameters of the system (e.g., flight heights and camera parameters). The required image scales and resolutions vary depending on the processes observed. For considering the different observation scales and site extents, we employed different flight plan designs for image acquisition.

3.1. Investigation of UAS flight parameters

With different flight altitudes and flight parameters, the image resolution taken by UAS will be different. For taking aerial photographs in higher resolution, flight tests of the UAS parameters were undertaken to achieve the optimal flight parameters, which would be appropriate for the system.

Flight heights (m)	Shooting interval (s)	Overlap (%)
300	4	63–36
	5	54–20
500	4	73–64
	5	67–55
800	4	86–75
	5	82–69

Table 1. The image overlaps taken at flight heights of 300, 500, and 800 m.

Figure 6. Images taken from three different flight altitudes: (a) 300 m, (b) 500 m, and (c) 800 m.

Firstly, different flight altitudes were chosen for the system, like 200, 300, 500, and 800 m. We performed the tests in the same environment: (a) infinite focal length, (b) exposure time of 1/1000 second, (c) ISO 200, (d) interval timer of shooting of 4 seconds (minimum shooting interval of Sigma DP2), (e) wind speed of 5 m/s, (f) length of the flight line of 8 km, (g) horizontal and vertical viewing angle of 60 and 45°, and (h) flying speed of 30 m/s. The achieved results showed that the image resolution taken at these flight altitudes meets our requirements. Nevertheless, at the flight altitude of 200 m, we only obtained oblique images during the UAS turns in the overshoot zone and missed some shots. Considering the minimum turn radius of the UAS and its average speed, this case represents almost the lower survey limit and should not be set during the flight. The overlap rates of images acquired at other three flight altitudes are shown in **Table 1**. It can be seen from **Table 1** that the image overlap rate depends on the shooting interval and the flight altitude and all of these overlap rates fulfill our splicing requirements.

After obtaining the image overlap rates at various flight altitudes, we carried out different tests for acquiring the camera parameters involving ISO sensitivity and exposure time at different types of weather (cloudy, dark cloud, cloudless, and fog). For ISO sensitivity, the auto control mode is more suitable for the first two types of weather than the fixed scale one. However, the fixed scale mode operates superior in the latter two types of weather. The reason is that the light in the foggy and cloudy weather conditions changes based on the clouds or fog states. For achieving the optimal camera exposure time, the UAS at different flight altitudes (i.e., 300, 500, and 800 m) with infinite focal length and the same ISO sensitivity is tested. Six types of exposure time were set during the flight, including 1/250, 1/320, 1/500, 1/640, 1/700, and 1/1000 second. The obtained findings demonstrated that the clarity and resolution of images, which were taken at flight altitudes of 300 and 500 m, were lower than those taken at flight altitude of 800 m (**Figure 6**). Besides, because of high-speed movement of the camera during the flight, long exposure time would blur images, especially for 1/250, 1/320, and 1/500 second. Thus, in a real field test, only flight altitude of 800 m and exposure time with 1/1000 second are set for images taken.

3.2. Data analysis

The image resolution can be obtained by a formula referring to the 1:500, 1:1000, and 1:2000 aerial photography standard [19], based on the relation of focus length, flight altitude, and resolution. The formula can be expressed as follows:

$$f = H \times C / A \tag{1}$$

where f is the focus length, H denotes the flight altitude, C represents the CCD size, and A indicates the ground covering area computed by A = pixel number × resolution.

For images taken at flight altitude of 800 m, fixed focus model was proposed during shooting. The focus length was 41 mm, with image sensor size of 20.7 × 13.8 mm^2 and image size of 2640 × 1760 pixels. Substituting data into Eq. (1), the obtained resolution is 0.06 m, and the real test result is higher than the theoretical value, which is between 0.1 and 0.12 m.

3.3. Regional panoramic image of Yuxi County

After determining the UAS flight parameters, we selected Yuxi County for flying test for acquiring the panoramic image with high resolution and verifying our UAS performance. Yuxi County has a northeast-southwest distance of 8 km as well as a northwest-southeast distance of 6 km. The total area for mapping is approximately 42 km^2. Based on the overlap requirement, we need at least 25 lines at flight altitude of 800 m. To obtain more auxiliary data, 30 lines are finally set for flight, and the total distance is 200 km as well. Herein, 1711 images (3 MB per image) with data volume of 20 GB were captured in three flights. We excluded images which exceeded a predefined threshold for maximum deviance from a nadir position during image acquiring or deviated from flight altitude, and a total of 1610 images were left. We then used Pix4UAV software for rapidly splicing these images. **Figure 7** showed the automatically computed flight plan. During the triangulation process, the whole three-dimensional (3D) points and key-point observations for bundle block adjustment are 387,784 and 1,268,824, respectively, while the mean value of re-projection error is 0.711432 pixel. The geotag localization variances, which were calculated by Pix4UAV, were 0.251682, 0.258808, and 0.374265 m for X-, Y-, and Z-directions, respectively.

Eventually, through the use of point cloud densification in Pix4UAV software, we acquired the panoramic image of Yuxi County with high resolution shown in **Figure 8**. One can clearly distinguish the violin strings of Nie Er Square, which is the landmark of Yuxi County, by magnifying the Square step by step.

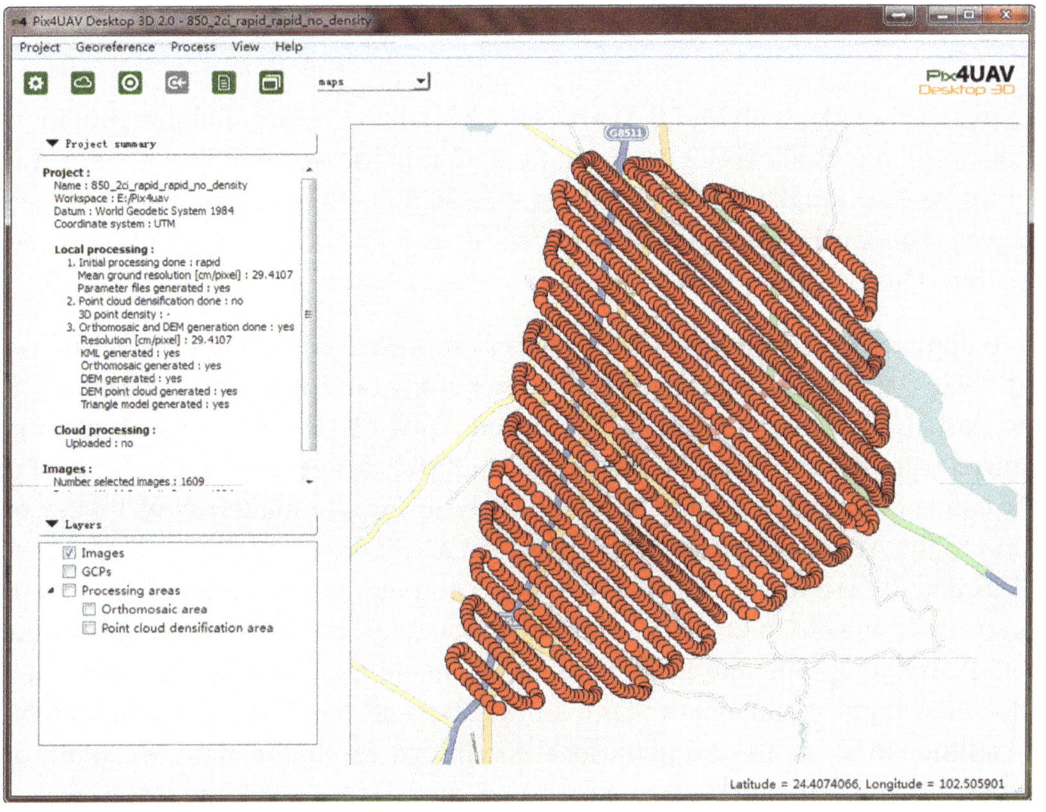

Figure 7. The flight line automatically calculated by the Pix4UAV software.

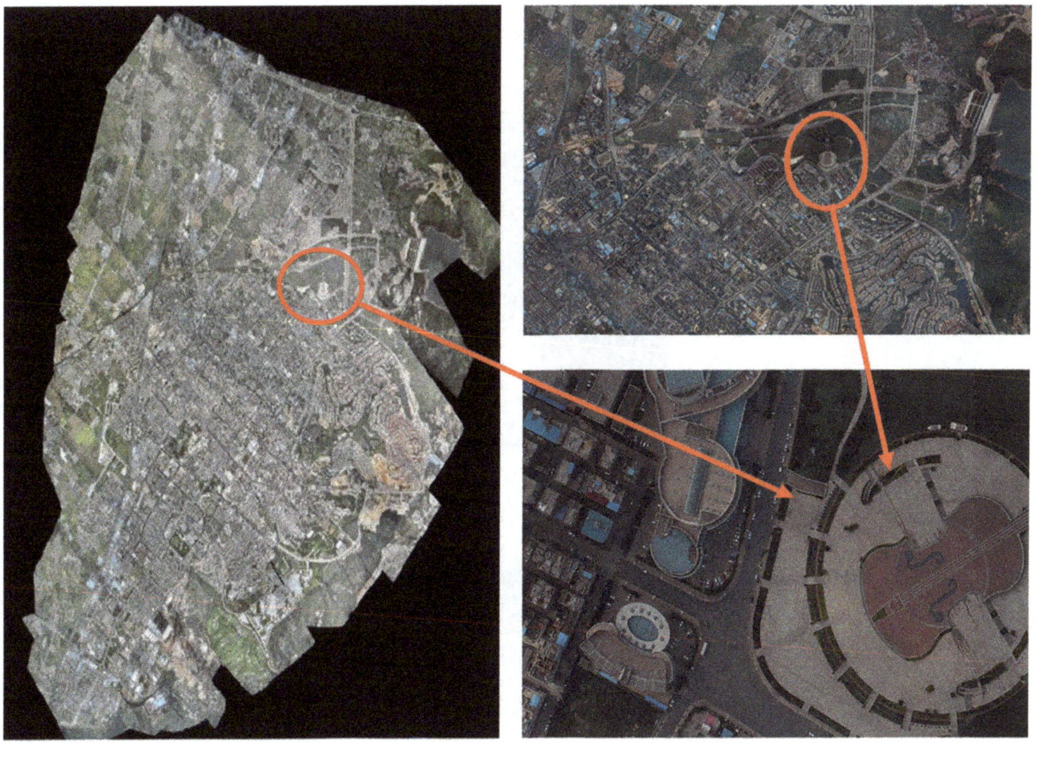

Figure 8. Panorama and the landmark of Yuxi county (Nie Er Square).

4. Real field application to Lushan earthquake

A destructive earthquake with Ms7.0/Mw6.6 struck Lushan County, Sichuan province, China, at 08:02 Beijing Time (00:02 UTC) on April 20, 2013 [1]. The location of the main shock provided by China Earthquake Network Center was 30.3°N, 103.0°E, and the focal depth was 13 km as well. This earthquake caused 193 fatalities and 25 missing, in addition to more than 10,000 injuries. Direct economic losses were estimated to more than 10 billion RMB.

In order to apply the presented system to this earthquake for obtaining its performance, we departed Beijing at 15:00 Beijing Time on the same day, and it took us more than 1 day to reach the Lushan field headquarter (12:00 pm on April 22). Based on the disaster information acquired from the earthquake emergency team, we designed a flight plan and committed to the China Earthquake Administration for censoring. It took us about 1 week to obtain permission to fly. After the plan was approved, an aerial flight was immediately carried out on April 29 and 30. We totally spent 9 hours on completing this designed plan, which was from Longmen Township to Lushan County with the total roundtrip distance of about 40 km (**Figure 9(a)**). The longitude and latitude of waypoints of the flight line are listed in **Table 2**. We used similar flight parameters obtained from the field test in Yuxi County. Here, we set the flight altitude to 600 m according the local conditions. From this flight, we totally acquired 1101 images for 25.51 km^2 affected area mapping, and the data volume stored was approximately 9.16 GB. It took us 2 hours to splice these images through Pix4UAV software, and the obtained result is displayed in **Figure 9(b)**.

For the Lushan earthquake, due to the failure to obtain flight permission timely, we only took the flight after the end of the earthquake emergency response (EER), and the achieved result

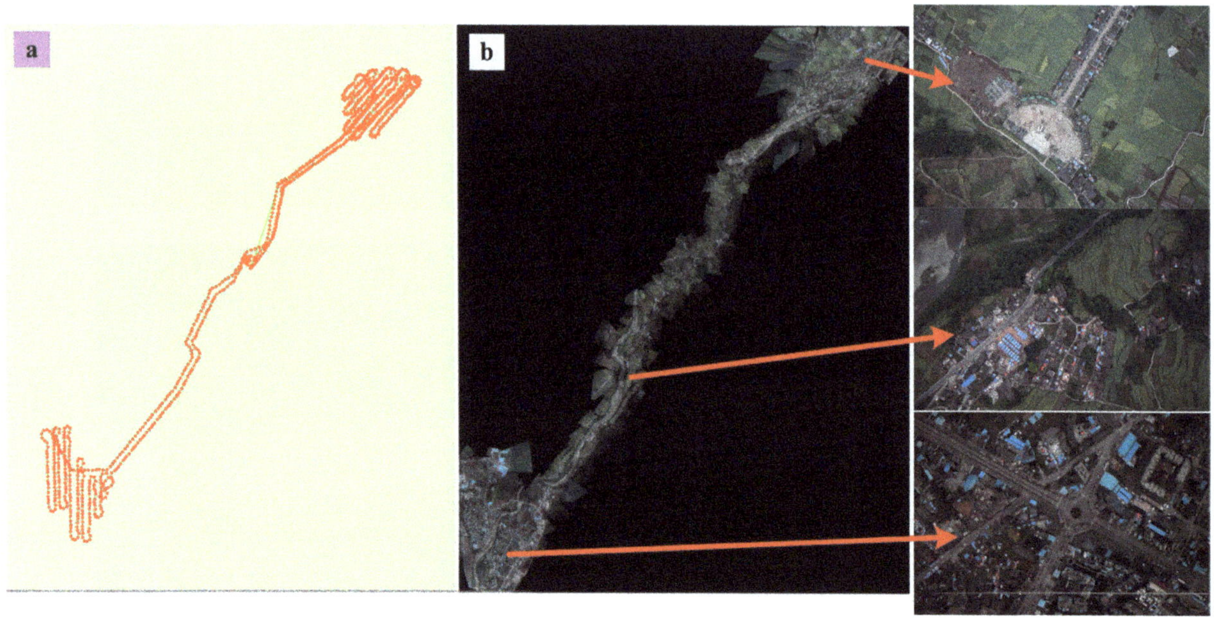

Figure 9. (a) The flight line and (b) the splicing results for Lushan earthquake.

No.	Long.	Lat.	No.	Long.	Lat.	No.	Long.	Lat.
1	102.877	30.153	22	102.866	30.193	43	102.819	30.270
2	102.927	30.158	23	102.863	30.199	44	103.040	30.281
3	102.875	30.161	24	102.865	30.202	45	102.799	30.290
4	102.920	30.162	25	102.958	30.204	46	103.044	30.291
5	102.886	30.162	26	102.852	30.207	47	102.791	30.292
6	102.893	30.164	27	102.857	30.208	48	103.037	30.297
7	102.933	30.164	28	102.850	30.211	49	103.038	30.300
8	102.868	30.167	29	102.846	30.212	50	102.790	30.304
9	102.942	30.165	30	102.955	30.215	51	102.786	30.306
10	102.897	30.168	31	102.959	30.215	52	103.035	30.310
11	102.915	30.170	32	102.961	30.219	53	102.783	30.316
12	102.894	30.173	33	102.965	30.224	54	103.045	30.321
13	102.870	30.174	34	102.970	30.224	55	102.772	30.330
14	102.948	30.177	35	102.968	30.229	56	102.778	30.332
15	102.868	30.178	36	102.974	30.232	57	102.786	30.340
16	102.900	30.178	37	102.986	30.233	58	102.795	30.342
17	102.955	30.181	38	102.999	30.243	59	102.803	30.349
18	102.863	30.182	39	103.014	30.249	60	102.807	30.363
19	102.863	30.188	40	102.827	30.252	61	102.812	30.371
20	102.956	30.190	41	103.023	30.264			
21	102.953	30.193	42	103.033	30.266			

Table 2. Longitude and latitude of the flight line waypoints.

was not used for emergency rescue and decision-making. However, it further attempted to test the combat capability of the presented UAS for EER.

5. Real earthquake emergency application during Ludian earthquake

Regarding the Ludian earthquake, it was the first time that the presented UAS was practically used for EER. This earthquake occurred at 16:30 on 3 August 2014 (Beijing Time), with the location of 27.1°N, 103.3°E, and focal depth of 12 km provided by the China Earthquake Network Center. According to the Chinese intensity scale, the maximum intensity of this earthquake was IX, similar to the modified Mercalli intensity IX, and this disaster caused 617 deaths, 112 missing, and 3143 injuries.

Due to serious damage to the traffic roads, it was impossible for the earthquake emergency team to enter the most difficult hit areas timely for emergency rescue and disaster assessment. Therefore, the headquarters decided to dispatch the UAS team to learn more about the road destruction. On the early morning of the 5th of August, we arrived at the field command and immediately began to design the flight route from Shaba to Longtoushan to Lehong town and to apply for air traffic control (ATC) clearances. At 12:00, under coordination of the Yunnan Bureau of Surveying, Mapping, and Geoinformation, we obtained the permission to fly. In

Figure 10. Panorama of the barrier lake.

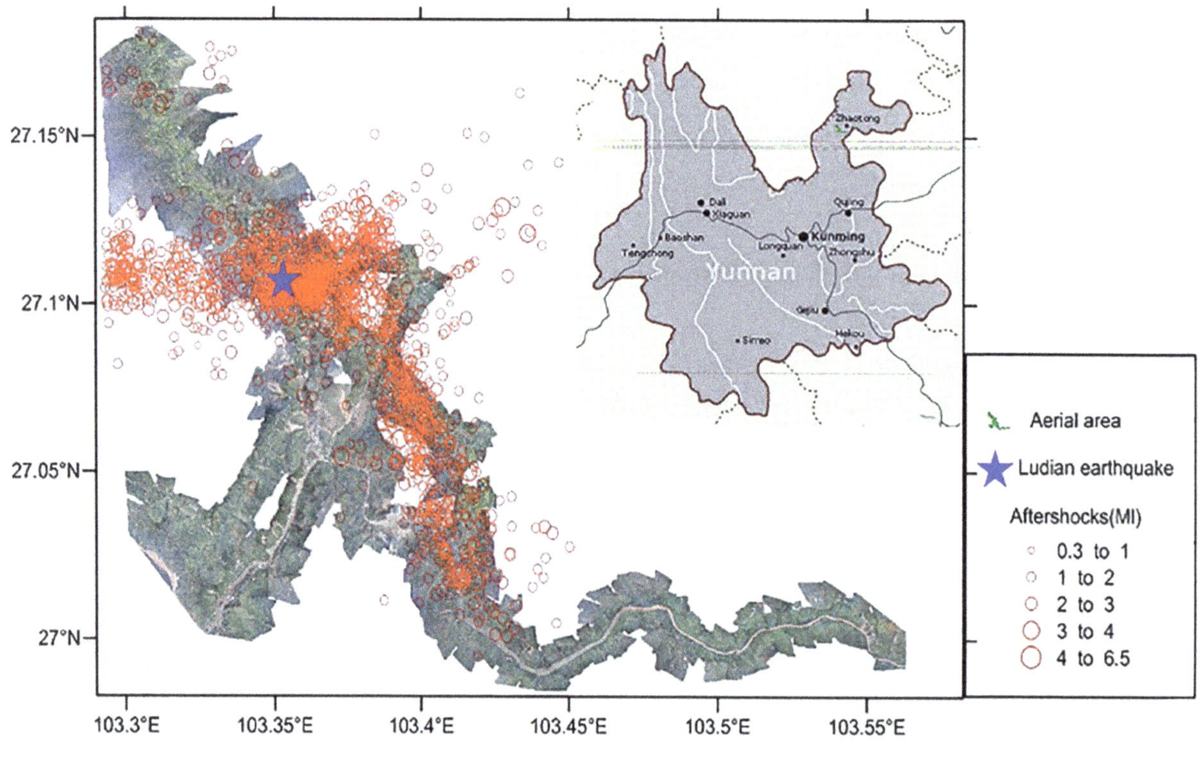

Figure 11. Coverage of the panorama image for the Ludian earthquake compared to the aftershock distribution of this earthquake. The inset represents the study region within Yunnan Province.

response to the situation that a barrier lake formed in the Niulan River because of a huge landslide was too dangerous, we immediately used Google Earth to design a flight plan along the river for this barrier lake. At 15:00, we departed from the field command and selected Li yard lied in Huodehong town as a landing point. It took us about 3 hours to complete the two flight plans. After returning to the headquarters (at 21:00), we spent 2 hours on splicing the images, and the panoramic image of the barrier lake was obtained (**Figure 10**).

On August 6 and 7, we designed another six flight plans and finally obtained aerial images of almost the whole extreme disaster areas (**Figure 11**), compared with the aftershock distribution. When the image is partially enlarged, we can clearly see different types of disaster caused by this earthquake, including landslides, road destruction, structural damages, and floods caused by barrier lakes.

6. Discussion

Although accuracy of the processing workflow cannot be compared with those with installed filed-measured GCPs, the results obtained from this workflow are very suitable for post-earthquake epicentral area mapping, and these data are much more accurate and detailed than traditional SRS data. The image resolution is ca. 10–15 cm, and it is approximately one order of magnitude better than those images taken by QuickBird or WorldView. Such commercially available imagery can be used to analyze larger areas in notable detail. However, UAS data obtained from flight altitude of about 800 m can provide the coverage of post-earthquake epicentral regions with much higher precision at different repeat rates, user-specified times, and lighting conditions.

Based on the authors' experience, the flight altitude should not be below 300 m with the same performance of the presented UAS; otherwise the accuracy of the initial exterior orientation values obtained from the in-flight GPS logs would be extremely low relative to the size of area covered by the image. When the initial position of the image using in the triangulation process is inaccurate, we cannot calculate the overlap rate with the neighboring images. Therefore, we cannot establish tie points and relative orientation between the images. In addition, larger image extents obtained at higher flight altitudes could reduce the relative images' mispositioning. Accordingly, the necessary initial triangulation process and subsequent refinement should be performed by using the Pix4UAV software.

To obtain results with a high level of details, image processing is rather work-intensive, and we need to improve it in the future. However, it may be beneficial when image processing requires manual input because the user can have more control over the process [20]. Recent techniques may lead to further facilitation, such as workflows developed based on structure from motion (SfM) techniques [21].

7. Conclusions

GIS techniques and Earth observation can be used to improve efforts dedicated in developing applicable disaster mitigation strategies. In addition, they can provide relevant agencies with crucial information to alleviate the impact caused by a disaster. However, due to technical and financial issues, the traditional use of aerial and satellite images for this task has been challenged.

More and more UAS are being used as photogrammetric platforms for civilian applications because of their ease of operation, their relatively low cost, and the emergence of low-cost imaging and navigation sensors with performances comparable with more expensive sensors. Based on their operational nature and cost of fabrication, we can adopt this technology to establish a low-cost mapping system.

In this chapter, we presented the application of an UAS to rapidly collect images of post-earthquake disaster. With the chosen fixed-wing UAS, we successfully performed data acquisition at different scales. The results obtained from the two disastrous earthquakes, especially the 2014 Ludian earthquake, demonstrated that the system plays a significant role on the processes of investigating and gathering information about a disaster in epicentral areas of the earthquake, like secondary disaster investigation, road detection, and rapid disaster assessment. The salvation headquarters can effectively use the obtained earthquake information to rapidly develop the relief measures to improve the emergency rescue efficiency.

As to the image analysis, we applied a photogrammetric workflow for coping with the high resolution of the obtained SFAPs, which were captured from relatively high flight altitudes (i.e., 800 m above ground). At these flight altitudes, the resolution of the acquired images still remains below 15 cm. Because we carried out the aerial survey in post-earthquake epicentral areas and it covered a very large area, there were no GCPs distributed in the mapping field. Instead, we applied direct georeferencing using the GPS log recorded by the UAS to create the image block. Among other information, the content of the log files includes the measured position (in X-, Y-, and Z-directions) by GPS for each image obtained as well as the information on tilt of the image axes (κ, ϕ, and ω). We used this information as initial values of exterior orientation for the images in the photogrammetric block fil, and directly analyzed the accuracy of these values through the Pix4UAV software. According to our analysis, the accuracy of these values highly relies on the camera alignment relative to the IMU coordinate system and the measurement precision of the GPS and IMU units. In addition, the accuracy may be influenced by a time-lag between triggering command and actual triggering of the camera. Therefore, these values can be expected to deviate up to several degrees and meters from the actual ones. We need further study to investigate these problems in the future.

Acknowledgements

This research was cofunded by the National Natural Science Foundation of China (41404048) and the Special Fund of the Institute of Geophysics, China Earthquake Administration (DQJB15C04).

Conflict of interest

The authors declare no conflict of interest.

Author details

Chaoyong Peng*, Zhiqiang Xu, Jiansi Yang, Yu Zheng, Weiping Wang, Sha Liu and Baofeng Tian

*Address all correspondence to: pengchaoyong@cea-igp.ac.cn

Institute of Geophysics, China Earthquake Administration, Beijing, China

References

[1] Peng CY, Yang JS, Chen Y, Zhu XY, Xu ZQ, Zheng Y, Jiang XD. Application of a threshold-based earthquake early warning method to the Mw6.6 Lushan earthquake, Sichuan, China. Seismological Research Letters. 2015;**86**(3):841-847

[2] Peng CY, Yang JS, Zheng Y, Zhu XY, Xu ZQ, Chen Y. New τ_c regression relationship derived from all P wave time windows for rapid magnitude estimation. Geophysical Research Letters. 2017;**44**:1724-1731. DOI: 10.1002/2016GL071672

[3] Peng CY, Chen Y, Chen QS, Yang JS, Wang HT, Zhu XY, Xu ZQ, Zheng Y. A new type of tri-axial accelerometers with high dynamic range MEMS for earthquake early warning. Computers & Geosciences. 2017;**100**:179-187. DOI: 10.1016/j.cageo.2017.01.001

[4] Boccardo P. New perspectives in emergency mapping. European Journal of Remote Sensing. 2013;**46**:571-582. DOI: 10.5721/EuJRS20134633

[5] Bendea H, Boccardo P, Dequal S, Tonolo FG, Marenchino D, Piras M. Low cost UAV for post-disaster assessment. In: Chen Jun, Jiang Jie, Ammatzia Peled, editors. Beijing, China: The International Archives of the Photogrammetry, Remote Sensing and Spatial Information Sciences; 3-11 July 2008; Vol. XXXVII, No. Part B8. 2008. p. 1373-1379

[6] Gaszczak A, Breckon TP, Han JW. Real-time people and vehicle detection from UAV imagery. In: Proc. SPIE 7878, Intelligent Robots and Computer Vision XXVIII: Algorithms and Techniques. January 24, 2011. 78780B. DOI: 10.1117/12.876663

[7] Al-Tahir R, Arthur M, Davis D. Low cost aerial mapping alternatives for natural disasters in the Caribbean. In: FIG Working Week 2011, Bridging the Gap Between Cultures; Marrakech, Morocco. May 18-22, 2011

[8] Al-Tahir R, Arthur M. Unmanned aerial mapping solution for small island developing states. In: Global Geospatial Conference 2012; Québec City, Canada. 9 p

[9] Meo R, Roglia E, Bottino A. The exploitation of data from remote and human sensors for environment monitoring in the SMAT project. Sensors (Basel). 2012;**12**(12):17504-17535. DOI: 10.3390/s121217504

[10] Watts AC, Ambrosia VG, Hinkley EA. Unmanned aircraft systems in remote sensing and scientific research: Classification and considerations of use. Remote Sensing. 2012;**4**:1671-1692. DOI: 10.3390/rs4061671

[11] Baiocchi V, Dominici D, Milone MV, Mormile M. Development of a software to plan UAVs stereoscopic flight: An application on post earthquake scenario in L'Aquila city. In: 13th International Conference of Computational Science and Its Applications (ICCSA 2013); Ho Chi Minh City, Vietnam. June 24-27, 2013. pp. 150-165. DOI: 10.1007/978-3-642-39649-6_11

[12] Baiocchi V, Dominici D, Mormile M. Unmanned aerial vehicle for post seismic and other hazard scenarios. WIT Transactions on the Built Environment. 2013;**134**:113-122

[13] Adams SM, Friedland CJ. A survey of unmanned aerial vehicle (UAV) usage for imagery collection in disaster research and management. In: Proceedings of the Ninth International Workshop on Remote Sensing for Disaster Response; Stanford, CA, USA. September 15-16, 2011

[14] d'Oleire-Oltmanns S, Marzolff I, Peter KD, Ries JB. Unmanned aerial vehicle (UAV) for monitoring soil erosion in Morocco. Remote Sensing. 2012;**4**:3390-3416. DOI: 10.3390/rs4113390

[15] Xu ZQ, Yang JS, Peng CY, Wu Y, Jiang XD, Li R, Zheng Y, Gao Y, Liu S, Tian BF. Development of an UAS for post-earthquake disaster surveying and its application in Ms7.0 Lushan earthquake, Sichuan, China. Computers & Geosciences. 2014;**68**:22-30. DOI: 10.1016/j.cageo.2014.04.001

[16] Pastor E, Lopez J, Royo P. UAV payload and mission control hardware/software architecture. IEEE Aerospace and Electronics Systems Magazine. 2017;**22**:3-8

[17] Pix4D. Pix4D Software Introduction [Internet]. 2010. Available from: https://pix4d.com/ [Accessed: February 1, 2018]

[18] Sigma DP2. The Sigma DP2: A Full Spec Compact Digital Camera with all the Power of DSLR [Internet]. 2009. Available from: http://www.sigma-dp.com/DP2/specification.html [Accessed: February 1, 2018]

[19] GB/T 6962-2005. 1:500 1:1 000 1:2 000 Aerial Photogrammetric Standard. Beijing: State Standardization Publishing House; 2005

[20] Hendrickx M, Gheyle W, Bonne J, Bourgeois J, de Wulf A, Goossens R. The use of stereoscopic images taken from a microdrone for the documentation of heritage – An example from the Tuekta burial mounds in the Russian Altay. Journal of Archaeological Science. 2011;**30**:2968-2978

[21] Turner D, Lucieer A, Watson C. An automated technique for generating georectified mosaics from ultra-high resolution unmanned aerial vehicle (UAV) imagery, based on structure from motion (SfM) point clouds. Remote Sensing. 2012;**4**:1392-1410

Permissions

All chapters in this book were first published in EFPERC, by InTech Open; hereby published with permission under the Creative Commons Attribution License or equivalent. Every chapter published in this book has been scrutinized by our experts. Their significance has been extensively debated. The topics covered herein carry significant findings which will fuel the growth of the discipline. They may even be implemented as practical applications or may be referred to as a beginning point for another development.

The contributors of this book come from diverse backgrounds, making this book a truly international effort. This book will bring forth new frontiers with its revolutionizing research information and detailed analysis of the nascent developments around the world.

We would like to thank all the contributing authors for lending their expertise to make the book truly unique. They have played a crucial role in the development of this book. Without their invaluable contributions this book wouldn't have been possible. They have made vital efforts to compile up to date information on the varied aspects of this subject to make this book a valuable addition to the collection of many professionals and students.

This book was conceptualized with the vision of imparting up-to-date information and advanced data in this field. To ensure the same, a matchless editorial board was set up. Every individual on the board went through rigorous rounds of assessment to prove their worth. After which they invested a large part of their time researching and compiling the most relevant data for our readers.

The editorial board has been involved in producing this book since its inception. They have spent rigorous hours researching and exploring the diverse topics which have resulted in the successful publishing of this book. They have passed on their knowledge of decades through this book. To expedite this challenging task, the publisher supported the team at every step. A small team of assistant editors was also appointed to further simplify the editing procedure and attain best results for the readers.

Apart from the editorial board, the designing team has also invested a significant amount of their time in understanding the subject and creating the most relevant covers. They scrutinized every image to scout for the most suitable representation of the subject and create an appropriate cover for the book.

The publishing team has been an ardent support to the editorial, designing and production team. Their endless efforts to recruit the best for this project, has resulted in the accomplishment of this book. They are a veteran in the field of academics and their pool of knowledge is as vast as their experience in printing. Their expertise and guidance has proved useful at every step. Their uncompromising quality standards have made this book an exceptional effort. Their encouragement from time to time has been an inspiration for everyone.

The publisher and the editorial board hope that this book will prove to be a valuable piece of knowledge for researchers, students, practitioners and scholars across the globe.

List of Contributors

Roberto Guidotti
Department of Civil and Environmental Engineering, MAE Center: Creating a Multi-hazard Approach to Engineering, University of Illinois, Urbana, IL, USA

Alberto Castellani
Structural Engineering Department, Politecnico di Milano, Milano, Italy

Comingstarful Marthong
National Institute of Technology Meghalaya, Shillong, India

Hideyuki O-tani
RIKEN Advanced Institute for Computational Science, Chuo-ku, Hyogo, Japan

Muneo Hori and Lalith Wijerathne
Earthquake Research Institute, University of Tokyo, Tokyo, Japan

Michihiro Ohori
Research Institute of Nuclear Engineering, University of Fukui, Kanawa, Tsuruga, Japan

Michaela Ibrion and Nicola Paltrinieri
Faculty of Engineering, Department of Mechanical and Industrial Engineering, Norwegian University of Science and Technology (NTNU), Trondheim, Norway

Sui Tung and Timothy Masterlark
Department of Geology and Geological Engineering, South Dakota School of Mines and Technology, Rapid City, SD, USA

Daniel Sai Huen Lo
Department of Civil Engineering, The University of Hong Kong, Hong Kong SAR

Li Jianbo and Liu Weihong
Institute of Earthquake Engineering, Dalian University of Technology, Dalian, China

Mei Li, Chong Yue, Chen Yu, Tao Xie and Xian Lu
China Earthquake Networks Center, China Earthquake Administration, Beijing, China

Wenxin Kong and Shu Song
China University of Geosciences, Beijing, China

Zuhair Hasan El-Isa
Department of Geology, University of Jordan, Amman, Jordan

Walter Salazar
Catholic University of El Salvador UNICAES, Santa Ana, El Salvador

Huai-feng Wang
School of Civil Engineering and Architecture, Xiamen University of Technology, Xiamen, China

Chaoyong Peng, Zhiqiang Xu, Jiansi Yang, Yu Zheng, Weiping Wang, Sha Liu and Baofeng Tian
Institute of Geophysics, China Earthquake Administration, Beijing, China

Index

CPSIA information can be obtained
at www.ICGtesting.com
Printed in the USA
LVHW060855030122
707669LV00002B/175

9 781639 891610